Pre-Calculus

for technical programs (PART 1)

A STUDENT ORIENTED TEACHING OR SELF-STUDY TEXT

Third Edition

Rudolf A. Zimmer
Fanshawe College

$\boxed{\mathcal{B}|C}$ PUBLISHING COMPANY, LONDON, ONTARIO

ISBN 0-9697446-9-2

Printed in Canada
10 9 8 7 6 5 4 3

B C PUBLISHING COMPANY
247 Millbank Dr.
London, Ontario, N6C 4V9

In memory of
Sharon Trafford

Acknowledgments

I wish to make a number of acknowledgments for the assistance in the preparation of this book.

Firstly, I should like to thank Bep Houghton for her considerable assistance in typing and editing the final manuscript.

Secondly, I should like to thank Sharon Trafford and Patricia Mosher for their work in typing the draft copies, Joanne Schuck for her assistance in preparing solutions and my son Rudy for his imaginative cartoons.

In addition, I should like to express my appreciation to Karen Buck, Don Forrest, Janet Gallagher, Beb Houghton, Eileen Hull, Rose McCabe, Wesley Prescod, John Vranch, Anneke Wall and Joseph Woodall who have extensive experience working with students on an individual basis. Their help and suggestions for improvements were most valuable.

Any acknowledgement would be incomplete without thanking all the students who were involved in the class testing of this material.

Contents

Preface to the Third Edition

Since users in general liked the text as is, the third edition is basically still the same book as the second edition with unit 22, Graphs of Functions, Exponential Growth and Decay added on.

Additional changes have been made in the following units:

Unit 3 has been substantially altered by the addition of frames dealing with more complex problems. Furthermore, frames have been added to discuss the mistakes students make when simplifying exponential expressions. Check point exercises have been added.

In Unit 10, examples 3 and 4 on pages 140 and 141 have been changed to result in equations of the form $ax^2 + ax + c = 0$, and in Check point 5 on page 141, question 4 has been changed accordingly.

Unit 12 has been substantially altered by the addition of complex algebraic fractions on pages 178 and 179. On page 181, Drill exercises numbered 39 to 44, and on page 182, Assignment exercises numbered 31 to 34 have been added.

In Unit 13, Drill exercises 1 and 2 in the second edition have been deleted, and the exercises numbered 22 to 24 on page 195 have been added. On page 196, the Assignment exercises numbered 16 to 19 have been added.

Some teachers requested a unit on graphing and I concurred that the text should include a discussion of graphs of the functions dealt with in the text. Furthermore, it was felt that exponential growth and decay problems would be an appropriate addition as applications to the work done in Units 20 and 21.

I wish to thank the following people for reviewing and/or suggesting alterations to the text.

Frank Harnadek, St. Clair College
Joseph MacPherson, University College of Cape Breton
Sherry May, Memorial University of Newfoundland
Wesley Prescod, Rose McCabe, John Vranch and Joseph Woodall, Fanshawe College

Rudolf A. Zimmer

To the Student

This Learning Material is designed to help you learn Mathematics topics more effectively and more easily than would be the case if you used traditional textbooks. The Learning Material consists of:

(a)   A self-study book,

(b)   A student manual.

1.   The Structure of the Book

(a)   The book, Pre-Calculus (Part 1), contains 22 units which are grouped into 5 Modules.

(b)   Each unit has the same organization which consists of:
(i)   A sequence of learning and exercise frames. The exercise frames are called *Check Points*.

(ii)   Drill Exercises.
(iii)   Assignment Exercises.

(c)   At the end of the book we have:
(i)   Solutions to Check Point Exercises.
(ii)   Answers to Drill Exercises.

2.   The Structure of the Student Manual

The Student Manual is divided into 2 parts.

Part 1   contains complete solutions to all Drill Exercises and answers to odd numbered Assignment Exercises.

Part 2   contains Module Self-Tests with answer keys and a review analysis. The review analysis will indicate to you the topics to which you should pay special attention in your review.

3.   Recommended study procedure

To benefit fully from this Learning Material, you must use it as follows:

(a)   Carefully study the contents of each learning frame and do each Check Point exercise in order. Immediately compare your answers to Check Point exercises with the solutions given at the end of the book. These exercises are designed so that you will become familiar with the skills needed to learn the material which follows.

(b)   Do all odd numbered exercises in the Drill section and check your answers against the answers located at the end of the book. If you have made mistakes or took a long time to answer these questions, go back and do the even numbered exercises of the Drill section for further practice. Complete solutions to all the Drill exercises are provided in the Student Manual.

(c) The Assignment Exercises should be attempted only after you have worked through the unit as described in (a) and (b) above.

Do all the odd numbered exercises in the Assignment without looking at the examples given in the unit. In other words, use the Assignment as a self-test. Check your answers against the answers located in the *Student Manual*. If you have wrong answers, see your instructor for assistance and then do the even numbered exercises of the Assignment section.

(d) After having worked through the assigned units of a given Module, write the Module Self-Test located in the Student Manual.

(e) Use Table 2 on the front page of each Module Self-Test to thoroughly review those parts of the test in which you scored below a passing mark. In reviewing these topics you may want to see your instructor for extra help.

(f) You should write all mathematical symbols very carefully. Carelessly written symbols are often misinterpreted or overlooked and this leads to unnecessary errors.

(g) You should use a proper notebook in which you do your exercises and make a summary which lists the rules and terms in each unit. The notebook should contain examples of solved problems. A carefully organized notebook will help you to review more efficiently and be more successful when writing tests.

Rudolf A. Zimmer

To the Instructor

Introduction

The development of this learning material is based on many years of practical experience at Fanshawe's Mathematics Learning Centre, which has been in operation since 1972.

Pre-Calculus (Part 1) for technical Programs, is designed as a companion text to the following series of learning packages:

1. Fundamental Mathematics,
2. Essential Mathematics,
3. Pre-Calculus (Part 2 and Part 3),
4. Sets, Relations and Functions.

Content and emphasis

Many students encounter difficulties in learning mathematics because they do not clearly understand concepts and/or do not possess skills which are prerequisites to what they are currently attempting to learn. Pre-Calculus (Part 1) reviews fundamental concepts of algebra and then introduces the student to topics of intermediate algebra.

The emphasis is on building the necessary mathematical skill for calculus and to foster independent study habits. Special attention has been paid to the following:

(a) The ability to identify what a student knows with the use of a diagnostic test.

(b) The use of a learning and exercise sequence which is to ensure that new terms, concepts and processes are well understood *before* they are used in solving more complex problems.

(c) Explanations which relate new concepts and skills to what a student already knows.

(d) The need for estimating and how to estimate answers when using a calculator.

(e) The pointing out of common errors and how to avoid making them.

Features of the learning materials

1. **Diagnostic test.**

   The Teacher's Manual contains a diagnostic test which has been developed over many years and has been extensively used to:

   (a) prescribe self-study programs for laboratory, homestudy, or correspondence course situations,

   (b) place students in appropriate classes,

   (c) identify to what extent prerequisite topics should be reviewed.

2. **Learning by doing**

Each unit is divided into a number of learning and exercise frames.

Exercise frames are called *Check Points* and are designed to provide a step by step learning and exercise experience. Hence, a student is encouraged to work and not just read through a unit. Complete solutions for Check Point exercises are provided.

3. **Competency through drill**

Each unit contains an extensive set of Drill Exercises to foster competency in the performance of prerequisite skills. A set of exercises for assignment purposes follows the drill exercises. The textbook is accompanied by a Student Manual which contains complete solutions to all Drill Exercises and answers to odd numbered Assignment Exercises.

4. **Self tests with review analysis**

The Student Manual also contains criterion referenced Module *Self-Tests*. Each self-test contains an analysis table referring students to units and frames dealing with those topics to which he/she should pay special attention in his/her review.

5. **Criterion referenced tests**

The Teacher's Manual contains a battery of criterion referenced tests for each Module. This evaluation method ensures that a student has adequate knowledge of the fundamentals required to learn topics covered in subsequent modules.

Use of the learning material

The material has been designed to be flexible. The following describes how the material has been and is currently used.

1. As a conventional classroom text, it performs the function of a text and a complete set of notes. This liberates the instructor from the routine drudgery of writing detailed notes on the chalkboard. Instead, it allows the teacher to deal more thoroughly with conceptual difficulties and applications by concentrating on the Socratic aspect of teaching.

2. As a self-study text used for learning laboratory courses, it allows students to learn at different levels and, hence, is suited for programs which have a continuous intake.

3. As a self-study text used for correspondence or homestudy courses, it has all the elements required for total individualized learning. Over the past years, several thousand students have participated in our homestudy programs, which involves visits to the Mathematics Learning Centre for both tutorial assistance and the writing of tests.

Fanshawe College

MODULE 1

Review of
Fundamental Algebra

Table of contents **Page**

UNIT 1

REAL NUMBERS AND PROPERTIES

Objectives: After having worked through this unit, the student should be capable of:

1.   understanding the following terms and their relationship:
     natural numbers
     integers
     rational numbers
     real numbers
     complex numbers;
2.   recognizing the properties of real numbers.

1

REAL NUMBERS AND THEIR BASIC PROPERTIES

Introduction:

Numbers are basic concepts in mathematics. Most students are familiar with the terms natural numbers, integers, rational numbers and real numbers. However, to learn the mathematical material discussed in this book, a student has to know what these terms mean, how these numbers relate to each other and what their individual properties are.

2

THE NATURAL NUMBERS

Natural numbers are used for counting. They are: 1, 2, 3, 4, 5, 6, 7, 8, 9, 10, 11, 12, 13, 14, ... The three dots at the end indicate "and so on" which means that in this sequence of numbers, the pattern of adding "1" to a number to obtain the next one continues. There is no last natural number.

3

SOME BASIC PROPERTIES OF NATURAL NUMBERS

1. Divisibility.

A natural number is said to be evenly divisible by another if the remainder is zero.

Example:

12 is evenly divisible by 2, 3, 4 and 6.

2. Prime and Composite Numbers.

A prime number is a natural number other than 1 that is evenly divisible only by itself and 1.

Examples:

2, 3, 5, 7, 11, 13, 17, 19, 23, 29, 31, 37, 41, 43, 47 are the prime numbers between 1 and 50.

Note:

1 is not a prime number and 2 is the only prime number which is even. Numbers which are not prime are called composite numbers.

3. Unique Factorization Property

Any natural number **greater than 1** can be written as a unique product of prime numbers.

Examples:

1. $12 = 2 \cdot 2 \cdot 3$ 2. $30 = 2 \cdot 3 \cdot 5$ 3. $51 = 3 \cdot 17$

4

PRIME FACTORIZATION

To find the prime factors of a natural number, we try to divide the number by consecutive prime numbers, starting with 2. If the number is divisible by a prime number, continue to divide the quotient by the same prime number until no longer possible, then try to divide by the next higher prime number 3, 5, 7, ...

Repeat the process until the quotient is 1.

Example:
Find the prime factorization of 360.

Solution:

prime factors

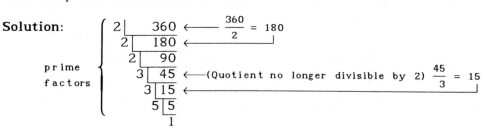

Hence, the prime factorization of 360 is 2 • 2 • 2 • 3 • 3 • 5.

Normally, we write this in exponential form $2^3 \cdot 3^2 \cdot 5$.

5

CHECK POINT 1

Find the prime factorization of the following numbers.
1.   18   2.   45   3.   610   4.   324

6

We can add and multiply any two natural numbers to obtain another natural number.

Examples:
1.   3 + 8 = 11   2.   4 × 3 = 12

However, we cannot subtract any two natural numbers and obtain another natural number.

Example:
5 − 7 = −2

Since −2 is not a natural number, we have to enlarge the collection of numbers in order to subtract.

7

THE INTEGERS

If we add the negatives of the natural numbers and zero to the natural numbers, we have

        ... −5, −4, −3, −2, −1, 0, 1, 2, 3, 4, 5, ...

a collection of numbers called the integers.

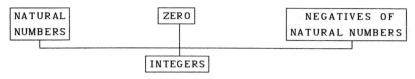

8

We can now add, subtract and multiply any two integers and obtain another integer. However, we cannot divide any two integers and obtain another integer.

Example:
The numbers −3 and 4 are integers, but $(-3) \div (4) = -\frac{3}{4}$ is not an integer.

Hence, in order to add, subtract, multiply and divide integers we have to enlarge the collection of numbers.

THE RATIONAL NUMBERS

If we add to the integers fractions, such as $\frac{3}{4}$, $-\frac{1}{2}$, $\frac{19}{21}$, etc.,
we obtain a collection of numbers called rational numbers.

Any rational number can be expressed as $\frac{a}{b}$, where a and b are integers and
$b \neq 0$.

Any integer can be written as a rational number with b = 1.

Examples: $-2 = -\frac{2}{1}$, $6 = \frac{6}{1}$, $210 = \frac{210}{1}$.

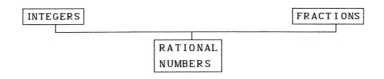

Solutions of many problems are not rational numbers. Given [triangle figure with sides h, 1, 1], find h.
Using the Pythagorean Theorem, we have $h = \sqrt{1^2 + 1^2} = \sqrt{2}$. It can be shown
that $\sqrt{2}$ cannot be written as $\frac{a}{b}$ where a and b are integers and $b \neq 0$.
Hence, $\sqrt{2}$ is not a rational number. Other familiar numbers which are not
rational are π and e, two symbols we find on scientific calculators.
Numbers such as $\sqrt{2}$, π and e are called irrational numbers.

THE REAL NUMBERS

If we combine the rational and irrational numbers, we obtain a collection
of numbers called the real numbers. A real number is any number that can be
represented by an infinite decimal.

The **rational numbers** can be represented by **repeating decimals**.

Examples:

1. $\frac{7}{4} = 1.750000\ldots$ 2. $9 = 9.00000\ldots$

3. $\frac{1}{7} = 0.142857142857\ldots$ 4. $\frac{2}{3} = 0.66666\ldots$

The **irrational numbers** can be represented as **nonrepeating decimals**.

Examples:

1. $\sqrt{2} = 1.4142135623\ldots$ 2. $\pi = 3.1415926536\ldots$ 3. $e = 2.718281\ldots$

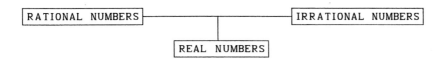

12

Many problems arise in Science and Technology that cannot be solved if we work only with real numbers. For example, when solving the equation $x^2 + 1 = 0$, we obtain the expression $x^2 = -1$. However, the square of any real number is always positive. Hence, the solution of $x^2 = -1$ cannot be a real number. To deal with such problems, the symbol j (or i) is introduced. We define $j = \sqrt{-1}$ and call it the imaginary unit. In general the square root of a negative number is expressed as the product of j and the square root of a positive number. Hence, if d is positive, $\sqrt{-d} = \sqrt{d}j$.

Numbers written with the imaginary unit j are called imaginary numbers.

Examples: $\sqrt{5}j$ and $\sqrt{49}j = 7j$ are imaginary numbers.

13

THE COMPLEX NUMBERS

If we add the imaginary numbers to the real numbers, we obtain a collection of numbers called Complex Numbers.

A complex number is an expression **a + bj** where a and b are real numbers.

The number **a** is called the **real part** and **bj** the **imaginary part** of the complex number.

Examples:
$2 + 3j$ and $\frac{3}{4} - 2\sqrt{2}j$ are complex numbers.

Any real number is a complex number since it may be written in complex form $a + bj$ where $b = 0$.

Example:
The real number 7 can be written as $7 + 0j$ in complex form.

Any imaginary number is a complex number since it may be written in complex form $a + bj$ where $a = 0$ and $b \neq 0$.

Example:
The imaginary number $\sqrt{5}j$ can be written as $0 + \sqrt{5}j$ in complex form.

14

TYPES OF NUMBERS AND THEIR RELATIONSHIP

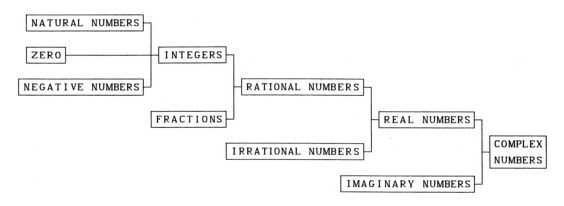

CHECK POINT 2

Categorize the following numbers as natural numbers, integers, rational numbers, irrational numbers, real numbers, imaginary numbers or complex numbers. Note that a number may belong to all or several categories.

$-\dfrac{1}{2}$,   5, $\dfrac{12}{13}$, $\sqrt{7}j$, $-\pi$, $\sqrt{\dfrac{9}{25}}$, $3 - 2j$,

0, -37,   e, $\dfrac{5}{3}j$,   1, $\sqrt{3} + \pi j$,   6.50.

BASIC PROPERTIES OF REAL NUMBERS

In this book we shall deal mainly with problems of combining and manipulating algebraic expressions which represent real numbers. The following are some of the basic properties of real numbers which we will use in our work with algebraic expressions. In what follows, when we use the term numbers, we mean real numbers.

1. COMMUTATIVE PROPERTY OF:

(a) Addition

The **sum** of two numbers is the same regardless of the order in which we add.

Example: $9 + 6 = 6 + 9$
                    $15 = 15$

If a and b are any two numbers, then $\boxed{a + b = b + a}$

(b) Multiplication

The **product** of two numbers is the same regardless of the order in which we multiply.

Example: $11 \cdot 5 = 5 \cdot 11$
                    $55 = 55$

If a and b are any two numbers, then $\boxed{a \cdot b = b \cdot a}$

Note: Subtraction and Division are **not commutative**.
Examples:

(i) $9 - 6 \neq 6 - 9$
         $3 \neq 6 - 9$

(ii) $18 \div 6 \neq 6 \div 18$
          $\dfrac{18}{6} \neq \dfrac{6}{18}$
          $3 \neq \dfrac{1}{3}$

The symbol $\neq$ means "does not equal".

2. ASSOCIATIVE PROPERTY OF:

(a) Addition

The **sum** of three or more numbers is the same regardless of how we group when we add.

Example:
$6 + (19 + 25) = (6 + 19) + 25$
$44 = 25 + 25$
$50 = 50$

If a, b and c represent numbers, then $\boxed{a + (b + c) = (a + b) + c}$

Note: The operations inside brackets are performed first.

(b) Multiplication

The **product** of three or more numbers is the same regardless of how we group when we multiply.

Example:

$$2 \times (5 \times 8) = (2 \times 5) \times 8$$
$$2 \times 40 = 10 \times 8$$
$$80 = 80$$

If a, b and c represent numbers, then $\boxed{a \times (b \times c) = (a \times b) \times c}$

Note: Subtraction and Division are **not associative**.

Examples:

(i) $15 - (7 - 2) \neq (15 - 7) - 2$ (ii) $16 \div (8 \div 2) \neq (16 \div 8) \div 2$

$15 - 5 \neq 8 - 2$ $16 \div 4 \neq 2 \div 2$

$10 \neq 6$ $4 \neq 1$

3. DISTRIBUTIVE PROPERTY OF MULTIPLICATION WITH RESPECT TO ADDITION

The product of a number "a" and the sum "b + c" is the same as the sum of "a × b" and "a × c".

Example:

$$3(5 + 2) = (3 \times 5) + (3 \times 2)$$
$$3(7) = 15 + 6$$
$$21 = 21$$

If a, b and c are any three numbers, then $\boxed{a(b + c) = (a \times b) + (a \times c)}$

Notes:

1. We distribute the **factor** "a" over the **terms** "b" and "c".

2. To illustrate this property, we use the fact that
$$3(5 + 2) = (5 + 2) + (5 + 2) + (5 + 2)$$

Using the associative and commutative properties, we can rearrange the addition so that
$$(5 + 2) + (5 + 2) + (5 + 2) = (5 + 5 + 5) + (2 + 2 + 2)$$
$$= (3 \times 5) + (3 \times 2)$$

Hence, $3(5 + 2) = (3 \times 5) + (3 \times 2)$

SPECIAL PROPERTIES OF ZERO AND ONE

1. Identity elements with respect to:

(a) Addition

The **sum** of zero and any number is the original number.

Examples: $0 + 9 = 9$
$16 + 0 = 16$

If a is any number, then $\boxed{a + 0 = 0 + a = a}$

Note: **Zero** is called the **identity** element of **addition**.

(b) Multiplication

The **product** of the number "1" and any number is the original number.

Examples: $1 \times 5 = 5$
$18 \times 1 = 18$

If a is any number, then $\boxed{a \times 1 = 1 \times a = a}$

Note: The number 1 is called the **identity** element of **multiplication**.

2. Using zero in multiplication and division.

Examples:

(a) The product of zero and any number is zero.

$$4 \times 0 = 0 + 0 + 0 + 0$$
$$= 0$$

If a is any number, then $\boxed{a \times 0 = 0}$

(b) The quotient of zero divided by any number, except zero, is zero.

$$\frac{0}{16} = 0$$

If a is any number, except zero, then $\boxed{\dfrac{0}{a} = 0}$

(c) The quotient of any number "b" divided by zero is **undefined**.

$$\boxed{\dfrac{b}{0} \text{ is undefined}}$$

Reason: If $\dfrac{18}{0}$ = ? (some number), then ? (some number) $\times$ 0 = 18.

This contradicts (a) above which says that the product of zero and **any** number is always zero.

$\boxed{\textbf{Never divide by zero.}}$

CHECK POINT 3

I. True or False. (Use properties – do not evaluate.)

1. $236 + 52.698 = 52.698 + 236$

2. $\dfrac{0}{185}$ is undefined.

3. $\sqrt{2} \times 1 = \sqrt{2}$

4. $15.296 \times (18.960 \times 5) = (15.296 \times 18.960) \times 5$

5. $296 \div 299 = 299 \div 296$

6. $8\pi \div 0 = 0$

7. $31(\pi + 5.63) = (31 \times \pi) + (31 \times 5.63)$

8. $56.984 - 57.981 = 57.981 - 56.984$

9. $1.5 + 0 = 1.5$

10. $42 \times 0 = 42$

II. Evaluate the following using the Distributive Property.

11. $7(5 + 9)$ 12. $8(9 + 10)$

GEOMETRIC REPRESENTATION OF REAL NUMBERS

The real numbers can be represented geometrically as points on a line. We call this line the real number line. To construct it, we draw a line horizontally and select at random a point on the line to represent zero. This point is called the origin. Next, we mark off points left and right of the origin such that the distance between points is the same. We assign positive integers to points on the right and negative integers to points on the left of the origin.

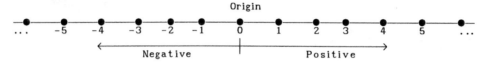

Rational numbers which are not integers and all irrational numbers are located as points between the integers.

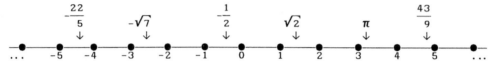

Note:

Every real number can be represented as one and only one point on the line and every point on the line can be represented by one and only one real number.

ORDER RELATIONS

Given any two distinct numbers a and b on the real number line, one of the following two situations occurs.

(i)   If a is to the right of b, then a is greater than b. We symbolize "greater than" by > and write a > b.

   Example: The statement 12 > 5 means 12 is greater than 5.

(ii)   If a is to the left of b, then a is less than b. We symbolize "less than" by < and write a < b.

   Example: The statement -2 < 9 means -2 is less than 9.

Statements such as 12 > 5 and -2 < 9 are called **inequalities**.

There are many instances when we need to compare two real numbers. The basic properties of real numbers stated previously do not provide us with the means to do so. However, using the fact that the real number line preserves order, we can see that given any two real numbers a and b, one and only one of the following is true:

$$a < b, \qquad a = b, \qquad a > b.$$

Examples:

Comparing the following pairs of real numbers

$\frac{1}{2}$ and $\frac{6}{12}$,   1 and $\sqrt{2}$, $-\sqrt{7}$ and -5,   we have

$\frac{1}{2} = \frac{6}{12}$, $1 < \sqrt{2}$, $-\sqrt{7} > -5$.

CHECK POINT 4

I. Construct a real number line and show the approximate locations of the following numbers.

$-\dfrac{6}{9}$, π, $\dfrac{11}{15}$, $\sqrt{3}$, $-\dfrac{\pi}{2}$,   3.5, -4.75, $-e$, 2π.

II. Compare the given pairs of real numbers by using one of the symbols <,   =,   >.

1. 0 and -9   2. 6 and 2π 3. $-\sqrt{2}$ and -1.6   4. $\dfrac{1}{3}$ and $\sqrt{\dfrac{1}{9}}$

DRILL EXERCISES

1. Given the natural numbers
    2, 9, 4, 5, 12, 48, 13, 15, 29, 1, 37, 51, 45, 7, 26, 11, 39, 63,
    100, 23, 81, 31,
    identify the prime numbers.

Prime factorize the following numbers.

2. 8   3. 26   4. 100   5. 94   6. 250

7. 115   8. 144   9. 730   10. 1452   11. 1925

12. Categorize the following numbers as integers, natural, rational, irrational, real, imaginary or complex numbers. Note that a number may belong to all or several categories.

-6, $\sqrt{3}$,   256, $\sqrt{\dfrac{1}{4}}$, -2π, $-\dfrac{11}{13}$,   j,   0, $1 - 2j$,

$5 + 0j$, $0 - \sqrt{2}j$, -2670,   1.75,   0.666..., $-\dfrac{1}{3}$.

True or False. (Use properties – do **not** evaluate.)

13. $\dfrac{0}{2.75}$ is undefined.   14. $3\pi(2.5 + 1.3) = (3\pi \times 2.5) + (3\pi \times 1.3)$

15. $24.327 \times (7.934 \times 10) = (24.327 \times 7.934) \times 10$

16. $3\sqrt{2} + \dfrac{\pi}{2} = \dfrac{\pi}{2} + 3\sqrt{2}$ 17. $2.5 - \sqrt{7} = \sqrt{7} - 2.5$

18. $0 \times 1 = 1$ 19. $\dfrac{-2\pi}{0} = 0$ 20. $0 + \dfrac{3}{4} = \dfrac{3}{4}$

21. Construct a real number line and show the approximate location of the following numbers.

$\dfrac{6}{5}$, $\sqrt{2}$, -2π,   5.25, $-\sqrt{5}$, $\dfrac{\pi}{6}$, -1.75.

Compare the given pairs of real numbers by using one of the following symbols <,   =,   >.

22. $\sqrt{2}$ and 1.2   23. $-\dfrac{\pi}{2}$ and -1.75   24. e and 2.5

25. $\sqrt{\dfrac{1}{16}}$ and 0.25   26. 0 and 0.1   27. -6 and -0.75

28. 0.06 and -1.5   29. $\dfrac{1}{5}$ and 0.2   30. -150 and -2

ASSIGNMENT EXERCISES

1. Given the natural numbers 3, 16, 8, 17, 21, 19, 51, 47, 69, identify the prime numbers.

Prime factorize the following numbers.

2. 12
3. 63
4. 95
5. 248

6. 495
7. 539
8. 738
9. 1573

10. Categorize the following numbers as integers, natural, rational, irrational, real, imaginary or complex numbers. Note that a number may belong to all or several categories.

$$-\frac{1}{9}, \quad \sqrt{5}, \quad 17, \quad 3\pi, \quad 6\sqrt{2}j, \quad -8, \quad \sqrt{\frac{81}{4}}, \quad -920, \quad 6 - j,$$

$$\frac{7}{13}, \quad 9.25, \quad 0.333...$$

True or False. (Use properties - do **not** evaluate.)

11. $26.7(3.2 + \pi) = (26.7 \times 3.2) + (26.7 \times \pi)$

12. $36.2 \div 0 = 0$

13. $30.41 - (18.2 - 10.59) = (30.41 - 18.2) - 10.59$

14. $\frac{3\pi}{4} \times \sqrt{2} = \sqrt{2} \times \frac{3\pi}{4}$

15. $7.5 \times 0 = 0$

16. Construct a real number line and show the approximate location of the following numbers.

$$\sqrt{3}, \quad 4.25, \quad -\frac{\pi}{5}, \quad \frac{4}{3}, \quad -\sqrt{7}, \quad -\pi, \quad 2.5.$$

Compare the given pairs of real numbers by using one of the following symbols <, =, >.

17. 2e and 5.5
18. $\sqrt{\frac{25}{81}}$ and $\frac{9}{5}$
19. -1.5 and $-\frac{\pi}{3}$

20. $-\sqrt{7}$ and -2.9
21. 0.125 and $\frac{1}{8}$
22. 3e and 8.0

UNIT 2

SIGNED NUMBERS, VARIABLE EXPRESSIONS

Objectives: After having worked through this unit, the student should be capable of:

1.   adding and subtracting signed numbers;
2.   multiplying and dividing signed numbers;
3.   evaluating expressions involving exponents and the four basic operations of addition, subtraction, multiplication and division;
4.   evaluating variable expressions.

1

SIGNED NUMBERS

If we attach signs such as a minus or a plus to a real number, we call it a signed number. Numbers which are preceded by a plus sign are called positive and those preceded by a minus sign are called negative numbers.

Zero is neither positive nor negative.

In this unit, we shall review how to add, subtract, multiply and divide signed numbers.

2

The negative of a negative number is a positive number, that is, for all real numbers "a"   -(-a) = a.

Examples:

1.   -(-5) = 5   2.   -(-π) = π

3

THE TWO PARTS OF A SIGNED NUMBER

A signed number has two parts.

The sign is one part.
The magnitude or size of the number is the other part.

While the positive numbers are to the right, the negative numbers are to the left of the origin. Hence, the sign indicates the **direction** of the number from the origin.

The numerical size or magnitude of the signed number tells us **how far** or the **distance** the number is from the origin.

Example:

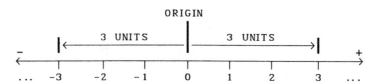

-3 and +3 are the same distance from the origin, but in opposite directions.

4

ABSOLUTE VALUE

The absolute value of a number is the distance between the origin and the number. We use the symbol |x| to denote the absolute value of the number x.

$$|x| = \begin{cases} x \text{ if } x > 0 \\ x \text{ if } x = 0 \\ -x \text{ if } x < 0 \end{cases}$$

Examples:

1. |+3| = 3 and |-3| = -(-3) = 3.   Hence, |+3| = |-3|.

Note: Both +3 and -3 are 3 units from the origin.

2. |-16| = -(-16) = 16 and |+16| = 16.

To find the absolute value of a number, we ignore its sign.

Examples:

1. |-π| = π   2. |+5| = 5   3. |-216| = 216

5

ADDITION OF SIGNED NUMBERS

When adding signed numbers, we have two situations, namely, the numbers added have:

1. the **same** signs: **Examples:**   (+3) + (+11),   (-4) + (-2)

2. **opposite** signs: **Examples:**   (-6) + (+8),   (+7) + (-1)

6

RULE: ADDING NUMBERS WHICH HAVE THE SAME SIGNS

1. Add the absolute value of the two numbers.
2. Place the common sign in front of the sum.

Examples:

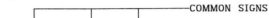

1. (-81) + (-5) = -86   since |-81| + |-5| = 86

2. (+125) + (+20) = +145   since |+125| + |+20| = 145

7

RULE: ADDING NUMBERS WHICH HAVE OPPOSITE SIGNS

1. Find the absolute values of the two numbers, then subtract the smaller from the larger.

2. Place the sign of the number which has the larger absolute value in front of the difference.

Examples:

SIGN OF THE NUMBER HAVING
THE LARGER ABSOLUTE VALUE

1.   (+16) + (-24) = -8; |-24| - |+16| = 24 - 16
                                              = 8

SIGN OF THE NUMBER HAVING
THE LARGER ABSOLUTE VALUE

2.   (+35) + (-20) = + 15; |+35| - |-20| = 35 - 20
                                              = 15

8

The sum of any signed number and zero is the signed number.

Examples:

1. (+6) + 0 = +6
2. 0 + (-12) = -12   **Remember:** Zero is the identity element of addition.

9

SUBTRACTION OF SIGNED NUMBERS

Method:

Step 1.   Change the number to be subtracted to its opposite.
Step 2.   Change the subtraction sign to an addition sign.
Step 3.   Use the rules of adding signed numbers to evaluate the expression.

$$\overset{\displaystyle \ulcorner\text{change to}\urcorner}{\underset{\text{subtraction}}{\underset{\llcorner\text{change to}\lrcorner}{\overset{\text{subtrahend }b \qquad\qquad \text{opposite of } b}{(a) - (b) = (a) + (-b)}}}}$$

Example:

Horizontal Form	Vertical Form

1. $(+8) - (+2) = (+8) + (-2)$

   $= +6$

Subtract: $\begin{array}{r} +8 \\ +2 \\ \hline \end{array}$ or Add: $\begin{array}{r} +8 \\ -2 \\ \hline +6 \end{array}$

CHECK POINT 1

True or False.

1. $|-6| = |6|$ 2. $|0| < |-9|$ 3. $|-7| = -7$ 4. $|+2| > |-1|$

Add or subtract as indicated.

5. $(+25) + (-10)$ 6. $(-5) - (+10)$ 7. $(-6) + (-12)$ 8. $(-8) - (-15)$

ADDITION AND SUBTRACTION OF MORE THAN TWO TERMS

Method:
   Change all subtractions to additions.

An Agreement
   When **all** terms in an expression are **added,** we can remove the addition signs with the understanding that the signed numbers are to be added.

Example:

The expression	$(-8) + (-6) + (+9) + (+6) + (-18)$
can be written as	$-8 - 6 + 9 + 6 - 18.$
Conversely,	$16 - 2 + 3 - 19 - 5 - 3 + 10$
means the sum of	$+16, -2, +3, -19, -5, -3$ and $+10.$

Example:
   Evaluate $(+9) - (-3) + (-4) - (+12)$

Solution:
   Step 1. $(+9) - (-3) + (-4) - (+12) = (+9) + (+3) + (-4) + (-12)$

   Step 2.
$$= 9 + 3 - 4 - 12$$
$$= 12 - 4 - 12$$
$$= 8 - 12$$
$$= -4$$

14

CHECK POINT 2

Evaluate the following expressions.
1. (+16) – (–12) – (+4) + (–16)   **2.**   (–10) + (–5) – (+2) + (+7) – (–9)

3. (–3) – (–19) + (–25) – (+6) – (–8)   **4.**   16 – 2 + 3 – 19 – 5 – 3 + 10

15

MULTIPLICATION OF SIGNED NUMBERS

When multiplying signed numbers, we can have two situations.

Namely, the numbers multiplied have:
1. opposite signs;   examples: (+3)(–4) or (–6)(+2)

2. identical signs;   examples: (+5)(+3) or (–7)(–3)

16

MULTIPLICATION WITH ZERO AS A FACTOR

We have seen earlier that:

(a)   zero is neither positive nor negative,
(b)   zero times any number is zero.

Examples:
1.   (0)(–6) = 0   2.   (+9)(0) = 0

17

THE PRODUCT OF TWO NUMBERS WITH OPPOSITE SIGNS IS NEGATIVE

Examples:

1.   (–18)(+5) = –90   2.   (+7)(–7) = –49

3.   6(–10) = –60   4.   (–1)(+5) = –5

18

THE PRODUCT OF TWO NUMBERS WITH IDENTICAL SIGNS IS POSITIVE

Examples:
1.   (+10)(+20) = +200   2.   (–1)(–8) = +8   3.   (–5)(–5) = +25

19

If a product has an:
1. **even** number of negative factors, the product is **positive**;
2. **odd** number of negative factors, the product is **negative**.

20

Example:
Evaluate   (+3)(–3)(–4)(+2)

Solution: (+3)(–3)(–4)(+2) = (–9)(–4)(+2)
                              = (+36)(+2)
                              = +72

21

CHECK POINT 3

Evaluate the following expressions.
1. (–5)(+2)(+3)(–1)   2.   (–6)(–5)(–2)

3. (+3)(–4)(–2)(+4)(–1)   4.   (+1)(–16)(–1)(+2)(–10)(–2)

NEGATIVE NUMBERS RAISED TO AN EXPONENT

We recall that $2^4 = 2 \times 2 \times 2 \times 2$. The number "2" is called the base and "4" is called its exponent.

Beware:

Exponents apply only to **one symbol**, namely, the symbol on their immediate left unless there are grouping symbols indicating otherwise.

Example:

$$-2^4 = (-1)2^4$$
$$= (-1)(2)(2)(2)(2)$$
$$= -16$$

Here the exponent applies only to the number "2" **not** to the "−" sign.

but

$$(-2)^4 = (-2)(-2)(-2)(-2)$$
$$= 16$$

Here the exponent applies to the number "−2" as indicated by the parentheses.

DIVISION OF SIGNED NUMBERS

When dividing signed numbers, we can have two situations.
The dividend and divisor may have:

1. opposite signs; examples: $(-30) \div (+10)$; $(+6) \div (-2)$
2. identical signs; examples: $(-25) \div (-5)$; $(+16) \div (+4)$

THE QUOTIENT OF TWO NUMBERS WITH OPPOSITE SIGNS IS NEGATIVE

Examples:

1. $(-10) \div (+5) = -2$; $\dfrac{-10}{+5} = -2$

2. $(+27) \div (-3) = -9$; $\dfrac{+27}{-3} = -9$

THE QUOTIENT OF TWO NUMBERS WITH IDENTICAL SIGNS IS POSITIVE

Examples:

1. $(-100) \div (-20) = +5$; $\dfrac{-100}{-20} = +5$

2. $(+6) \div (+3) = +2$; $\dfrac{+6}{+3} = +2$

3. $(-1) \div (-1) = +1$; $\dfrac{-1}{-1} = +1$

Note:

Zero divided by any number, whether positive or negative, will always be zero. However, **any number divided by zero is undefined**.

$\dfrac{0}{-14} = 0$ $\dfrac{0}{+25} = 0$ $\dfrac{+64}{0}$ is undefined $\dfrac{-23}{0}$ is undefined

CHECK POINT 4

Evaluate the following.

1.　$(-3)^4$　　　　2.　-6^2　　　　3.　$(-2)^3$　　　　4.　-1^6

Divide the following.

5.　$(-1) \div (+1)$　　6.　$(+10) \div (-5)$　　7.　$(-6) \div (-3)$

8.　$\dfrac{-1}{-1}$　　　　9.　$\dfrac{0}{-7}$　　　　10.　$\dfrac{-45}{-9}$　　　　11.　$\dfrac{-110}{0}$

REVIEW: ORDER OF OPERATIONS RULES

Step 1. Evaluate all exponential expressions

Step 2. Multiply or divide, whichever comes first, from left to right.

Step 3. Add or subtract, whichever comes first, from left to right.

USE OF GROUPING SYMBOLS

If we want to evaluate an expression in a different order than indicated by the order of operations rules, we use grouping symbols.

These symbols are normally referred to as　　(a)　parentheses
　　　　　　　　　　　　　　　　　　　　　　　　　(b)　brackets
　　　　　　　　　　　　　　　　　　　　　　　　　(c)　braces

and are normally denoted by　　(a)　　( )
　　　　　　　　　　　　　　　　　(b)　　[ ]
　　　　　　　　　　　　　　　　　(c)　　{ }

EVALUATING EXPRESSIONS CONTAINING A SERIES OF GROUPINGS

Method:

Step 1. Evaluate the expression inside the innermost parentheses ( ).

Step 2. Evaluate the expression inside the second grouping indicated by the symbol [ ].

Step 3. Evaluate the expression inside the third grouping indicated by the symbol { }.

EVALUATING EXPRESSIONS CONTAINING A SEQUENCE OF OPERATIONS

To evaluate expressions containing a sequence of operations, we use the following procedure:

(a)　Perform operations within grouping symbols starting with the innermost group.

(b)　To evaluate expressions inside a group or expressions where all grouping symbols have been removed, we follow the rules of the order of operations.

31

Example 1:
Evaluate $(-16)[4^2 \div (-2)^3]$

Solution:
$$(-16)[4^2 \div (-2)^3] = (-16)[16 \div (-8)]$$
$$= (-16)[-2]$$
$$= +32$$

32

Example 2:
Evaluate $27 \div (-3) + 3\{2(-3)^2 - [(-2)^3 - 3(4)]\}$

Solution:
$$27 \div (-3) + 3\{2(-3)^2 - [(-2)^3 - 3(4)]\} = 27 \div (-3) + 3\{2(9) - [-8 - 12]\}$$
$$= 27 \div (-3) + 3\{2(9) - [-20]\}$$
$$= 27 \div (-3) + 3\{18 + [+20]\}$$
$$= 27 \div (-3) + 3\{38\}$$
$$= -9 + 114$$
$$= 105$$

33

CHECK POINT 5

Evaluate the following expressions.

1. $-24 \div [(3)(2)]$
2. $(3)\{5^2 + 3[4^2 - 2^2]\}$
3. $\{6 - 3^2(-4) + [2(5) - 2^3]\} \div (0 - 2)$
4. $[(-3)(-4)(0)] \div 2^2 + 6$

34

VARIABLE EXPRESSIONS

Introduction

In arithmetic we represent quantities explicitly by numbers such as 7, 0, and 25. In algebra we use, in addition to these numbers, various letters such as x, y, z, m, n, p to designate general quantities.

Since the symbol x is frequently used to represent a general quantity, we shall now indicate multiplication by the symbols "•" or "( )( )" instead of $\times$.

35

Suppose a meal at a restaurant costs you $8.00.

If you take people out for dinner to this restaurant, it would cost:
    (8)(1)   dollars for one person,
    (8)(3)   dollars for three persons,
    (8)(12)   dollars for twelve persons,
    (8)(x)   dollars for x persons.

36

CONSTANTS

A constant is a symbol which represents only one number. For example, in the expression (8)(x) above, the number 8 is a constant; that is, the price of the dinner does not vary.

VARIABLES

A variable is a symbol which can represent different numbers. For example, in the expression (8)(x) in the previous frame, the symbol x can represent such numbers as one, three, six, ten, twelve; that is, x can represent any number of persons.

Examples:

1. In the formula for calculating the area of a circle, $A = \pi \cdot r^2,$ the symbol "π" is a constant **approximately** equal to 3.14; the symbol "r" is a variable which can represent the lengths of the radii of different circles.

2. In the formula for calculating the perimeter of a rectangle, $P = 2 \cdot L + 2 \cdot W$ the symbol "2" is a constant; the symbols "L" and "W" are variables representing the lengths and widths of various rectangles.

3. If symbols such as a, b, and c represent only one specific number, then they are constants.

SYMBOLS INDICATING OPERATIONS

In algebra, we use basically the same symbols as in arithmetic to indicate the operations of addition, subtraction, multiplication and division. However, there is a notation in algebra indicating multiplication which is different from the notations we have seen so far.

ALGEBRAIC NOTATION INDICATING MULTIPLICATION
Examples:
1. 8x   indicates 8 is multiplied by x.

2. xy   indicates x is multiplied by y.

3. x(y + z)   indicates x is multiplied by the sum y + z.

4. mnt   indicates the **product** of the factors m, n and t.

VARIABLE EXPRESSIONS

If we combine variables and constants, using the operations of addition, subtraction and multiplication, we obtain variable expressions.

Examples:

1. x + 5	2. x + y	3. $2\pi r$
4. 8x	5. 2L + 2W	6. 5(y + 3t)

42

EVALUATING VARIABLE EXPRESSIONS

Depending on the value we assign to a variable or the variables, a variable expression will supply us with a specific value.

Examples:

If x = 5,   y = -2,   z = 12, then

1.   8x = (8)(5)   2.   xy = (5)(-2)   3.   x(x + z) = (5)[(-2) + 12]

  = 40   = -10   = 5(10)

                                                     = 50

4.   6xy + 2xz = 6(5)(-2) + 2(5)(12)

           = 30(-2) + 10(12)

           = -60 + 120

           = 60

43

Note:

When evaluating variable expressions such as 3x, xyz, and so on, we must indicate the intended multiplication by the use of symbols when replacing the variable by the assigned number.

Examples:

Multiplication Symbol

1.   If x = 2, then 3x = 3(2)   or   3 • 2

2.   If x = 4, y = 6, z = -2, then xyz = (4)(6)(-2)   or   4 • 6 • (-2)

44

AN AGREEMENT ON SUBSTITUTION

Evaluating an expression, such as 6xy + 2xz, where a variable occurs several times, we must substitute the **same** value wherever the variable occurs.

Examples:

1.   If x = -4, y = 3,   then   x(x + y) + 7xy = (-4)[(-4) + 3] + 7(-4)(3)

2.   If x = 13, y = 10,   then   x(x + y) + 7xy = (13)(13 + 10) + 7(13)(10)

45

CHECK POINT 6

If x = 5,   y = -3,   z = 2, evaluate the following expressions.

1. 2x + 4   2.   x - 3y   3.   xyz   4.   3xy + 5xz   5.   z(xz + 3y) + 6

DRILL EXERCISES

Find the sum of each of the given expressions.

1.   (-18) + (-7)   2.   (+36) + (+11)   3.   0 + (+17)

4.   (+8) + (-3)   5.   (-10) + (+1)   6.   (-15) + 0

7.   (-2) + (+7) + (-3) + (+5)   8.   (+11) + (-7) + (-16) + (+16) + (-11)

9.   (+7) + (-3) + (-9) + (-11) + (+6)

DRILL EXERCISES (continued)

Evaluate the following expressions.

10. $(+11) - (+30)$

11. $(+16) - (+10)$

12. $(-15) - (+5)$

13. $0 - (-7)$

14. $(+2) - 0$

15. $(+14) - (-14)$

16. $(-8) - (-12)$

17. $(-16) - (-9)$

18. $(-14) - (+7) - (-11) + (-14)$

19. $(-11) + (-3) - (+1) + (+9) - (-4)$

20. $(+2) + (-19) - (-20) - (+13) - (-12)$

21. $6 - 2 + 3 - 10 + 3$

22. $11 - 14 + 6 - 9 + 2$

Multiply as indicated.

23. $(+6)(-9)$

24. $(-13)(0)$

25. $(-5)(-7)$

26. $(-8)(+4)$

27. $(+7)(+6)$

28. $(+33)(-1)$

29. $(-4)(+1)$

30. $(-7)(-2)(-3)$

31. $(-4)(+6)(-2)(+3)$

32. $(+4)(-1)(0)(+6)$

33. $(-1)(+2)(-1)(-3)(+4)(-2)$

Evaluate the following exponential expressions.

34. $(-2)^4$

35. $(-4)^4$

36. -3^2

37. $(-1)^{251}$

38. -5^2

Divide the following.

39. $(+20) \div (-4)$

40. $\dfrac{+17}{-1}$

41. $\dfrac{-100}{-10}$

42. $\dfrac{-120}{0}$

43. $(-60) \div (-5)$

Evaluate the following expressions.

44. $(-22)[(-2)^4 \div 4]$

45. $2\{4[5^2 - 3^2] + (-4)^3\}$

46. $\{(-5) + 2^2(-3) - 10^2 \div 5 + (-3)] \div (2^3 - 4^2)$

47. $[-4^2 \div 2 + (2)(6) - (-6)] \div (-1) \cdot (2)$

If $x = 4$, $y = 6$, and $z = -2$, evaluate the following expressions.

48. $2z + 6$

49. $-x^2yz$

50. $x - 4z$

51. $5xz^2 + 2yz$

52. $y(xy + 2z) - 3$

ASSIGNMENT EXERCISES

Find the sum of each of the given expressions.

1.   (+23) + (+8)   2.   (−50) + (−14)

3.   (+35) + (−30)   4.   (+11) + (−27)

5.   0 + (−17)   6.   (+24) + 0

7.   (+9) + (−7) + (+13) + (−7)   8.   (+4) + (−17) + (+21) + (+17) + (−21)

9.   (−2) + (−4) + (+7) + (−9) + (+6)

Evaluate the following expressions.

10.   (−14) − (−17)   11.   (−7) − (+11)

12.   0 − (+4)   13.   (+6) − (+10)

14.   (+17) − (−17)   15.   (−13) − (−6)

16.   (+17) − (−4) − (+17) + (−15)   17.   (−21) − (+6) + (−4) − (−10) + (+25)

18.   (−8) + (+2) − (−5) − (−6) + (−5)

19.   12 − 10 + 8 + 2 − 6 − 7   20.   1 − 12 − 4 + 6 + 7 − 3

Multiply as indicated.

21.   (−3)(−7)   22.   (0)(−25)

23.   (−7)(+9)   24.   (−1)(+29)

25.   (+1)(−16)   26.   (−8)(−9)

27.   (+6)(−10)   28.   (+3)(−2)(−4)(+2)

29.   (+9)(+2)(−3)   30.   (−2)(−4)(−5)(−1)(+2)

31.   (+2)(+10)(−3)(+1)(−2)(−5)

Evaluate the following exponential expressions.

32. -7^2 33. $(-1)^{168}$ 34. $(-3)^5$ 35. $(-2)^4$ 36. -2^6

Divide the following.

37.   (−42) ÷ (−6)   38. $\dfrac{0}{-7}$ 39.   (−63) ÷ (+7)

40. $\dfrac{-14}{+1}$ 41. $\dfrac{+20}{+5}$

Evaluate the following expressions.

42.   48 ÷ [(4)(−3)]

43. $2[(-7)^2 - 2^2] + (-4)^3 \div (16)(-3)$

44. $[(-7)^2 - 4^3 \div 2^3 - (-7)(-3)] \div (3^2 - 2^2)$

45. $[(-25) + (-4)(3) + 3^2 - 7] \div (-2 - 3)$

If   a = −1,   b = 3 and   c = −4, evaluate the following expressions.

46.   3b − 7   47. $-a^2bc$

48.   c − 5a   49.   6ab − 2ac

50. $a(bc - 6a)^2 + 4$

UNIT 3

EXPONENTIAL LAWS, SCIENTIFIC NOTATION

Objectives: After having worked through this unit, the student should be capable of:

1.   simplifying expressions by using the Laws of Exponents;
2.   evaluating exponential expressions;
3.   changing numbers in ordinary notation to their scientific notation and vice versa;
4.   using scientific notation to multiply and divide very large and small numbers;
5.   using the calculator to evaluate expressions involving very large and small numbers.

1

EXPONENTIAL NOTATION

You may recall that in a product the numbers being multiplied are called factors.

Examples:

1. $5 \cdot 16 \cdot 32$ 5, 16 and 32 are factors.

2. $4 \cdot 4 \cdot 4$ 4 is a factor used three times.

When the same number is used as a factor many times, we employ a shorter notation to indicate the sequence of multiplication.

$$4 \cdot 4 \cdot 4 = 4^3$$

In the expression 4^3 the factor 4 is called the **base**. The number of times the base is used as a factor is called the **exponent** or the **power**.

Note:

The word power is also used to refer to the whole exponential expression.

2

EXPONENTIAL EXPRESSIONS

When we use a variable or a variable expression several times as a factor, we use exponential notation to indicate the sequence of multiplication. Variable expressions written in exponential notation are called exponential expressions.

Examples:

1. $\overbrace{x \cdot x \cdot x \cdot x \cdot x \cdot x}^{\text{Product}} = x^6$;   Note: $6x = \overbrace{x + x + x + x + x + x}^{\text{Sum}}$

$\therefore 6x \neq x^6$.

2. $(x + y)(x + y)(x + y) = (x + y)^3$ Here the base is $x + y$ and the exponent is 3.

3

Note:

We normally do not write the exponent "1".
When no exponent is indicated, we understand the exponent to be "1".

Examples:

1. x means x^1 and $-x$ means $-x^1$

2. $y + 7$ means $(y + 7)^1$

3. $x^2 + xy$ means $(x^2 + xy)^1$

4

LAWS OF EXPONENTS

The ability to work with exponential expressions is very important; particularly for people in technology and the sciences.

Hence, in the following frames, we shall review the properties of exponential expressions which are often referred to as the "Laws of Exponents".

These Laws will help us to combine and simplify exponential expressions. You should note that these Laws apply only when we **multiply** or **divide** exponential expressions.

5

MULTIPLYING EXPONENTIAL EXPRESSIONS

Let us look at some examples.

1. $x^5 \cdot x^6 = x \cdot x \cdot x \cdot x \cdot x \cdot x \cdot x \cdot x \cdot x \cdot x \cdot x = x^{11}$

Short Cut: add exponents $\quad x^5 \cdot x^6 = x^{5+6} = x^{11}$

2. $6^3 \cdot 8^2 = 6 \cdot 6 \cdot 6 \cdot 8 \cdot 8 \neq (6 \cdot 8)^5 = (6 \cdot 8)(6 \cdot 8)(6 \cdot 8)(6 \cdot 8)(6 \cdot 8)$

$\qquad$ 3 factors $\quad$ 2 factors
$\qquad$ of "6" $\qquad$ of "8"

Here we **cannot** add the exponents because the factors (bases) are **not the same**.

6

MULTIPLICATION LAW

We can see from the above examples that when we **multiply** exponential expressions which have the **same base**, we can simplify by **adding the exponents**. In general, if x is any number and m and n are positive integers, then:

3.1 $\quad \boxed{x^m \cdot x^n = x^{m+n}}$

7

Examples:

1. $x^{12} \cdot x^{26} = x^{12+26} = x^{38}$ $\qquad$ The base is x.

2. $(y + 4)^2(y + 4) = (y + 4)^{2+1} = (y + 4)^3$ $\qquad$ The base is (y + 4).

3. $(2ab)^4(2ab)^3 = (2ab)^{4+3} = (2ab)^7$ $\qquad$ The base is 2ab.

The law **3.1** above only applies when we **multiply** exponential expressions which have the **same base**.

Beware: $x^2 + x^4 \neq x^6$, since $x^2 + x^4 = x \cdot x + x \cdot x \cdot x \cdot x$, we do **not** have six **factors** of x.

8

DIVIDING EXPONENTIAL EXPRESSIONS

Depending on the exponential expressions involved, we may be able to simplify expressions which indicate division.

Examples: $\qquad\qquad\qquad$ **Short Cut**: Subtract exponents

1. $\dfrac{y^6}{y^4} = \dfrac{\cancel{y} \cdot \cancel{y} \cdot \cancel{y} \cdot \cancel{y} \cdot y \cdot y}{\cancel{y} \cdot \cancel{y} \cdot \cancel{y} \cdot \cancel{y}} = y^2$ $\qquad$ $\dfrac{y^6}{y^4} = y^{6-4} = y^2$

2. $\dfrac{y^3}{x^2} = \dfrac{y \cdot y \cdot y}{x \cdot x}$ $\qquad$ Since the bases are **not the same**, the expression cannot be simplified.

Expressions such as $y^4 - y^3$, $(ab)^2 + (ab)$ cannot be combined using the Laws of Exponents.

9

DIVISION LAW

The above examples show that we can simplify the division of exponential expressions by **subtracting exponents** if they have the **same** base.

In general, if x is any number **except zero**, and m and n are positive integers, then:

3.2
$$\frac{x^m}{x^n} = x^{m-n}; \quad x \neq 0$$

10

Examples:

1. $\dfrac{10^{25}}{10^{20}} = 10^{25-20} = 10^5$

2. $\dfrac{x^7}{x^5} = x^{7-5} = x^2$

3. $(2ab)^4 \div (2ab)^3 = (2ab)^{4-3} = (2ab)$

4. $(2x - 7)^7 \div (2x - 7)^2 = (2x - 7)^{7-2} = (2x - 7)^5$

11

CHECK POINT 1

Use Exponential Laws to simplify, if possible, the following expressions. (Assume none of the divisors is zero.)

1. $10^3 \cdot 10^{21}$

2. $\dfrac{8^{12}}{8^5}$

3. $x^4 \cdot y^6$

4. $(5xy)^2(5xy)^7$

5. $a^7 \div a^5$

6. $(y + 2)^3 \div (y - 1)^2$

7. $(x + 2)(x + 2)^5$

8. $a^2 + a^4$

9. $(2a + 3)^5 \div (2a + 3)^4$

10. $x^8 - x^3$

11. $(3x)^9 \div (3x)^4$

12

RAISING EXPONENTIAL EXPRESSIONS TO AN EXPONENT

Example:

$$(3^2)^4 = \overbrace{3^2 \cdot 3^2 \cdot 3^2 \cdot 3^2}^{2 \text{ four times}} = 3 \cdot 3 \cdot 3 \cdot 3 \cdot 3 \cdot 3 \cdot 3 \cdot 3 = 3^8$$

Short Cut: multiply exponents $\qquad (3^2)^4 = 3^{2 \cdot 4} = 3^8$

Exponential expressions raised to an exponent can be simplified by multiplying the exponents.

In general, if x is any number **except zero**, and n and k are positive integers, then

3.3 $\boxed{(x^n)^k = x^{nk}}$

Examples:

1. $(10^2)^8 = 10^{2 \cdot 8} = 10^{16}$

2. $(a^4)^6 = a^{4 \cdot 6} = a^{24}$

EXAMPLES INVOLVING PRODUCTS OF EXPONENTIAL EXPRESSIONS.

The Exponential Law 3.3 can be extended to products and quotients of exponential expressions.

1. $(2^4 \cdot 3^5)^2 = (2^4 \cdot 3^5)(2^4 \cdot 3^5) = 2^4 \cdot 2^4 \cdot 3^5 \cdot 3^5 = 2^8 \cdot 3^{10}$

Short Cut: multiply exponents $(2^4 \cdot 3^5)^2 = 2^{4 \cdot 2} \cdot 3^{5 \cdot 2} = 2^8 \cdot 3^{10}$

2. $(2ab)^3 = (2ab)(2ab)(2ab) = 2 \cdot a \cdot b \cdot 2 \cdot a \cdot b \cdot 2 \cdot a \cdot b = 2^3 a^3 b^3$

Short Cut: multiply exponents $(2ab)^3 = 2^{1 \cdot 3} \cdot a^{1 \cdot 3} \cdot b^{1 \cdot 3} = 2^3 a^3 b^3$

Remember: $2 = 2^1$, $a = a^1$, $b = b^1$.

In general, if n, m and k are positive integers, then:

3.4 $\boxed{(x^n y^m)^k = x^{nk} y^{mk}}$

Examples:

1. $(4x^6 y^3)^5 = 4^{1 \cdot 5} \cdot x^{6 \cdot 5} \cdot y^{3 \cdot 5} = 4^5 x^{30} y^{15}$

2. $(nm^3 t^3)^2 = n^2 m^6 t^6$

QUOTIENTS RAISED TO AN EXPONENT

Examples: $\left(\dfrac{x^2}{y}\right)^3 = \left(\dfrac{x^2}{y}\right)\left(\dfrac{x^2}{y}\right)\left(\dfrac{x^2}{y}\right) = \dfrac{x^2 \cdot x^2 \cdot x^2}{y \cdot y \cdot y} = \dfrac{x^6}{y^3}$

Short Cut: multiply exponents $\left(\dfrac{x^2}{y}\right)^3 = \dfrac{x^{2 \cdot 3}}{y^{1 \cdot 3}} = \dfrac{x^6}{y^3}$

In general, if n, m and k are positive integers, then:

3.5 $\boxed{\left(\dfrac{x^n}{y^m}\right)^k = \dfrac{x^{n \cdot k}}{y^{m \cdot k}} , \ y \neq 0}$

Examples:

1. $\left(\dfrac{1}{5}\right)^3 = \dfrac{1^{1 \cdot 3}}{5^{1 \cdot 3}} = \dfrac{1^3}{5^3} = \dfrac{1}{125}$

2. $\left(\dfrac{3xy^4}{2z^3}\right)^2 = \dfrac{3^2 x^2 y^8}{2^2 z^6} = \dfrac{9x^2 y^8}{4z^6}$

CHECK POINT 2

Use the exponential laws to simplify the following expressions. Assume none of the denominators are zero.

1. $(10^3)^4$
2. $\left(\dfrac{1}{2}\right)^5$
3. $(3x)^4$

4. $\left(\dfrac{a^2}{b^4}\right)^3$
5. $\left(\dfrac{x^2}{3y}\right)^3$
6. $\left(\dfrac{2n^3 m}{p^2 t^2}\right)^2$

ZERO POWER

If we use the division law $\dfrac{x^m}{x^n} = x^{m-n}$ and set $m = n$, we obtain a zero power or exponent.

Examples:

1. $\dfrac{10^3}{10^3} = 10^{3-3} = 10^0$, and $\dfrac{10^3}{10^3} = \dfrac{\overset{1}{\cancel{10}} \cdot \overset{1}{\cancel{10}} \cdot \overset{1}{\cancel{10}}}{\underset{1}{\cancel{10}} \cdot \underset{1}{\cancel{10}} \cdot \underset{1}{\cancel{10}}} = 1$

2. $x^8 \div x^8 = x^{8-8} = x^0$, and $\dfrac{x^8}{x^8} = \dfrac{\overset{1}{\cancel{x}} \cdot \overset{1}{\cancel{x}} \cdot \overset{1}{\cancel{x}} \cdot \overset{1}{\cancel{x}} \cdot \overset{1}{\cancel{x}} \cdot \overset{1}{\cancel{x}} \cdot \overset{1}{\cancel{x}} \cdot \overset{1}{\cancel{x}}}{\underset{1}{\cancel{x}} \cdot \underset{1}{\cancel{x}} \cdot \underset{1}{\cancel{x}} \cdot \underset{1}{\cancel{x}} \cdot \underset{1}{\cancel{x}} \cdot \underset{1}{\cancel{x}} \cdot \underset{1}{\cancel{x}} \cdot \underset{1}{\cancel{x}}} = 1$

3. $\dfrac{(3ab)^2}{(3ab)^2} = (3ab)^{2-2} = (3ab)^0$, and $\dfrac{(3ab)^2}{(3ab)^2} = \dfrac{\overset{1}{\cancel{(3ab)}}\,\overset{1}{\cancel{(3ab)}}}{\underset{1}{\cancel{(3ab)}}\,\underset{1}{\cancel{(3ab)}}} = 1$

We can see that in each case above when the exponent is zero, the expression is equal to 1.

In general, for any number x **except zero,**

3.6 $\boxed{x^0 = 1}$

Note: $x \neq 0$, since $x^0 = \dfrac{x^n}{x^n}$, remember, **never divide by zero.**

Examples:

1. $(3y)^0 = 1$

2. $10(xy + 2z)^0 = 10 \cdot 1 = 10$

3. $3y^0 = 3 \cdot 1 = 3$ Remember, the exponent o **applies only** to the base y.

CHECK POINT 3

Simplify the following expressions, if possible.

1. 10^0
2. $7x^0$
3. $(3x)^0$
4. $x^3(5x^2 + 3y^4 z)^0$

NEGATIVE EXPONENTS

If we use the division property $\dfrac{x^m}{x^n} = x^{m-n}$ where n is larger than m, we obtain a negative exponent.

Examples:

1. $\dfrac{10^4}{10^6} = 10^{4-6} = 10^{-2}$, and $\dfrac{10^4}{10^6} = \dfrac{\overset{1}{\cancel{10}} \cdot \overset{1}{\cancel{10}} \cdot \overset{1}{\cancel{10}} \cdot \overset{1}{\cancel{10}}}{\underset{1}{\cancel{10}} \cdot \underset{1}{\cancel{10}} \cdot \underset{1}{\cancel{10}} \cdot \underset{1}{\cancel{10}} \cdot 10 \cdot 10} = \dfrac{1}{10^2}$

   Hence, $10^{-2} = \dfrac{1}{10^2}$.

2. $\dfrac{x^2}{x^5} = x^{2-5} = x^{-3}$, and $\dfrac{x^2}{x^5} = \dfrac{\overset{1}{\cancel{x}} \cdot \overset{1}{\cancel{x}}}{\underset{1}{\cancel{x}} \cdot \underset{1}{\cancel{x}} \cdot x \cdot x \cdot x} = \dfrac{1}{x^3}$

   Hence, $x^{-3} = \dfrac{1}{x^3}$.

Note: $\dfrac{1}{5^{-3}} = \dfrac{1}{\frac{1}{5^3}} = 1 \times \dfrac{5^3}{1} = 5^3$. Hence, $\dfrac{1}{5^{-3}} = 5^3$.

In general, if x is any number **except zero** and n is an integer, then

3.7 $\boxed{x^{-n} = \dfrac{1}{x^n} \text{ and } \dfrac{1}{x^{-n}} = x^n}$

EVALUATING EXPRESSIONS WITH NEGATIVE EXPONENTS

When evaluating an expression with negative exponents, convert the expression so that it has positive exponents.

Examples:

1. $2^{-5} = \dfrac{1}{2^5} = \dfrac{1}{32}$ 2. $10^{-2} = \dfrac{1}{10^2} = \dfrac{1}{100}$

Note:

The properties of exponents also apply to negative exponents. Normally, we use the exponential properties **first** to simplify expressions, then we convert negative exponents to positive ones.

$\left(\dfrac{2^3}{5}\right)^{-2} = \dfrac{2^{-6}}{5^{-2}} = \dfrac{1}{2^6} \cdot 5^2 = \dfrac{25}{64}$ $\left(\dfrac{1}{2}\right)^{-6} = \dfrac{1^{-6}}{2^{-6}} = \dfrac{1}{1^6} \cdot 2^6 = 64$

CHECK POINT 4

Evaluate the following expressions.

1. 2^{-4} 2. 10^{-3} 3. $\left(\dfrac{1}{2^2}\right)^{-2}$ 4. $\left(\dfrac{2}{3}\right)^{-3}$

CHANGING NEGATIVE EXPONENTS TO POSITIVE EXPONENTS

When simplifying exponential expressions which cannot be evaluated, the answers are normally written with positive exponents.

Examples:

1. $\dfrac{x^2}{x^9} = x^{2-9} = x^{-7} = \dfrac{1}{x^7}$

2. $a^3 b^{-2} = a^3 \cdot \dfrac{1}{b^2} = \dfrac{a^3}{b^2}$

3. $\dfrac{m^3}{n^{-7}} = m^3 \cdot \dfrac{1}{n^{-7}} = m^3 n^7$

4. $2x^0 y^4 z^{-6} = 2 \cdot 1 \cdot y^4 \cdot \dfrac{1}{z^6} = \dfrac{2y^4}{z^6}$

5. $\dfrac{x^{-4}}{y^5} = x^{-4} \cdot \dfrac{1}{y^5} = \dfrac{1}{x^4} \cdot \dfrac{1}{y^5} = \dfrac{1}{x^4 y^5}$

CHECK POINT 5

Write the following expressions so that they have no negative or zero exponents. Assume none of the denominators is zero.

1. $x^2 y^{-3}$ 2. $\dfrac{5a^2}{b^{-4}}$ 3. $2m^3 n^{-4}$ 4. $\dfrac{a^{-5}}{b^6}$ 5. $7x^3 y^{-8} z^0$ 6. $9m^0 n^{-6} t^{-2}$

SHORTCUT: CHANGING NEGATIVE EXPONENTS TO POSITIVE EXPONENTS

Given the expression $\dfrac{x^{-2} \cdot y^3}{z^4}$, the negative exponent -2 of x in the numerator can be changed to positive 2 in the denominator by moving the factor x^{-2} from the numerator to the denominator.

This can be justified by multiplying $\dfrac{x^{-2} \cdot y^3}{z^4}$ by $\dfrac{x^2}{x^2}$ as follows

$$\dfrac{x^{-2} \cdot y^3}{z^4} = \dfrac{x^2}{x^2} \cdot \dfrac{x^{-2} \cdot y^3}{z^4} = \dfrac{x^0 y^3}{x^2 z^4} = \dfrac{1 \cdot y^3}{x^2 z^4}$$

Hence, $\dfrac{x^{-2} \cdot y^3}{z^4} = \dfrac{y^3}{x^2 z^4}$

Using a similar argument, we can change a negative exponent in the denominator to a positive exponent in the numerator by moving the factor with the negative exponent to the numerator.

$$\dfrac{b^3}{a^{-5} \cdot c} = \dfrac{a^5}{a^5} \cdot \dfrac{b^3}{a^{-5} \cdot c} = \dfrac{a^5 b^3}{a^0 c} = \dfrac{a^5 b^3}{1 \cdot c}$$

Hence, $\dfrac{b^3}{a^{-5} \cdot c} = \dfrac{a^5 \cdot b^3}{c}$

Example 1:

Write the following expressions so that they have no negative or zero exponents.

(a) $\dfrac{4x^{-3}y}{5^{-1}z^2} = \dfrac{(4)(5)y}{x^3z^2}$ Move the factors with negative exponents across the fraction bar.

$\qquad = \dfrac{20y}{x^3z^2}$

(b) $\dfrac{8(x^2-y)^{-2}}{z^{-3}} = \dfrac{8z^3}{(x^2-y)^2}$ **Note:** The exponent of 8 is not -2, but 1. $8 = 8^1$, the -2 exponent applies only to the base $(x^2 - y)$.

32

Example 2:

Write the expression $\left(\dfrac{-3a^{-2}b^3}{2c^{-4}}\right)^{-4}$ so that it has no negative or zero exponent.

Solution

Step 1. Use the exponential property to remove the brackets. Multiply the exponent of each factor by -4.

$$\left(\dfrac{-3a^{-2}b^3}{2c^{-4}}\right)^{-4} = \dfrac{(-3)^{-4}a^8b^{-12}}{2^{-4}c^{16}}$$

Step 2. Change negative exponents to positive exponents by moving factors with negative exponents across the fraction bar.

$$\dfrac{(-3)^{-4}a^8b^{-12}}{2^{-4}c^{16}} = \dfrac{2^4a^8}{(-3)^4b^{12}c^{16}} = \dfrac{16a^8}{81b^{12}c^{16}}$$

33

WORD OF CAUTION

Many students make unnecessary mistakes when using the exponential properties. Reading the following should help you avoid making some of the most common errors.

1. $3x^{-4} = \dfrac{3}{x^4}$ The exponent applies only to x, not to 3.

         Remember, $3 = 3^1$.   **No!** $\boxed{3x^{-4} = \dfrac{1}{3x^4}}$ Common Error.

$(3x)^{-4} = \dfrac{1}{(3x)^4} = \dfrac{1}{3^4x^4}$ The use of brackets indicates that now the exponent applies to 3x.

34

2. Be especially careful when dealing with grouping symbols, additive terms and exponents.

For example:

(a) $(1 + 2)^2 = 3^2 = 9$ (Recall order of operation, Unit 2, frame 35)

but $1^2 + 2^2 = 1 + 4 = 5$

Hence, **No!** $\boxed{(1 + 2)^2 = 1^2 + 2^2}$ **Similarly** **No!** $\boxed{(x + y)^a = x^a + y^a}$

          Common Error   Common Error

(b) $(a - b)^{-3} = \dfrac{1}{(a-b)^3}$ **No!** $\boxed{(a - b)^{-3} = a^{-3} - b^{-3}}$

                                             Common Error

3. $(5x^2)^3 = (5^1x^2)^3 = 5^3x^6$

Remember when no exponent is shown, we understand the exponent to be "1".

No! $\boxed{(5x^2)^3 = 5x^6}$ **Common Error.**

In general, when using the exponential properties to simplify expressions keep in mind what the exponent means.

Visualize: $(5x^2)^3 = 5x^2 \cdot 5x^2 \cdot 5x^2 = 5^3x^6$

$$m^5 \div m^2 = \frac{\cancel{m} \cdot \cancel{m} \cdot m \cdot m \cdot m}{\underset{1}{\cancel{m}} \cdot \underset{1}{\cancel{m}}} = m^{5-2} = m^3$$

CHECK POINT 6

Write the following expressions so that they have no negative or zero exponents. Assume none of the denominators is zero.

1. $\dfrac{5x^{-3}}{yz^2}$ 2. $9(a - b)^{-4}$ 3. $\dfrac{2^{-3}m^2n^{-1}}{3^{-1}t^3}$ 4. $\left(\dfrac{-7x^2y^{-3}}{2^{-3}z^4}\right)^2$ 5. $\left(\dfrac{3^{-1}r^2t^{-3}}{-2sz^{-1}}\right)^{-3}$

SIMPLIFYING AND EVALUATING EXPONENTIAL EXPRESSIONS

Before evaluating exponential expressions, we normally write them in their simplest form without negative exponents. To do this, we use the Laws of Exponents discussed previously.

Example 1:
Evaluate $(3x^{-2})(4x^7)$ when $x = 2$.

Solution:

Step 1. Simplify, using the Laws of Exponents.

$(3x^{-2})(4x^7) = 12x^{-2+7} = 12x^5$ (Multiplication Law **3.1**)

Step 2. Substitute $x = 2$ and evaluate.

$12x^5 = 12(2)^5 = 12(32) = 384$

Example 2:

Evaluate $\dfrac{(m^3n^{-5})^0m^2}{m^6n^{-2}}$ when $m = 3$, $n = 8$.

Solution:

Step 1. Simplify, using the Laws of Exponents.

$$\frac{(m^3n^{-5})^0m^2}{m^6n^{-2}} = \frac{1 \cdot m^2}{m^6n^{-2}} = \frac{n^2}{m^4}$$ (Use Laws **3.6**, **3.7** and **3.2**)

Step 2. Substitute $m = 3$, $n = 8$ and evaluate. $\dfrac{n^2}{m^4} = \dfrac{8^2}{3^4} = \dfrac{64}{81}$

Example 3:
Evaluate $4(a^2b^{-4}c)^3$ when $a = 2$, $b = -1$, $c = 10$.

Solution:

Step 1. Simplify $4(a^2b^{-4}c)^3 = 4a^6b^{-12}c^3 = \dfrac{4a^6c^3}{b^{12}}$ (Use Laws **3.4** and **3.7**)

Step 2. Substitute $a = 2$, $b = -1$, $c = 10$ and evaluate.

$\dfrac{4a^6c^3}{b^{12}} = \dfrac{4(2)^6(10)^3}{(-1)^{12}} = \dfrac{4(64)(1000)}{1} = 256\ 000$

40

Example 4:

Evaluate $\left(\dfrac{4a^{-1}}{3b^3}\right)^{-2}$ when $a = 4$, $b = 2$.

Solution :

Step 1. Simplify $\left(\dfrac{4a^{-1}}{3b^3}\right)^{-2} = \dfrac{4^{-2}a^2}{3^{-2}b^{-6}} = \dfrac{3^2a^2b^6}{4^2}$ (Use Laws **3.5** and **3.7**)

Step 2. Substitute $a = 4$, $b = 2$ and evaluate.

$\dfrac{3^2a^2b^6}{4^2} = \dfrac{3^2 \cdot 4^2 \cdot 2^6}{4^2} = 9 \times 64 = 576$

41

CHECK POINT 7

1. Evaluate $(2a^7)(5a^{-10})$ when $a = 3$.

2. Evaluate $x^0(x^{-2}y^3z)^2$ when $x = 3$, $y = 2$, $z = 9$.

3. Evaluate $\left(\dfrac{m^{-1}n^2}{2r^5}\right)^{-3}$ when $m = 3$, $n = 2$, $r = 1$.

4. Evaluate $2x(x^{-3}y^2)^{-1}(3x^{-4}y^3z^{-1})^0$ when $x = 3$, $y = (-5)$, $z = 6$.

5. Evaluate $\left(\dfrac{x^{-2}y}{y^{-3}x^{-5}}\right)^{-2}$ when $x = (-1)$, $y = (-2)$.

6. Evaluate $\left(\dfrac{3r^2t^{-1}}{r^{-3}t^4}\right)\left(\dfrac{4r^4}{9t^7}\right)^{-1}$ when $r = (-7)$, $t = 4$.

42

USING NEGATIVE EXPONENTS TO WRITE A QUOTIENT AS A PRODUCT

In some situations it may be desirable to use negative exponents to write a quotient, which involves exponential expressions, as a product.

Examples:

1. $\dfrac{b}{10^5}$ may be written as $b \times 10^{-5}$.

2. $\dfrac{p}{(1 + i)^n}$ may be written as $p(1 + i)^{-n}$.

3. $\dfrac{MF^2Q^3}{T^2L^3}$ may be written as $MF^2Q^3T^{-2}L^{-3}$.

4. $\dfrac{N}{e^{kt}}$ may be written as Ne^{-kt}.

35

43

CHECK POINT 8

Write the following fractions in product form using negative exponents.

1. $\dfrac{2.56}{10^3}$
2. $\dfrac{I}{R^2}$
3. $\dfrac{g}{4\pi^2 f^2}$

44

WORKING WITH VERY LARGE AND VERY SMALL NUMBERS

Knowing how to work with exponents is very helpful when working with very large and small numbers.

For example, to find the product $3\ 296\ 000 \times 9\ 170\ 000$ or the quotient $81\ 400\ 000 \div 0.000\ 642$ would be awkward without a calculator and impossible with some calculators unless we changed the numbers to Scientific Notation.

45

SCIENTIFIC NOTATION

We say that a number is in scientific notation when it is written as a product of a number between 1 and 10 and the appropriate power of ten.

Examples:

Number		Number in Scientific Notation
123	=	1.23×10^2
0.008	=	8×10^{-3}
0.000 041	=	4.1×10^{-5}
2.84	=	2.84×10^0

46

CHANGING ORDINARY NUMBERS TO THEIR SCIENTIFIC NOTATION

Method:

Step 1. Write the original number as a number between 1 and 10 by moving the decimal point.

Step 2. Multiply the number in Step 1 by 10 raised to an exponent which we determine as follows:

    (a) If the original number is 1 or larger, the exponent is positive and is equal to the number of places we moved the decimal point.

    (b) If the original number is less than 1, the exponent is negative and is equal to the number of places we moved the decimal point.

47

Example 1:

Change 673.5 to scientific notation.

Solution:

Step 1. Writing 673.5 as a number between 1 and 10, we have **6.735**.

Step 2. Multiply by the appropriate power of ten.

$$673.5 = 6.735 \times 10^2$$

2 places

Original number is larger than 1. Hence, the exponent is positive.

48

Example 2:
Change   0.000 075   to scientific notation.

Solution:
Step 1.   Writing   0.000 075   as a number between   1 and 10, we have **7.5.**

Step 2.   Multiply by the appropriate power of ten.

$0.000\ 07\ 5 = 7.5 \times 10^{-5}$ Original number is less than 1.
Hence, the exponent is negative.
5 places

49

Example 3:

$9\ 170\ 000 = 9\ 170\ 000 = 9.17 \times 10^{6}$
6 places

Note:
We normally discard zeros between the decimal point of the original **whole** number and the first non-zero digit.

50

CHANGING NUMBERS IN SCIENTIFIC NOTATION TO ORDINARY NOTATION
When we change numbers from scientific notation to ordinary notation, we proceed as follows.

1.   A positive exponent of 10 indicates that we are to **multiply** by that power of 10, thus the result will be a larger number. Hence, we move the decimal point to the right by the same number as the exponent.

Example:
Change 4.25×10^{5} to ordinary notation.

Solution:
$4.25 \times 10^{5} = 425\ 000$
5 places

51

2.   A negative exponent of 10 indicates that we are to **divide** by the corresponding positive power of 10, thus, the result will be a smaller number. Hence, we move the decimal point to the left by the same number as the exponent.

Example:
Change 1.478×10^{-3} to ordinary notation.

Solution:
$1.478 \times 10^{-3} = 0.001\ 478$
3 places

52

CHECK POINT 9
I   Change each of the following to scientific notation.

1.   801.43   2.   0.000 91   3.   7.21   4.   81 400 000

II   Change each of the following to ordinary notation.

5. 3.045×10^{-2} 6. 1.09×10^{4} 7. 5×10^{-6} 8. 7×10^{3}

COMPUTATIONS WITH NUMBERS IN SCIENTIFIC NOTATION

Computations with very large and small numbers is considerably simplified by changing these numbers to scientific notation.

Example 1:

Find the product 150 000 × 4 000 000.

Solution:

Step 1.   Convert numbers to scientific notation.

$$150\ 000 \times 4\ 000\ 000 = (1.5 \times 10^5) \times (4 \times 10^6)$$

Step 2.   Group the powers of ten and multiply using the properties of exponents.

$$(1.5 \times 10^5) \times (4 \times 10^6) = 1.5 \times 4 \times 10^5 \times 10^6$$

$$= 6 \times 10^{11}$$

$$= 600\ 000\ 000\ 000$$

Example 2:

Find the quotient $\dfrac{0.000\ 000\ 48}{0.000\ 12}$.

Solution:

Step 1.   Convert numbers to scientific notation.

$$\frac{0.000\ 000\ 48}{0.000\ 12} = \frac{4.8 \times 10^{-7}}{1.2 \times 10^{-4}}$$

Step 2.   Divide, using the properties of exponents.

$$\frac{4.8 \times 10^{-7}}{1.2 \times 10^{-4}} = \frac{4.8}{1.2} \times 10^{-7+4}$$

$$= 4 \times 10^{-3}$$

$$= 0.004$$

USING CALCULATORS AND SCIENTIFIC NOTATION

Generally, most scientific calculators have the capacity to compute directly with numbers in scientific notation.

For this purpose, the calculator has a button with either one of the following symbols:

$$\boxed{\text{EE}}, \quad \boxed{\text{EXP}} \quad \text{or} \quad \boxed{\text{EEX}}.$$

In the following discussion, we shall use the symbol $\boxed{\text{EE}}$.

Example 1:

Enter 1.275×10^7 into a scientific calculator.

Solution:

Follow the given sequence.

Enter: 1.275

Press: EE

Enter: 7 (the exponent)

Display: $\boxed{1.275 \quad 07}$ or $\boxed{1.275 \times 10^{07}}$

57

Example 2:
Enter 9.567×10^{-5} into a scientific calculator.

Solution:
Enter: 9.567

Press: EE

Enter: 5 (the exponent without the sign)

Press: +/−   (changes 5 to −5)

Display: $\boxed{9.567 \quad -05}$ or $\boxed{9.567 \times 10^{-05}}$

58

COMPUTATIONS WITH NUMBERS IN SCIENTIFIC NOTATION USING SCIENTIFIC CALCULATORS

Example 1:
Evaluate $(5.25 \times 10^{12})(1.12 \times 10^{-3})$.

Solution:
Enter: (5.25×10^{12}) ⟶ 5.25   EE   12

Display: $\boxed{5.25 \quad 12}$

Press: x

Enter: 1.12×10^{-3} ⟶ 1.12   EE   3   +/−

Display: $\boxed{1.12 \quad -03}$

Press: =

Display: $\boxed{5.88 \quad 09}$

Hence, $(5.25 \times 10^{12})(1.12 \times 10^{-3}) = 5.88 \times 10^{9}$.

59

Example 2:
Evaluate $(9.341 \times 10^{-7}) \div (3.059 \times 10^{13})$.

Solution:
Enter: (9.341×10^{-7}) ⟶ 9.341   EE   7   +/−

Display: $\boxed{9.341 \quad -07}$

Press: ÷

Enter: (3.059×10^{13}) ⟶ 3.059   EE   13

Display: $\boxed{3.059 \quad 13}$

Press: $\boxed{=}$

Display: $\boxed{3.053\ 612\ 3\quad -20}$ (floating decimal point)

Hence, $(9.341 \times 10^{-7}) \div (3.059 \times 10^{13}) = 3.053\ 612\ 3 \times 10^{-20}.$

60

CHECK POINT 10
Use scientific notation to perform the indicated operations.

1. $250\ 000 \times 30\ 000\ 000$

2. $\dfrac{0.0048}{0.000\ 016}$

3. $\dfrac{0.002 \times 0.000\ 015}{0.000\ 3 \times 0.04}$

4. $\dfrac{125\ 000 \times 400}{2\ 000\ 000 \times 10\ 000}$

DRILL EXERCISES

Simplify the following expressions by using the properties of exponents. Assume none of the denominators is zero.

1. 13^0

2. $a^7 \cdot a^{14}$

3. $(6y)^0$

4. $(2b)^5$

5. $y^{13} \div y^9$

6. $(x + y)^2 (x + y)^3$

7. $(2ab)^3 (2ab)^2$

8. $(3x + 5)^2 \div (3x + 5)$

9. $\dfrac{10^{11}}{10^9}$

10. $(x - 2)(x - 2)^6$

11. $\left(\dfrac{x^3}{a^6}\right)^5$

12. $\left(\dfrac{-1}{x}\right)^8$

13. $\left(\dfrac{3m^2 n}{2s^3 t^3}\right)^3$

14. $\dfrac{11^4 \cdot 11^{12}}{11^{14}}$

15. $(xy)^{15} \div (xy)^3$

16. $23y^0$

17. $(y - 3)^5 \div (y - 3)^2$

18. $(2a^2 b)^4 (3ab)^2$

19. $x^7 (3x^2 y^2 + 2z^2)^0$

Evaluate the following expressions.

20. 4^{-2}

21. $\left(\dfrac{1}{3^3}\right)^{-2}$

22. $\left(\dfrac{4}{5}\right)^{-3}$

Write the following expressions so that they have no negative or zero exponents.

23. $a^{-4} b^6$

24. $\dfrac{2^3 x^3}{2^5 y^{-6}}$

25. $3m^{-7} n^{-6}$

26. $\dfrac{p^{-3}}{q^2}$

27. $2^3 m^{-3} n^0 s^{-4}$

28. $5x^0 y^{-7} z^2$

29. $\dfrac{10^{-5} \cdot 10^{11}}{10^4}$

30. $\dfrac{10^{-12} \cdot 10^4}{10^{-9} \cdot 10^3}$

31. Evaluate $3(xy^{-2} z^{-3})^2$ when $x = 5$, $y = 3$, $z = (-1)$.

32. Evaluate $\dfrac{4m^{-5}}{7n^{-2}}$ when $m = 2$, $n = 3$.

DRILL EXERCISES (continued)

33. Evaluate $\dfrac{x^9 y^{-3}}{x^3 (x^4 y^{-2})^0}$ when $x = 2$, $y = 5$.

34. Evaluate $\left(\dfrac{2xy^2 z^{-1}}{3x^2 y^{-3}}\right)^{-2}$ when $x = 5$, $y = (-1)$, $z = (-3)$.

Write the following fractions in product form using negative exponents.

35. $\dfrac{3.74}{10^5}$ 　　　　 36. $\dfrac{KmM}{r^2}$ 　　　　 37. $\dfrac{V}{Re^{\frac{t}{RC}}}$

Change each of the following to scientific notation.

38. 3.75 　　　　 39. 0.000 04 　　　　 40. 0.31

41. 0.063 4 　　　　 42. 7 053 210 　　　　 43. 615.27

Change the following to ordinary notation.

44. 2.03×10^5 　　　　 45. 4×10^6 　　　　 46. 3×10^{-2}

47. 7.102×10^{-3} 　　　　 48. 5.2×10^7 　　　　 49. 9×10^{-5}

Use scientific notation to perform the indicated operations.

50. $5\ 300 \times 800\ 000$ 　　　　 51. $70\ 000 \times 0.000\ 000\ 003$

52. $\dfrac{0.000\ 93}{0.000\ 000\ 031}$ 　　　　 53. $\dfrac{0.003 \times 70\ 000}{5600 \times 0.002\ 7}$

54. $\dfrac{0.000\ 008\ 1 \times 600}{0.000\ 09 \times 1\ 200\ 000}$

Use a calculator to perform the indicated calculations.

55. $0.023\ 641 \times 50\ 800$ 　　　　 56. $2\ 314\ 000 \div 711\ 200$

57. $\dfrac{1500 \times 2.54}{645.16 \times 38\ 100\ 000}$ 　　　　 58. $\dfrac{0.005 \times 820}{128\ 000 \times 0.000\ 125}$

ASSIGNMENT EXERCISES

Simplify the following expressions by using the properties of exponents. Assume none of the denominators is zero.

1. $m^6 \cdot m^8$ 　　　 2. $(5x)^0$ 　　　 3. $13a^0$

4. $x^9 \div x^4$ 　　　 5. $(4x^3)^2$ 　　　 6. $(a + 4)^3 (a + 4)$

7. $(2xy)^4 (2xy)$ 　　　 8. $\left(\dfrac{y^3}{4x^4}\right)^3$ 　　　 9. $\left(\dfrac{-1}{2y}\right)^4$

10. $\dfrac{10^6}{10^3}$ 　　　 11. $a^3(2a^2 + 4bc^4)^0$ 　　　 12. $(2x^2 y)^3 (3xy)^2$

13. $(a - 2)^6 \div (a - 2)^2$ 　　　　 14. $(5c)^2 \div (5c)$

15. $\left(\dfrac{3x^5 y}{a^3 b^2}\right)^4$ 　　　　 16. $\dfrac{13^4 \cdot 13^{11}}{13^6 \cdot 13^7}$

ASSIGNMENT EXERCISES (continued)

Evaluate the following expressions.

17. 5^{-3}

18. $\left(\dfrac{1}{2}\right)^{-5}$

19. $\left(\dfrac{1}{2^3}\right)^{-2}$

20. $\dfrac{10^{10}}{10^3 \cdot 10^4}$

21. $\dfrac{10^{-4} \cdot 10^{-5}}{10^2 \cdot 10^{-10}}$

Write the following expressions so that they have no negative or zero exponents.

22. $x^{-3}y^{-2}$

23. $\dfrac{7^2 a^{-3}}{b^4}$

24. $\dfrac{x^7}{y^{-2}}$

25. $2^{-3}r^5 s^{-6}$

26. $3^2 x^0 y^2 z^{-3}$

27. $2a^{-1}b^0 c$

28. Evaluate $(3x^{-11})(7x^9)$ when $x = 5$.

29. Evaluate $(a^{-3}b^2 c)^2 (a^2 b^2)^0$ when $a = 4$, $b = 2$, $c = 7$.

30. Evaluate $\left(\dfrac{3r^3 s^{-1}}{t^{-4}}\right)^{-2}$ when $r = 2$, $s = 4$, $t = -1$.

31. Evaluate $(4a^{-4}b^{-1}c^0)(-3a^6 b^{-4}c^2)^{-1}$ when $a = (-2)$, $b = 4$, $c = 3$.

Write the following fractions in product form using negative exponents.

32. $\dfrac{x}{10^4}$

33. $\dfrac{\mu I}{2\pi d}$

34. $\dfrac{R}{\delta T^4}$

Change each of the following to scientific notation.

35. 215 040 000

36. 0.000 093

37. 75.042

38. 0.371

39. 0.005 76

40. 2.93

Change each of the following to ordinary notation.

41. 1.15×10^2

42. 8.001×10^{-4}

43. 1×10^{-2}

44. 3.6×10^9

45. 8×10^4

46. 2×10^{-7}

Use scientific notation to perform the indicated operations.

47. $4000 \times 2\ 500\ 000$

48. $60\ 000 \times 0.000\ 009$

49. $\dfrac{0.000\ 000\ 135}{0.004\ 500}$

50. $\dfrac{640 \times 270\ 000}{5\ 400\ 000 \times 100}$

51. $\dfrac{0.008 \times 21\ 000\ 000}{14\ 000 \times 3.20}$

Use a calculator to perform the indicated calculations.

52. $0.009\ 74 \div 0.000\ 038$

53. $249\ 100 \times 0.001\ 658$

54. $\dfrac{0.0437 \times 756\ 000}{0.002\ 913 \times 5276}$

55. $\dfrac{2\ 300\ 000 \times 0.006}{9800 \times 0.000\ 632}$

UNIT 4

ADDITION AND SUBTRACTION OF ALGEBRAIC EXPRESSIONS

Objectives: After having worked through this unit, the student should be capable of:

1. understanding the concepts of algebraic expressions, terms, like terms;
2. adding and subtracting like terms;
3. removing a grouping symbol which is preceded by a plus or a minus sign;
4. simplifying algebraic expressions by removing grouping symbols and collecting like terms;
5. vertically adding and subtracting algebraic expressions.

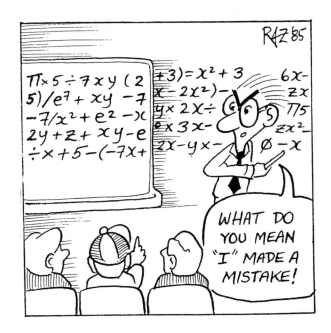

1

ALGEBRAIC EXPRESSIONS

Expressions such as

$$8x, \quad x + 9, \quad 5a - 3b, \quad 9(m + 2n), \quad 2x^2 - xy + 6$$

were introduced previously. We referred to them as variable expressions. They are also called algebraic expressions and are made up of constants, variables, symbols of operations, and grouping symbols. The most basic algebraic expression is called a term.

2

TERMS

A term is a **product** or **quotient** of numbers and variables.
A number or a variable by itself is also considered to be a term.

Examples:

1. $3x^2yz$ is a term.

2. x is a term.

3. $\dfrac{m^6 n^3}{5}$ is a term.

4. 7 is a term.

3

Complex algebraic expressions are made up of terms.

$3x^2 + 9xy + 21$ is an algebraic expression made up of the terms $3x^2$, $9xy$, and 21.

When an algebraic expression involves subtraction of terms, we have to rewrite the subtraction as an equivalent addition, in order to identify the terms.

Example:

$$a^2b^2 - 5a^2b - ab^2 + ab - 7 = a^2b^2 + (-5a^2b) + (-ab^2) + ab + (-7)$$

Hence, the terms of $a^2b^2 - 5a^2b - ab^2 + ab - 7$

are: a^2b^2, $-5a^2b$, $-ab^2$, ab and -7.

4

REARRANGING TERMS

It is sometimes necessary or desirable to rearrange the order in which terms appear in an expression.

For example, we may want to rearrange the terms in
$$5x - 7 + x^3 - 8x^2 - 6x^4$$
such that the exponents appear in descending order.

Since the commutative property does not apply to subtractions, we rewrite the subtractions as equivalent additions,
$$5x + (-7) + x^3 + (-8x^2) + (-6x^4).$$

Now, we can switch the terms in any order we like.

Hence, $5x + (-7) + x^3 + (-8x^2) + (-6x^4) = (-6x^4) + x^3 + (-8x^2) + 5x + (-7)$
$$= -6x^4 + x^3 - 8x^2 + 5x - 7.$$

5

CHECK POINT 1

I.   Identify the terms in the following algebraic expressions.

1. $5 - x^3 + 2x - 6x^2$ 2. $2a^3b^3 - ab - 3a^4b^4 + 12 - 8a^2b^2$

II.   Rearrange the terms in the above expressions so that the exponents appear in descending order.

6

NUMERICAL COEFFICIENTS

The number or the constant factor of the term is called the numerical coefficient .

Examples:

1.   "8" is the numerical coefficient of the term 8x.

2.   "-4" is the numerical coefficient of the term $-4x^6y^3$.

3.   "2" is the numerical coefficient of the term 2r.

4.   "π" is the numerical coefficient of the term πr^2.

7

Note:

Any constant term, such as   -5   in the expression

$$x^4 + 2x^3 - 7x^2 + 3x - 5$$

is considered to be the numerical coefficient of $-5x^0$.   ($x^0 = 1$, $x \neq 0$)

8

Note:

If a term has only variables as factors, the numerical coefficient is "1" or "-1".

Examples:

1.   The numerical coefficient of the terms x, ab^2, and y^3 is "1",

since $x = (1)(x)$, $ab^2 = (1)(ab^2)$, and $y^3 = (1)(y^3)$.

2.   The numerical coefficient of the terms $-y$, $-t^2z$, and $-a^3b^2c$ is "-1",

since $-y = (-1)(y)$, $-t^2z = (-1)t^2z$, and $-a^3b^2c = (-1)a^3b^2c$.

9

LIKE TERMS

Terms which differ only in their numerical coefficient and/or the order of their variables are called **like terms**.

10

Examples:

| **Like Terms** | **Not Like Terms** |

1. $8\underline{x}$ and $3\underline{x}$; $5\underline{x}$ and $5\underline{y}$

Identical. Variable factors are not identical.

2. $-\underline{x^2y}$ and $16\underline{x^2y}$; $6\underline{x^2y}$ and $3\underline{xy^2}$

Identical. Powers of variable factors are not identical.

3. $\pi\underline{xy^2z^3}$ and $-\underline{xy^2z^3}$; $-7\underline{xyz^3}$ and $9\underline{xyt^3}$

Variable factors and their powers are identical. Variable factors are not identical.

11

Note:

The variable factors of a term are normally written in alphabetical order.

Examples:

1. $5a^2b\,c^3$ 2. $-9mn^2t^3$ 3. xy^2z 4. p^2q^4r

12

CHECK POINT 2

I. Identify the numerical coefficients of each term in the following expressions.

    1. $3m^2t - 2mt^2 + 4m - n$ 2. $x^2 - xy - y^2$

II. Identify the like terms in the following expressions.

    3. $ab^2 - 2ab^3 + 6ab^2 - a^3b$

    4. $x^2 + 4x^2y - 3x^2y^2 - 3x^2 + xy^2 - 5x^2y$

    5. $7mn^2t^3 - 7m^2nt^3 + mn^2t^3 + 8mn^3t^2 - 2n^2t^3$

13

ADDITION AND SUBTRACTION OF LIKE TERMS

Only like terms of algebraic expressions can be combined by addition or subtraction.

Example:

Let us find the sum of $2x$ and $7x$.

If we reverse the distributive property $a(b + c) = ab + ac$

               so that $ab + ac = a(b + c)$

and use the commutative property of multiplication,

               we have: $ba + ca = (b + c)a$.

If $b = 2$, $a = x$, $c = 7$, then $2x + 7x = (2 + 7)x = 9x$

The distributive property can be extended to more than two terms.

Example: $-6xy + 14xy - 3xy = (-6 + 14 - 3)xy$

                        $= 5xy$

Hence, to combine like terms, we add the numerical coefficients of the like terms. This is similar to adding oranges, bananas and apples.

Example:

1. $8x + 3x = 11x$ is similar to: 8 oranges + 3 oranges = 11 oranges.

2. $9a - 3b - 5a + 10b = 4a + 7b$ is similar to:

    9 apples − 3 bananas − 5 apples + 10 bananas = 4 apples + 7 bananas.

Note:

$2a + 3b \neq 5ab$, just as:

2 apples + 3 bananas $\neq$ | 5 apples times bananas | .
                              meaningless

Thus, **unlike terms cannot be combined.**

When we simplify expressions with many terms, we underline or check the terms as we mentally add them to ensure that we don't overlook one.

Normally, we also **arrange the terms** so that:

1. the **variables** are in **alphabetical order** and/or
2. the **exponents** are in a **descending order**.

Example:

Simplify the following expression by combining like terms.

$2xy^2 + x^4y^2 + 5x^2y^2 - 3xy^2 + 4y + 3x^2y^2 - 6y$

Solution:

$$2xy^2 + x^4y^2 + 5x^2y^2 - 3xy^2 + 4y + 3x^2y^2 - 6y$$

$$= x^4y^2 + 2xy^2 + 5x^2y^2 - 3xy^2 + 4y + 3x^2y^2 - 6y$$

$$= x^4y^2 + 8x^2y^2 + 2xy^2 - 3xy^2 + 4y - 6y$$

$$= x^4y^2 + 8x^2y^2 - xy^2 + 4y - 6y$$

$$= x^4y^2 + 8x^2y^2 - xy^2 - 2y$$

MORE EXAMPLES

1. $-4b^2 + 3a - 2b + 8b^2 - 7a + 3a^2 = 3a^2 + 4b^2 - 4a - 2b$

2. $y - 6y^3 + 3y - 4y^2 + 2y^3 - 10y = -4y^3 - 4y^2 - 6y$

3. $m^3n^2 + 4m^2n^3 - pq^2 + t$ (This expression contains no like terms and hence cannot be simplified.)

CHECK POINT 3

Simplify the following expressions by combining like terms.

1. $-xy + 2x^2 + 10xy - y - x^2 + 3y$

2. $4a^3 - 3a^4 - 2a^3 + 5ab + 5a^4 - 3ab$

3. $2m^4nt^3 - 3mn + 6m^4nt^3 - 9mn - n + 2mn$

4. $6a + 2b + 3c$

RULES FOR REMOVING GROUPING SYMBOLS

In algebra we often encounter expressions where certain terms are grouped. In order to simplify these expressions by adding like terms, we have to remove the grouping symbols.

If a grouping symbol is preceded by a:

1. plus sign,   remove the grouping symbol and the sign preceding it, then write terms with their original signs;

2. minus sign,   change the sign of each term inside the grouping symbol to its **opposite**, then remove the grouping symbol and omit the minus sign preceding it.

REMOVING GROUPING SYMBOLS AND COLLECTING LIKE TERMS

Simplify the following expressions by removing the grouping symbols and collecting like terms.

Example:

$(-x^2 + 2xy) - (-y^2 + 2xy + 3x^2) + (4x^2 - 4xy + 2y^2)$

Solution:

$(-x^2 + 2xy) - (-y^2 + 2xy + 3x^2) + (4x^2 - 4xy + 2y^2)$

Step 1. $= -x^2 + 2xy + y^2 - 2xy - 3x^2 + 4x^2 - 4xy + 2y^2$

Step 2. $= 3y^2 - 4xy$

REMOVING A SEQUENCE OF GROUPING SYMBOLS INDICATING GROUPS WITHIN GROUPS

When removing grouping symbols indicating groups within groups, we remove the symbol of the **innermost group first**. Then, we remove the symbol of the next innermost group and so on.

Simplify the following expression by removing the grouping symbols and collecting like terms.

Example 1:

$4a - [2ab - (3a - 4ab) + a] - ab$

Solution:

$$4a - [2ab - (3a - 4ab) + a] - ab$$

Step 1. $= 4a - [2ab - 3a + 4ab + a] - ab$

Step 2. $= 4a - 2ab + 3a - 4ab - a - ab$

Step 3. $= 6a - 7ab$

24

Example 2:

$$-5x - \{4y - [(4x - 3y) - (6x - 2y)] - 3xy\}$$

Solution:

$$-5x - \{4y - [(4x - 3y) - (6x - 2y)] - 3xy\}$$

Step 1. $= -5x - \{4y - [4x - 3y - 6x + 2y] - 3xy\}$

Step 2. $= -5x - \{4y - 4x + 3y + 6x - 2y - 3xy\}$

Step 3. $= -5x - 4y + 4x - 3y - 6x + 2y + 3xy$

Step 4. $= -7x + 3xy - 5y$

25

CHECK POINT 4

Simplify the following expressions by removing the grouping symbols and collecting like terms.

1. $(y^2 - y) - (4y + 6)$

2. $-(a^2 - 4a + 1) + (3a^2 - a + 2) - (-a^2 + 3a)$

3. $4pq - [6p + (3pq - 4)]$

4. $m^2n^2 - [3m^2n - (-m + n - 1)]$

5. $x^2 - \{4x - [2x - (6x - 3)] - 3x^2 + 9\}$

26

VERTICAL ADDITION AND SUBTRACTION OF ALGEBRAIC EXPRESSIONS

If we arrange like terms in columns, we can add or subtract algebraic expressions vertically.

Example:

Add $(3x^2 - 5) + (-x + 6) + (5x^2 + 6x - 2)$

Solution:

Step 1.   Arrange the expressions vertically so that like terms are in the same column.

$$
\begin{array}{rrr}
3x^2 & & -\ 5 \\
& -\ x & +\ 6 \\
5x^2 & +\ 6x & -\ 2 \\
\end{array}
$$

Step 2.   Add the like terms:

$$
\begin{array}{rrr}
3x^2 & & -\ 5 \\
& -\ x & +\ 6 \\
5x^2 & +\ 6x & -\ 2 \\
\hline
8x^2 & +\ 5x & -\ 1 \\
\end{array}
$$

REVIEW: VERTICAL SUBTRACTION OF SIGNED NUMBERS

Example: Subtract: $\underbrace{\begin{array}{r} -6 \\ -9 \end{array}}_{\text{Change}}$ Add: $\begin{array}{r} -6 \\ +9 \end{array}$

We will use the same principle to vertically subtract algebraic expressions.

Example:
Subtract: $(7a^2 - 5a + 1) - (3a^2 - a + 4)$

Solution:

Step 1.   Arrange expressions vertically so that like terms are in the same column.

$$7a^2 - 5a + 1$$

Subtract: $3a^2 - a + 4$

Step 2.   **Change the sign of each term in the subtrahend** (expression to be subtracted) and then **add the like terms**.

$$\begin{array}{r} 7a^2 - 5a + 1 \\ \text{Add:} \quad -3a^2 + a - 4 \\ \hline 4a^2 - 4a - 3 \end{array}$$

CHECK POINT 5

Vertically add or subtract the following algebraic expressions.

1. $(a - 3) + (4a^2 + 6a) + (3a^2 - 9a - 4)$

2. $(3y^2 - 6y + 4) - (y^2 + 2y + 8)$

DRILL EXERCISES

I   Simplify the following expressions by combining like terms.

1. $2a^2 - 2b^2 - 18ab + 12b^2 - 2ab + 6a^2$

2. $x^3 + 3x^2 - 4x^2 - 12xy - 2x^3 + 9xy$

3. $3x^2yz + 2xz - 5x^2yz - x - 3xz$

4. $m + 3m^2 - 4p - 5p$

DRILL EXERCISES (continued)

II   Remove the grouping symbols in the following expressions.

5. $-(r + 3a)$

6. $(a^3b^3 + 3ab) + (-2b + a)$

7. $4x^2 + (2x - 7)$

8. $m^2 - (3m - 2n)$

9. $(x - 3xy) + (-x^2 + 7z - y)$

10. $p^3 - q^3 - (-p^2q^2 - pq + 2p - 7)$

III   Simplify the following expressions by removing the grouping symbols and collecting like terms.

11. $(a^3 + 2a^2) - (a^2 + 3a - 2)$

12. $-(-2x^2 - 3x + 2) - (x^2 + 2x) + (4x^2 + x - 3)$

13. $3mn - [-7n + (4mn - 2)]$

14. $-6x - [2x^2y - (-x^2 + y - 3)]$

15. $p^2 - \{4p - [-2p - (3p^2 - 4)] + 5p^2 - 8\}$

IV   Vertically add or subtract the following algebraic expressions.

16. $(2x^2 - 3x) + (x - 5) + (4x^2 + 5x + 6)$

17. $(2y^2 + 9y - 5) - (5y^2 - 6y - 7)$

18. $(a^3 + 6a^2 + 3) - (-a^3 + 4a - 1)$

19. $(a^4 - 5a^3 - 3a) + (2a^2 - 4a + 1) + (a^4 - a^2 - 2)$

20. $(3x^4 - 5x^2 - 2x + 3) - (x^3 - 2x^2 - 3)$

ASSIGNMENT EXERCISES

I   Simplify the following expressions by combining like terms.

1. $9mn + 6m^2 - 12mn + 3n^2 + m^2 - 4n^2$

2. $3a^5 - 2a^2 - 2a^5 + 7a^2b + a^2 - 4a^2b$

3. $x^2y^3 - xy^2 - xy^3 + x^3 - x^2y^2 + x^2y$

4. $rst - r^2s - rs^2 + 3rst^2 + 2rs^2 + 4rst$

51

ASSIGNMENT EXERCISES (continued)

II Remove the grouping symbols in the following expressions.

5. $(x^3y^3 + xy) + (-y + 3x)$

6. $-(-2m + 3n)$

7. $p^3 - (-p^2 + pq - q - 4)$

8. $3a^2 + (2b - 5)$

9. $(x^3 - 2xy) + (-3x^2 + y^2 - 4x)$

10. $-(a^2b - ab^2) - (a^2b^2 - ab - b + 6)$

III Simplify the following expressions by removing the grouping symbols and collecting like terms.

11. $(x^2 - x) - (-3x + 4)$

12. $-(m^2 - 5m + 2) + (2m^2 + m - 3) - (-4m^2 + 4m)$

13. $7ab - [2a^2 + (-3ab + 6)]$

14. $-4xy - [-2xy^2 - (x - y + 2)]$

15. $y^3 - \{-7y^2 - [5y - (-2y^2 + 6) - 6y] + 7\}$

IV Vertically add or subtract the following algebraic expressions.

16. $(3y^2 - 7y + 9) - (6y^2 + y - 4)$

17. $(6x - 2) + (3x^2 + 4x) + (x^2 - 7x + 3)$

18. $(a^3 + 2a + 1) - (a^3 + 3a^2 + 9)$

UNIT 5

SOLVING LINEAR EQUATIONS AND FORMULAS

Objectives: After having worked through this unit, the student should be capable of:

1.   distinguishing between an identity and a conditional equation;
2.   solving a linear equation in one unknown;
3.   solving a simple formula for specific variables of the formula.

1

ALGEBRAIC EQUATIONS

An equation is a statement that two quantities or algebraic expressions are equal.

If an equation has only one variable to the first power, we call it a **linear equation in one unknown**.

Examples:

1. $x + 1 = 9$ is a linear equation.

2. $\dfrac{3x}{4} + 2(x - 5) = 7x + 2$ is a linear equation.

3. $x^2 + 2x + 1 = 9$ is **not** a linear equation since the highest power of x is two.

2

IDENTITIES

Algebraic equations which are true, no matter what values we substitute for the variables, are called **identities**.

Examples:

1. $y + 2 = 2 + y$ Note: Both sides of the equation are the
2. $3x + x + 6 = 4x + 6$ same, just written in a different form.
4. $2(W + L) = 2W + 2L$

3

CONDITIONAL EQUATIONS

Algebraic equations which are true only if we substitute certain values for the variables are called **conditional equations**.

Examples:

1. $x + 1 = 9$ is only true under the condition that $x = 8$.
2. $2y - 4 = 16$ is only true under the condition that $y = 10$.

4

SOLUTIONS OR ROOTS OF EQUATIONS

The values of the variables which make an algebraic equation true are called **solutions** or **roots** of the equation.

Examples:

1. The number "8" is a solution of $x + 1 = 9$,

                       since $8 + 1 = 9$ is true.

2. The number "5" is **not** a solution of $x + 1 = 9$,

                       since $5 + 1 = 9$ is false.

3. The number "10" is a solution of $2y - 4 = 16$,

                       since $2(10) - 4 = 16$ is true.

4. The number "8" is **not** a solution of $2y - 4 = 16$,

                       since $2(8) - 4 = 16$ is false.

5

CHECK POINT 1

I   Check if the number "10" is a solution of the following equations.

1. $2x = 30$ 2. $x + 1 = 9$ 3. $2(y - 10) = 0$

4. $10(3y - 30) = 10$ 5. $3x + x + 6 = 4x + 6$

II   Check the following equations and indicate whether they are identities or conditional equations.

6. $2y + 1 = 1 + 2y$ 7. $x - 5 = 1$ 8. $x^2 + 1 = x^2 + 1$

9. $10y - 10 = 0$ 10. $3x + 6 = 3(x + 2)$

6

SOLVING EQUATIONS

To solve an equation is to find all the solutions of the equation. First, we shall learn how to solve linear equations in one unknown. In order to accomplish this task, we shall have to use the following important properties of equality.

7

PROPERTIES OF EQUALITY

1. Symmetry Property

The two sides of an equation can be interchanged without destroying the equality.

Example: If $5x - 4 = 3 - x$, then $3 - x = 5x - 4$.

2. Addition Property

Equal quantities may be **added to both sides** of an equation without changing the equality or solution.

Example: If $3x - 4 = 9$, then $3x - 4 + 4 = 9 + 4$ or $3x = 13$.

3. Subtraction Property

Equal quantities may be **subtracted from both sides** of an equation without changing the equality or solution.

Example: If $y + 8 = 12$, then $y + 8 - 8 = 12 - 8$ or $y = 4$.

4. Multiplication Property

Both sides of an equation may be **multiplied** by the **same quantity, except zero**, without changing the equality or solution.

Example: If $\frac{w}{5} = 3$, then $5 \cdot \frac{w}{5} = 3 \cdot 5$ or $w = 15$.

5. Division Property

Both sides of an equation may be **divided** by the **same quantity, except zero**, without changing the equality or solution.

Example: If $3m = 21$, then $\frac{3m}{3} = \frac{21}{3}$ or $m = 7$.

CHECK POINT 2

Check which of the following statements are true and which are false. For those statements which are true, identify which of the above five properties is being used.

1. If $\frac{1}{2}x = 6$, then $2(\frac{1}{2}x) = 6 + 2$.

2. If $x - 4 = 10$, then $x - 4 + 4 = 10 + 4$ or $x = 14$.

3. If $9w = 18$, then $\frac{9w}{9} = \frac{18}{9}$ or $w = 2$.

4. If $4y + 2 = 26$, then $4y + 2 - 2 = 26 - 2$ or $4y = 24$.

5. If $3z = 5$, then $z = 5 - 3$.

6. If $\frac{x}{6} = 10$, then $6 \cdot \frac{x}{6} = 6 \cdot 10$ or $x = 60$.

7. If $2(W + L) = 2W + 2L$, then $2W + 2L = 2(W + L)$.

SOLVING LINEAR EQUATIONS IN ONE UNKNOWN

To solve linear equations in one unknown, we use the following procedures:

Method:

Step 1.   Remove grouping symbols, if any.

Step 2.   Combine like terms on each side of the equation.

Step 3.   Use the addition and subtraction properties to obtain an equation so that the terms containing the variable are on the left and the numbers on the right side.

Step 4.   Divide or multiply both sides so that the coefficient of the variable is one, that is, we obtain $x = a$ (a is a number).

Step 5.   Check the solution by substituting it for the variable in the original equation.

Example 1:

Solve $5x - 2 = 8 + 3x$.

Solution:

Steps 1 and 2 are not needed.

Step 3.   Obtain an equation so that the terms with the variable are on the left and the numbers on the right side.

$5x - 2 = 8 + 3x$	To eliminate -2 on the left side,
$5x - 2 + 2 = 8 + 3x + 2$	we add 2 to both sides.
$5x + 0 = 10 + 3x$	To eliminate $3x$ on the right side,
$5x - 3x = 10 + 3x - 3x$	we subtract $3x$ from both sides.
$2x = 10 + 0$	

Step 4.   Divide both sides by the coefficient of the variable.

$\frac{2x}{2} = \frac{10}{2}$ Hence, $x = 5$.

Step 5. Check the solution.

Substituting $x = 5$ into $5x - 2 = 8 + 3x$ (original equation),
we have $5(5) - 2 = 8 + (3)(5)$
$23 = 23$ is true.

Hence, 5 is the solution of the equation $5x - 2 = 8 + 3x$.

11

Example 2:
Solve $6(y - 2) - 5y = 2(y + 3) - 3(2y + 1)$

Solution:
Step 1. Remove grouping symbols. $6(y - 2) - 5y = 2(y + 3) - 3(2y + 1)$
$6y - 12 - 5y = 2y + 6 - 6y - 3$

Step 2. Collect like terms on each side. $y - 12 = -4y + 3$

Step 3. Obtain an equation so that the terms with the variable are on the left and the numbers on the right side.

$y - 12 + 12 = -4y + 3 + 12$

$y = -4y + 15$

$y + 4y = -4y + 4y + 15$

$5y = 15$

Step 4. Divide both sides by the coefficient of the variable.
$$\frac{5y}{5} = \frac{15}{5}$$
Hence, $y = 3$.

Step 5. Check the solution.

Substituting $y = 3$ into $6(y - 2) - 5y = 2(y + 3) - 3(2y + 1)$

we have $6(3 - 2) - 5(3) = 2(3 + 3) - 3[2(3) + 1]$

$6 - 15 = 12 - 21$

$-9 = -9$ is true.

12

The process of solving equations can be shortened considerably once we understand what it involves.

Example 1: Solve $3x - 5 = 4$

Discussion:
To eliminate -5 on the left we would **add** 5 to both sides.
$3x - 5 + 5 = 4 + 5$
$3x = 4 + 5$

The same result can be achieved by "moving" -5 to the opposite side of the equal sign and changing its sign to **plus**.

$3x - 5 = 4$

$3x = 4 + 5$

Example 2:

Solve $9w = 3w + 12$.

Discussion:

To eliminate $3w$ on the right we would **subtract** $3w$ from both sides.

$9w - 3w = 3w + 12 - 3w$

$9w - 3w = 12$

Again, the same result can be achieved by "moving" $3w$ to the opposite side of the equal sign and changing its sign to **minus**.

$9w = 3w + 12$

$9w - 3w = 12$

TRANSPOSITION

When we "move" a term to the opposite side of the equation and **change its sign**, we say that we **transpose** the term. This is a short cut which has the same effect as adding terms to or subtracting terms from both sides of the equation.

Example 1:

Solve $6x - 7 = x + 3$.

Solution:

$6x - 7 = x + 3$ Transpose terms.

$6x - x = 7 + 3$

$5x = 10$

$x = 2$

Check: $6(2) - 7 = 2 + 3$ is true.

Example 2:

Solve $3y + 2(y - 4) = 5(y - 1) + 2y + 3$

Solution:

$3y + 2(y - 4) = 5(y - 1) + 2y + 3$

$3y + 2y - 8 = 5y - 5 + 2y + 3$

$5y - 8 = 7y - 2$

$5y - 7y = 8 - 2$

$-2y = 6$

$\dfrac{-2y}{-2} = \dfrac{6}{-2}$

$y = -3$

Check: $3(-3) + 2(-3 - 4) = 5(-3 - 1) + 2(-3) + 3$

$-9 + 2(-7) = 5(-4) + 2(-3) + 3$

$-9 - 14 = -20 - 6 + 3$

$-23 = -23$ is true.

17

Note:

1. Do not leave a solution in the form $-x = 7$
    To change the sign of x, simply multiply both sides by -1.

    Hence, $(-1)(-x) = (-1)(7)$
    $$x = -7.$$

2. To solve $5 + 7 = x + 3x$, we can use the symmetry property and write:
    $x + 3x = 5 + 7$ instead of transposing terms.

18

CHECK POINT 3

Solve the following equations using the transposition of terms or the symmetry property.

1. $3x + 5 = 2x + 12$ 2. $18 = 4x + 26$

3. $-x + 4 = 9$ 4. $3 + 12 = 3x + 2x$

5. $-3(w - 1) = -2(w - 4)$ 6. $8y - 6(2y + 5) = 12 + 2(y + 3)$

19

REVIEW:

THE LOWEST COMMON DENOMINATOR (LCD)

The lowest common denominator of a group of denominators is the smallest number into which all the denominators divide evenly.

Examples:

1. The LCD of $\frac{3}{4}$ and $\frac{5}{6}$ is 12.

2. The LCD of $\frac{x}{2y}$ and $\frac{7}{5y}$ is 10y.

20

In algebra, as in arithmetic, an expression is prime if it can be factored only into 1 and itself.

Examples:

The following expressions are prime: x,   3y - 1,   b,   7x + 9.

Each has no factors other than 1 and itself.

21

FINDING THE LCD BY INSPECTION

In many practical problems, we can find the LCD by simply inspecting the given denominators.

This is particularly the case in the following two situations:

1. The denominators are prime or have no common factors.
2. One denominator is a factor of the other denominator.

22

If the denominators are prime or have no common factors, the LCD is simply the product of the two denominators.

Examples:

1. The LCD of $\dfrac{3}{5}$ and $\dfrac{4}{7}$ is $5 \cdot 7 = 35$.

2. The LCD of $\dfrac{5x}{y}$ and $\dfrac{x^2}{z}$ is yz.

3. The LCD of $\dfrac{3}{R-1}$ and $\dfrac{5P}{R+3}$ is $(R-1)(R+3)$.

4. The LCD of $\dfrac{5}{2a}$ and $\dfrac{7}{3b}$ is 6ab.

23

If one denominator is a factor of the other denominator, the larger denominator is the LCD.

Examples:

1. The LCD of $\dfrac{2}{3}$ and $\dfrac{5}{12}$ is 12, since 3 is a factor of 12.

2. The LCD of $\dfrac{a}{x}$ and $\dfrac{b}{x^2}$ is x^2, since x is a factor of x^2.

3. The LCD of $\dfrac{P}{8R^3}$ and $\dfrac{V}{4R^2}$ is $8R^3$, since $4R^2$ is a factor of $8R^3$.

4. The LCD of $\dfrac{1}{(x-1)^2}$ and $\dfrac{x^3}{x-1}$ is $(x-1)^2$, since $(x-1)$ is a factor of $(x-1)^2$.

24

CHECK POINT 4

Find the LCD of the following groups of fractions by inspection.

1. $\dfrac{2}{3}, \dfrac{5}{6}, \dfrac{1}{18}$

2. $\dfrac{1}{x}, \dfrac{3}{y}$

3. $\dfrac{7}{n+1}, \dfrac{a}{m-1}$

4. $\dfrac{z}{3y}, \dfrac{9a}{5b}$

5. $\dfrac{1}{(p+1)^2}, \dfrac{p}{p+1}$

6. $\dfrac{1}{4R}, \dfrac{a}{8R^2}$

25

FRACTIONAL EQUATIONS

Equations which contain fractions are referred to as fractional equations.

To solve fractional equations, we first change the equation to one which no longer contains fractions.

This process is generally referred to as clearing an equation of fractions. The resulting equation may be a linear or quadratic equation which we then solve in the usual manner. In this unit we shall deal only with problems involving linear equations.

CLEARING AN EQUATION OF FRACTIONS

Rule:

To clear an equation of fractions, multiply each term of the equation by the lowest common denominator of all fractions in the equation.

Example:

Clear the following equation of fractions. $\dfrac{1}{2} + \dfrac{3x}{4} = \dfrac{5x}{6}$

Solution:

1. The LCD of $\dfrac{1}{2}$, $\dfrac{3x}{4}$ and $\dfrac{5x}{6}$ is 12.

2. $12\left(\dfrac{1}{2} + \dfrac{3x}{4}\right) = 12 \cdot \dfrac{5x}{6}$ (Multiply both sides of $\dfrac{1}{2} + \dfrac{3x}{4} = \dfrac{5x}{6}$ by the LCD.)

   $12 \cdot \dfrac{1}{2} + 12 \cdot \dfrac{3x}{4} = 12 \cdot \dfrac{5x}{6}$

   $6 + 9x = 10x$

SOLVING FRACTIONAL EQUATIONS

Method:

Step 1. Find the LCD of all fractions in the equation.

Step 2. Multiply both sides of the equation by the LCD and simplify.

Step 3. Solve the resulting non-fractional equations.

Step 4. Check the solutions.

CLEARING OF FRACTIONS RESULTING IN LINEAR EQUATIONS WITH NO FRACTIONAL COEFFICIENTS

Example 1: Solve $\dfrac{x}{6} = \dfrac{1}{8}$.

Solution:

Step 1. The LCD is 24.

Step 2. Multiply both sides of the equation by 24.

   $24 \cdot \dfrac{x}{6} = 24 \cdot \dfrac{1}{8}$

   $4x = 3$

Step 3. Solve the non-fractional equation.

   $4x = 3$

   $x = \dfrac{3}{4}$

Step 4. Check the solution.

   L.H.S. (Left hand side of the equation) $= \dfrac{\frac{3}{4}}{6} = \dfrac{3}{4} \cdot \dfrac{1}{6} = \dfrac{3}{24} = \dfrac{1}{8}$.

   R.H.S. (Right hand side of the equation) $= \dfrac{1}{8}$.

Hence, the solution is $x = \dfrac{3}{4}$.

Example 2: Solve $4 - \dfrac{a}{5} = \dfrac{3}{10}$.

Solution:

Step 1. The LCD is 10.

Step 2. Multiply both sides of the equation by 10.

$$10\left(4 - \frac{a}{5}\right) = 10 \cdot \frac{3}{10}$$

$$10 \cdot 4 - 10 \cdot \frac{a}{5} = 10 \cdot \frac{3}{10}$$

$$40 - 2a = 3$$

Step 3. Solve the non-fractional equation.

$$40 - 2a = 3$$

$$-2a = -37$$

$$a = \frac{37}{2}$$

Step 4. Check the solution as shown in the previous example.

Example 3: Solve $\dfrac{V}{6} - \dfrac{V-3}{4} = \dfrac{1}{2}$.

Solution:

Step 1. The LCD is 12.

Step 2. Multiply both sides of the equation by the LCD.

$$12\left(\frac{V}{6} - \frac{V-3}{4}\right) = 12 \cdot \frac{1}{2}$$

$$12 \cdot \frac{V}{6} - 12\left(\frac{V-3}{4}\right) = 12 \cdot \frac{1}{2}$$

$$2V - 3(V-3) = 6$$

Step 3. Solve the non-fractional equation.

$$2V - 3(V-3) = 6$$
$$2V - 3V + 9 = 6$$
$$-V = -3$$
$$V = 3$$

Step 4. Check the solution.

CHECK POINT 5

Solve the following equations.

1. $\dfrac{2m}{3} - \dfrac{1}{4} = \dfrac{1}{6}$

2. $\dfrac{R}{3} - \dfrac{R}{4} = \dfrac{2}{3}$

3. $\dfrac{2M}{3} - \dfrac{M+1}{2} = \dfrac{1}{3}$

LITERAL EQUATIONS

The equations we have solved previously have all contained only one letter representing a variable.

An equation which contains two or more **different** letters representing variables or constants is called a literal equation.

Examples:

1. $y = mx + b$ is a literal equation where m and b represent constants, and y and x represent variables.

2. $P = 2L + 2W$ is a literal equation where P, L and W represent variables.

FORMULA

A formula is a literal equation which expresses a relationship between two or more mathematical or physical quantities. Formulas are extensively used in business, engineering and science.

Example:

The formula $A = LW$ expresses the relationship between the area, A, the length, L, and the width, W, of a rectangle.

SOLVING LITERAL EQUATIONS AND FORMULAS

In working out problems, we are often confronted with the task of solving a literal equation or a formula for one of the variables in terms of the other variables.

Previously, we have discussed methods of solving equations in one variable. Literal equations and formulas are solved for specific variables in terms of the others in a similar way.

EXAMPLES OF SOLVING SIMPLE LITERAL EQUATIONS AND FORMULAS

Example 1:

Solve $mx = y - b$ for y.

Solution:

Step 1.   Transpose the terms so that y is on the left side and all other terms are on the right side of the equations.

$$mx = y - b$$

$$-y = -mx - b$$

Step 2.   Multiply both sides by -1.

$$(-1)(-y) = (-1)(-mx - b)$$

$$y = mx + b$$

Hence, expressing y in terms of the other variables, we write $y = mx + b$.

Example 2:

Solve $L = \dfrac{V}{HW}$ for W.

Solution:

Step 1. Multiply both sides by W to eliminate W from the denominator.

$$L = \frac{V}{HW}$$

$$WL = \frac{V}{HW} \cdot W$$

Step 2. Divide both sides by L.

$$WL = \frac{V}{H}$$

$$\frac{WL}{L} = \frac{\frac{V}{H}}{L} \qquad \left(\frac{\frac{V}{H}}{L} = \frac{\frac{V}{H}}{\frac{L}{1}} = \frac{V}{H} \cdot \frac{1}{L} = \frac{V}{HL} \right)$$

Therefore, $W = \dfrac{V}{HL}$.

Example 3:

Solve $Q = \dfrac{kAT(t_2 - t_1)}{d}$ for A. (Heat Conduction)

Solution:

Step 1. Multiply both sides by d to eliminate the denominator.

$$dQ = \frac{kAT(t_2 - t_1)}{d} \cdot d$$

Step 2. Divide both sides by $kT(t_2 - t_1)$ to isolate A.

$$dQ = kAT(t_2 - t_1)$$

$$\frac{dQ}{kT(t_2 - t_1)} = \frac{kAT(t_2 - t_1)}{kT(t_2 - t_1)}$$

(Use symmetry property)

Therefore, $A = \dfrac{dQ}{kT(t_2 - t_1)}$.

Example 4:

Solve $K = F + \dfrac{I}{i}$ for i. (Business: Capitalized Cost)

Solution:

Step 1. Transpose:

$$K - F = \frac{I}{i}.$$

Step 2.   Clear the equation of fractions.

$$i(k - F) = \frac{I}{i} \cdot i$$

Step 3.   Divide both sides by K - F.

$$\frac{i(K - F)}{K - F} = \frac{I}{K - F}$$

Therefore, $i = \dfrac{I}{K - F}$.

Example 5:

Solve $Z = \dfrac{X - u}{\sigma}$ for X.   (Statistics: Standard Score)

Solution:

Step 1.   Clear the fraction.

$$\sigma \cdot Z = \frac{(X - u)\sigma}{\sigma}$$

$$-X = -u - \sigma Z$$

Step 2.   Transpose.

$$-X = -u - \sigma Z$$

Step 3.   Multiply both sides by –1.

$$(-1)(-X) = (-1)(-u - \sigma Z)$$
$$X = u + \sigma Z$$

Therefore, $X = u + \sigma Z$.

CHECK POINT 6

1. Solve $V_1 P_1 = V_2 P_2$ for V_2.   Chemistry: Boyle's Law

2. Solve $E = \dfrac{MV^2}{R}$ for R.   Physics: Motion in a Circle Centripetal Force.

3. Solve $Q = C(T - t)$ for C.   Thermodynamics: Heat Change

4. Solve $\overline{X} = u + \sigma Z$ for Z.   Statistics

5. Solve $P = 2L + 2W$ for L.   Perimeter of Rectangle

6. Solve $T = \dfrac{1}{a} + t$ for a.   Temperature Conversion

---

DRILL EXERCISES

I   Check the following equations and indicate whether they are identities or conditional equations.

1. $3x^2 - 2 = 3x^2 - 2$ 2. $3x - 9 = 0$ 3. $y - 8 = -2$

4. $4x - 6 = 2(2x - 3)$ 5. $2x^2 - 1 = 2(x^2 - 1) + 1$

II   Solve the following equations.

6. $4y = 28$ 7. $16 = 2x + 4$

8. $5w + 2 = w + 10$ 9. $8x - 9 = 4x - 7$

10. $7w - 8 - 4w = 20 - w + 4$ 11. $6y + 3(y + 1) = y - 2(y - 3) + 7$

12. $-3y - 3 = 2$ 13. $-3(m - 1) + m = 2(m - 1) - 3$

14. $9x - 5(3x + 4) = 14 + 2(x - 1)$

15. $2(t - 3) - 2(3t - 4) = 2 + 4(t - 6)$

16. $\frac{1}{4} m = 7$ 17. $\frac{2}{3} p = \frac{1}{2}$

18. $\frac{5}{7} V - \frac{1}{14} = 0$ 19. $\frac{5}{8} = \frac{2}{3} R - \frac{1}{4}$

20. $\frac{3x}{5} - \frac{2x + 3}{2} = \frac{1}{4}$

Evaluate the given formulas for the indicated values.

21. $G = \frac{K - H}{K}$ $(K = 10, H = 8)$

22. $F = C - \frac{AB}{X}$ $(C = 10, A = 100, B = 0, X = 1000)$

23. $V_2 = \left(\frac{2m_1}{m_1 + m_2} \right) V_1$ $(m_1 = 20, m_2 = 10, V_1 = 27)$ (Physics)

24. $S = P(1 + rt)$ $(P = 1000, r = 0.16, t = \frac{60}{360})$ (Business)

25. $C = \frac{5}{9}(F - 32)$ $(F = 20°)$

26. $d = \frac{i}{1 + in}$ $(i = 0.015, n = 24)$ (Business)

Solve the following equations as indicated.

27.   Solve $y - 2x = 7$ for x.   28.   Solve $15x = 2y - 26$ for y.

29.   Solve $\frac{1}{2}x + \frac{13}{6}y = 4$ for y.   30.   Solve $y = \frac{3}{4}x - \frac{2}{3}$ for x.

31.   Solve $\frac{5}{2}x = -4y + 1$ for y.

DRILL EXERCISES (continued)

III Solve the following formulas as indicated.

32. Solve $C = 2\pi r$ for r.

33. Solve $\dfrac{V_1}{V_2} = \dfrac{P_1}{P_2}$ for P_2.

34. Solve $S = \dfrac{QV}{Ib}$ for V.

35. Solve $A = P(1 + i)^n$ for P.

36. Solve $T = \dfrac{M(g + a)}{2}$ for M.

37. Solve $F = \dfrac{Gm_1 m_2}{r_2}$ for m_1.

38. Solve $N = L - C$ for C.

39. Solve $D = P - FA$ for A.

40. Solve $P = S_1 + S_2 + S_3$ for S_2.

41. Solve $y = mx + b$ for m.

42. Solve $L = X - \dfrac{Z}{N}$ for Z.

43. Solve $V_1 = V_2 - \dfrac{n}{t}$ for t.

44. Solve $M = \dfrac{2T}{g + a}$ for a.

45. Solve $A = \dfrac{P - D}{F}$ for D.

ASSIGNMENT EXERCISES

Check the following equations and indicate whether they are identities or conditional equations.

I 1. $x + 4 = -1$

2. $x^3 + x + 1 = 1 + x^3 + x$

3. $-(x + 1) = -x - 1$

4. $15y + 30 = 0$

II Solve the following equations.

5. $3x = 27$

6. $19 = 3x - 2$

7. $7y + 3 = 5y - 11$

8. $4 + 2 = 6x + 2x$

9. $2y - 9 + 5y = 10 + 4y - 4$

10. $13 + 4(2m - 3) = 2(m - 3) + 7$

11. $-2w + 3 = 8$

12. $-4(y - 9) = -10(2 - y)$

13. $11x - 3(5x - 4) = 2 + 6(x + 5)$

14. $5(y + 7) - 3(2y + 4) = 2 - 4(y - 3)$

15. $6(2z - 3) - 4 = 3(5z - 7)$

16. $\dfrac{P}{7} = 1$

17. $\dfrac{3}{4}m = \dfrac{2}{3}$

18. $\dfrac{5R}{3} + \dfrac{1}{2} = 0$

19. $\dfrac{7}{9} = \dfrac{2}{5}M - \dfrac{1}{3}$

20. $\dfrac{3m}{2} - \dfrac{5m - 2}{3} = \dfrac{5}{12}$

21. $\dfrac{5}{4}p + \dfrac{2}{3} = 4$

ASSIGNMENT EXERCISES (continued)

Evaluate the given formulas for the indicated values.

22. $p = \dfrac{2wh}{S + 1}$ $\qquad$ (w = 5, h = 3, S = 4)

23. $t = \dfrac{V_2 - V_1}{n}$ $\qquad$ (V_2 = 20, V_1 = 10, n = 2)

24. $A = b - \dfrac{br}{100}$ $\qquad$ (b = 10, r = 0.5)

25. $M = \dfrac{2T}{a + g}$ $\qquad$ (T = 135, a = 3.7, g = 9.8)

26. $n = \dfrac{s - a}{d} + 1$ $\qquad$ (s = 18, a = 6, d = 3)

27. $b = \dfrac{v(s - t)}{at}$ $\qquad$ (v = 0.5, s = 10, t = 0.1, a = 5)

Solve the following equations as indicated.

28. Solve $y + x = 0$ for y.

29. Solve $3x - 4y = 12$ for y.

30. Solve $\dfrac{2}{3}x + \dfrac{1}{2}y = 1$ for x.

31. Solve $y = \dfrac{5}{8}x + \dfrac{1}{4}$ for x.

32. Solve $\dfrac{x}{3} = 2y - \dfrac{1}{2}$ for y.

III Solve the following formulas as indicated.

33. Solve $V = LWH$ for W.

34. Solve $n = \dfrac{PV}{RT}$ for V.

35. Solve $E = mc^2$ for m.

36. Solve $\dfrac{P_1 V_1}{T_1} = \dfrac{P_2 V_2}{T_2}$ for T_2.

37. Solve $n = \dfrac{I}{Pi}$ for P.

38. Solve $T = PG(a - b)$ for G.

39. Solve $V_2 = \left(\dfrac{2m_1}{m_1 + m_2}\right) V_0$ for V_0

40. Solve $I = S - P$ for P.

41. Solve $P = 2W + 2L$ for W.

42. Solve $A = b - \dfrac{br}{100}$ for r.

43. Solve $f = h - \dfrac{c}{n}$ for n.

44. Solve $u = \overline{x} - \sigma z$ for z.

45. Solve $t = \dfrac{V_2 - V_1}{n}$ for V_1.

46. Solve $a = \dfrac{2T - M_2}{M}$ for T.

47. Solve $s = \dfrac{c}{1 - p}$ for p.

48. Solve $p = \dfrac{2wh}{s + 1}$ for s.

SOLVING WORD PROBLEMS

Objectives: After having worked through this unit, the student should be capable of:

1. translating English phrases into algebraic expressions;
2. solving word problems involving one unknown;
3. expressing several unknowns in terms of one variable;
4. solving word problems involving several unknowns.

IDIOSYNCRATIC REACTION TO WORD PROBLEMS!

In the previous unit, we learned how to solve linear equations. Normally, these equations are the result of translating problems which are stated in English sentences into algebraic symbols.

Example:

The sentence

"**two times** a certain **number is** 12",

can be translated into the algebraic symbols

2n = 12, where n represents the "certain number".

Problems in business, engineering and science are often stated in English sentences and must be translated into algebraic equations in order to be solved.

CUE WORDS

To solve word problems, we first have to know how to translate English phrases into algebraic expressions. Certain words provide us with cues as to which mathematical operations are to be used. These words appear frequently in word phrases and we shall call them cue words.

CUE WORDS FOR ADDITION, SUBTRACTION AND MULTIPLICATION

Addition	**Subtraction**	**Multiplication**
plus	minus	times
more than	less than	product of
increased by	decreased by	multiplied by
sum of	difference of	twice
added to	subtracted from	doubled
enlarged by	reduced by	tripled, etc.

TRANSLATING ENGLISH PHRASES INTO ALGEBRAIC EXPRESSIONS

When translating English phrases into algebraic expressions, we translate words which represent:

1. unknown quantities into variables,
2. numbers and operations into their algebraic symbols.

Examples:

1. English phrase: The **sum of** a number and six.

    Algebraic expression: n + 6

2. English phrase: **Twice** a number.

    Algebraic expression: 2n.

3. English phrase: Three **times** a number **minus** four.

    Algebraic expression: 3n - 4.

6

CHECK POINT 1

Translate each of the following English phrases into algebraic expressions.

1. The sum of a number and ten.

2. Twelve times a number.

3. Four less than a number.

4. Twice a certain number increased by eight.

5. Twenty decreased by five times a number.

7

CUE WORDS FOR EQUALITY

The following words or phrases are frequently used to indicate equality:

equals
is, is the same as, is equal to,
the result is, what is left is,
we have, we obtain, we get.

8

CUE WORDS OR PHRASES POINTING TO THE UNKNOWN

The unknown in word problems represents the quantity we want to find. Hence, cue words or phrases pointing to an unknown are:

Find the ...
How much ...?
How long ...?
What is ...?
At what rate ...?

The unknown is normally found in the question part of the word problem.

9

Word problems can be very complex. Thus, competency in solving word problems requires a lot of practice. We will start with simple problems and gradually move to more complex ones in later units as we develop our algebraic skills.

10

SOLVING WORD PROBLEMS IN ONE UNKNOWN

Method:

Step 1. Represent the unknown by a letter.

Step 2. Translate the word statement into an equation.

Step 3. Solve the equation.

Step 4. Express the solution in terms of the original statement.

Step 5. Check the solution against the statement of the problem.

11

Example 1:

The sum of three times a number and 9 is 72. Find the number.

Solution:

Step 1. Let n = the number. (Find the ...)

Step 2. The sum of three times a number and 9 is 72.

$$3n + 9 = 72$$

Step 3. $3n + 9 = 72$
$3n = 63$
$n = 21$

Step 4. Hence, the number is 21.

Step 5. Check: "Three times 21 (is 63) and 9 is 72" is true.
Hence, 21 is correct.

12

Example 2:

If twice a certain number is decreased by 95, the result is –37. What is the number?

Solution:

Step 1. Let n = the number. (What is ...?)

Step 2. "Twice a certain number decreased by 95", $2n - 95$.

"the result **is** –37" $2n - 95 = -37$

Step 3. $2n - 95 = -37$
$2n = -37 + 95$
$2n = 58$
$n = 29$

Step 4. Hence, the number is 29.

Step 5. Check: "Twice 29 (is 58) decreased by 95 is –37" is true.
Hence, 29 is correct.

13

CHECK POINT 2

Solve the following word problems.

1. The sum of twice a number and 25 is 103. Find the number.

2. Three times a number decreased by 75 is –39. Find the number.

3. If we increase the product of five times a number by ten, the result is 45. Find the number.

14

EXPRESSING SEVERAL UNKNOWNS IN TERMS OF ONE VARIABLE
Normally, as word problems become more complex, the number of unknown quantities increases.

Example:

The sum of two numbers is 108. The second number is twice the first number plus 33. Find the numbers.

We shall now look at the techniques to express several unknowns in terms of one variable.

15

EXPRESSING ONE UNKNOWN IN TERMS OF ANOTHER
The statement

"the second number is twice the first number plus 33",

in the above example expresses one unknown in terms of the other. When solving problems where this type of statement occurs, it is important to identify the unknown which we use to describe the other unknowns.

In the above statement, we could evaluate the second number, **if we knew the first one**.

Hence, if we let n = the first number, then 2n + 33 = the second number.

16

To identify the unknown which describes the other, look for the one which we need in order to evaluate the other unknowns.

Examples:

In the following statements, represent one of the unknowns by x, then represent the other unknowns in terms of x.

1. "The larger number **is** three times **the smaller** less 7."

    If we knew the smaller number, we could evaluate the larger.

    Hence, let x = the smaller number;
    then 3x - 7 = the larger number.

17

2. "Of three circuits, the first carries three times the current of the second, the third carries 5 amperes less than twice the second."

    Here, if we knew how much the second circuit carried, we could evaluate what the first and third circuit carried.

    Hence, let x = the current of the second circuit,
    then   3x = the current of the first circuit,
    and   2x - 5 = the current of the third circuit.

18

3. "A water main separates into three branches. The second branch carries 500 litres per minute more than 7 times the number of litres per minute of the first. The third carries three times the sum of the first and second branch."

If we knew how much the first branch carries, we could calculate how much the other two carry.

Hence, let x = the number of litres per minute of the first branch,

then $7x + 500$ = the number of litres per minute of the second branch,

and $3[x + (7x + 500)]$ = the number of litres per minute of the third branch.

$\uparrow$ sum of the first and second branch)

19

CHECK POINT 3

In the following statements, represent one of the unknowns by x, then represent the other unknowns in terms of x.

1. The first number is three times the second number.

2. One number is six less than another number.

3. A wire is cut into three pieces. The first piece is 2 cm less than twice as long as the third. The second piece is 10 cm longer than the third piece.

4. Of three numbers, the first is 8 less than four times the second and the third is twice the sum of the first and second number.

20

SOLVING WORD PROBLEMS IN MORE THAN ONE UNKNOWN

Method:

Step 1. Represent the unknown which describes the others by a variable and, if possible, present the problem pictorially.

Step 2. Express the other unknowns in terms of the chosen variable.

Step 3. Analyze the word statements and translate them into an equation using the expressions obtained in Steps 1 and 2.

Step 4. Solve the equation.

Step 5. Evaluate the unknowns expressed in Step 2.

Step 6. Express and check the solutions in terms of the original statements.

21

Example 1:

We are given three numbers; the first is twice the second and the third is four times the second less 10. If the sum of the three numbers is 144, what are the three numbers?

Solution:

Step 1.   Let n = the second number.   (The second number describes the other two numbers.)

Step 2.   2n = the first number.   (The first is twice the second.)
          4n - 10 = the third number.   (The third is four times the second, less 10.)

Step 3.   "The **sum** of the three numbers **is** 144."

$$\boxed{\text{First number}} + \boxed{\text{second number}} + \boxed{\text{third number}} = 144$$
$$2n \quad + \quad n \quad + \quad (4n - 10) \ = 144$$

Step 4.   Solve   2n + n + 4n - 10 = 144
$$7n = 144 + 10$$
$$7n = 154$$
$$n = 22$$

Step 5.   If n = 22, then 2n = 44, and 4n - 10 = 4(22) - 10 = 78.

Step 6.   Hence, the first number is 44, the second number is 22 and the third number is 78.

44 + 22 + 78 = 144 is the correct sum of the three numbers.

22

Example 2:

An employee got a 15% increase in salary which brings his monthly pay to $1437.50. What was his salary before the increase?

Solution:

Step 1.   Let x = his pay before the increase.

Step 2. $x + \dfrac{15x}{100}$ = his present pay.

Step 3. $x + \dfrac{15x}{100} = 1437.50$

Step 4.   Solve the equation. $x + \dfrac{15x}{100} = 1437.50$

$$100(x + \frac{15x}{100}) = 100(1437.50)$$

$$100x + 100 \cdot \frac{15x}{100} = 143\ 750$$

$$115x = 143\ 750$$

$$x = 1250$$

Steps 5 and 6.
If x = 1250, then $1250 + \dfrac{15}{100} \cdot 1250 = 1437.50$.

Hence, his pay before the increase was $1250.00

Example 3:

A metal rod of length 42 cm is cut into two pieces.

The second piece is 14 cm less than $\frac{2}{5}$ the length of the first piece. Find the lengths of both pieces.

Solution:

Step 1.   Let x = the length of the first piece.

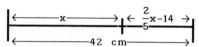

Step 2. $\frac{2}{5}$x – 14 = the length of the second piece.

Step 3.   "A metal rod of length 42 cm is cut into two pieces."
First piece + second piece = 42 cm.

$x + \frac{2}{5}x - 14 = 42.$

Step 4.   Solve the equation. $x + \frac{2}{5}x - 14 = 42$

$5(x + \frac{2}{5}x - 14) = 5 \cdot 42$

$5x + 2x - 70 = 210$

$7x = 280$

$x = 40$

Step 5.   Since the length of the first piece is 40 cm, therefore the length of the second piece is $\frac{2}{5}x - 14 = \frac{2}{5}(40) - 14 = 16 - 14 = 2$ cm.

Step 6.   Hence, the lengths of the two pieces are 40 cm and 2 cm.

Check: 40 + 2 = 42.

Example 4:

The perimeter of a triangle is 18 cm. If the second side is 5 cm more than half the first side, and the third side is two-thirds of the first side, find the lengths of the three sides.

Solution:

Step 1.   Let S = the length of the first side.

Step 2.   (a) $\frac{1}{2}$S + 5 = the length of the second side.

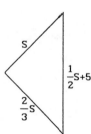

(b) $\frac{2}{3}$S = the length of the third side.

Step 3.   "The perimeter of a triangle is 18 cm."

First side + second side + third side = perimeter.

$S + (\frac{1}{2}S + 5) + \frac{2}{3}S = 18$

Step 4. Solve the equation.

$$S + \frac{1}{2}S + 5 + \frac{2}{3}S = 18 \quad \text{(LCD of 2 and 3 is 6.)}$$

$$6(S + \frac{1}{2}S + 5 + \frac{2}{3}S) = 6 \cdot 18$$

$$6S + 3S + 30 + 4S = 108$$

$$13S = 78$$

$$S = 6$$

Step 5. Since the length of the first side is 6, therefore the length of the second side is $\frac{1}{2}S + 5 = \frac{1}{2} \cdot 6 + 5 = 8$, and the length of the third side is $\frac{2}{3}S = \frac{2}{3} \cdot 6 = 4$.

Step 6. Hence, the lengths of the three sides of the triangle are 6, 8 and 4 cm.

Check: 6 + 8 + 4 = 18 cm.

25

CHECK POINT 4

Solve the following word problems.

1. Of three circuits, the second carries 12 amperes more than three times the first and the third carries twice the sum of the first and second. The three circuits carry together 84 amperes. Find the number of amperes each circuit carries.

2. The perimeter of a rectangle is 54 cm. If the width is 3 cm less than one-quarter of the length, find the length and width of the rectangle.

3. A wire which is 36 m long is to be cut into two pieces such that the smaller piece is 8 m less than one-third the larger. Find the length of each piece.

4. A salesman gets $375 per month plus 15% of his sales. Last month his pay was $4602. What were his sales last month?

26

THE SUM OF TWO UNKNOWN QUANTITIES

Sometimes we know the sum of two unknown quantities.

In this case, we can express one of the unknowns as the sum minus the other unknown.

Example:

"A collection of nickels and dimes has 30 coins."

If we let x = the number of nickels,
then $30 - x$ = the number of dimes.

Number of Nickels	Number of Dimes	Total Number of Coins
x	(30 - x)	30

Examples:

1. A 60 m rope is to be cut into two pieces.
   If we let x = the length of one piece,
   then   60 - x = the length of the other piece.

2. A 25 000 seat stadium has regular and box seats.
   If we let x = the number of regular seats,
   then   25 000 - x = the number of box seats.

CHECK POINT 5

For the following statements, express one of the unknowns as the sum minus the other unknown.

1. The sum of two numbers is 120.

2. A 350 cm wire is cut into two pieces.
   (Length of piece one, length of piece two.)

3. A collection of 75 coins consists of dimes and quarters.
   (Number of dimes, number of quarters.)

4. A developer bought two parcels of land. The total amount of land bought was 240 acres.
   (Number of acres in parcel I, number of acres in parcel II.)

5. The payment of principal and interest was $2350.00.
   (Amount of principal, amount of interest.)

ESTABLISHING EQUALITY

Both sides of an equation must represent the **same type** of quantity. As problems become more complex, it is necessary that we pay attention to the **type** of quantities we equate. A common error occurs when one equates the unknowns representing a certain type of quantity such as "the **number** of items" with another type of quantity such as "the **value** of the items."

Example:

A coin collection consisting of nickels and dimes is worth $2.10.
If there are 29 coins in the collection, find the number of nickels and the number of dimes.

Solution:

Step 1.   Let x = the number of nickels.

Step 2.   Then 29 - x = the number of dimes.

Step 3.   **Common Error**: $x + (29 - x) = 2.10$

$$\underbrace{\boxed{\underset{\text{Nickels}}{\downarrow} \qquad \underset{\text{Dimes}}{\downarrow}}}_{\text{Number of coins}} \neq \underset{\text{Value of 29 coins}}{\downarrow}$$

Different types of quantities

To solve this type of problem, we have to convert one type of quantity to the other type.

In this case, we know the value of a nickel and the value of a dime.

1 Nickel = 5 • 1 cents	1 Dime = 10 • 1 cents
3 Nickels = 5 • 3 cents	4 Dimes = 10 • 4 cents
7 Nickels = 5 • 7 cents	9 Dimes = 10 • 9 cents
x Nickels = 5 • x cents	x Dimes = 10 • x cents

Hence, the value of x nickels is 5x cents, and
the value of $29 - x$ dimes is $10(29 - x)$ cents.

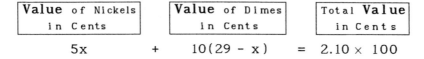

Value of Nickels in Cents		Value of Dimes in Cents		Total Value in Cents
5x	+	10(29 - x)	=	2.10 × 100

Example 1:

A coin collection worth $4.80 is made up of dimes and quarters. There are 6 fewer dimes than twice the number of quarters. Find the number of dimes and the number of quarters.

Solution:

Step 1.   Let y = the **number** of quarters.

Step 2.   2y - 6 = the **number** of dimes.

Step 3.

Value of Dimes in Cents		Value of Quarters in Cents		Total Value of coins in Cents
10(2y - 6)	+	25y	=	4.80 × 100

Step 4.   Solve $10(2y - 6) + 25y = 4.80 \times 100$

$$20y - 60 + 25y = 480$$

$$45y = 480 + 60$$

$$45y = 540$$

$$y = 12$$

Step 5.   If y = 12, then 2y - 6 = 2(12) - 6 = 18

Step 6.   Hence, there are 12 quarters and 18 dimes in the collection.

$10(18) + 25(12) = 180 + 300 = 480$ is correct.

Example 2:

A businessman has to invest $70 000. He is able to invest part at 12% and the rest at 15%. His investment earns him $9900 each year in interest. How much has he invested at 12% and how much at 15%?

Solution:

Step 1.   Let p = the amount invested at 15%. (We can choose p to be either the amount invested at 12% or at 15%.)

Step 2.   70 000 - p = the amount invested at 12%.

Step 3. "His investment earns him $9900 each year in interest."

Use the amounts p and (70 000 - p) invested to find expressions representing the amount of interest each earns.

Rate of interest for p is 15% and for (70 000 - p) is 12%.

Amount of Interest earned by p.	+	Amount of Interest earned by (70 000 - p)	=	Total Amount of Interest earned
$\frac{15}{100}p$	+	$\frac{12}{100}(70\ 000 - p)$	=	9900

Step 4. Solve $\frac{15}{100}p + \frac{12}{100}(70\ 000 - p) = 9900$

$$100 \cdot \frac{15}{100}p + 100 \cdot \frac{12}{100}(70\ 000 - p) = 100 \cdot 9900$$

$$15p + 12(70\ 000 - p) = 990\ 000$$

$$15p + 840\ 000 - 12p = 990\ 000$$

$$3p = 150\ 000$$

$$p = 50\ 000$$

Step 5. If p = 50 000, then 70 000 - p = 70 000 - 50 000 = 20 000.

Step 6. Hence, $50 000 has been invested at 15% and $20 000 has been invested at 12%.

Check:

Interest earned at 15% is $\frac{15}{100}(50\ 000) = \$7500$.

Interest earned at 12% is $\frac{12}{100}(20\ 000) = \$2400$.

Total interest earned on $70 000   = $9900.   This is correct.

34

MIXTURE PROBLEMS

Example 1:

How many litres of water must be added to 6 litres of a 30% acid solution to reduce it to a 20% solution?

Solution:

Step 1. Let x = the number of litres of water to be added.

Step 2. 6 + x = the number of litres in a new 20% solution.

Step 3.

Pure acid in 30% Solution	=	Pure acid in 20% Solution
30% of 6 litres	=	20% of (6 + x) litres
$\frac{30}{100}(6)$	=	$\frac{20}{100}(6 + x)$

Step 4. Solve the equation.

$$\frac{30}{100}(6) = \frac{20}{100}(6 + x)$$

$$180 = 20(6 + x)$$

$$9 = 6 + x$$

$$6 + x = 9$$

$$x = 3$$

Steps 5 and 6.   Hence, we have to add 3 litres.

Check:

$$30\% \text{ of } 6 = \frac{30}{100} \times 6 = 1.8 \text{ litres (pure acid)}$$

$$20\% \text{ of } 9 = \frac{20}{100} \times 9 = 1.8 \text{ litres (pure acid)}$$

Example 2:

One alloy contains 80% copper and another contains 50% copper. How much of each is required to obtain 280 kg of alloy containing 70% copper?

Solution:

Step 1.   Let x = the number of kg of the 80% alloy required.

Step 2.   280 - x = the number of kg of the 50% alloy required.

Step 3.

Pure copper in the 80% Alloy	+	Pure copper in the 50% Alloy	=	Pure copper in the 70% Alloy

$$80\% \text{ of } x \text{ kg} \quad + \quad 50\% \text{ of } (280 - x) \text{ kg} = 70\% \text{ of } 280 \text{ kg}$$

$$\frac{80}{100}x \quad + \quad \frac{50}{100}(280 - x) \quad = \frac{70}{100}(280)$$

Step 4.   Solve the equation.

$$\frac{80}{100}x + \frac{50}{100}(280 - x) = \frac{70}{100}(280)$$

$$80x + 50(280 - x) = 70(280)$$

$$80x + 14\ 000 - 50x = 19\ 600$$

$$30x = 5600$$

$$x = 186.67$$

Step 5.   If x = 186.67, then   280 - x = 280 - 186.67 = 93.33.

Step 6.   Hence, to obtain 280 kg of alloy containing 70% copper, we require 186.67 kg of the 80% alloy and 93.33 kg of the 50% alloy.

Check:

$$\frac{80}{100}(186.67) + \frac{50}{100}(93.33) = 149.34 + 46.66 = 196 \text{ kg}$$

$$\frac{70}{100}(280) = 196 \text{ kg}$$

EQUATIONS WITH DECIMALS

In the previous problem we have represented, for example, 80% as $\frac{80}{100}$ and 50% as $\frac{50}{100}$. We could have equally well represented these percentages as 0.80 and 0.50 in which case we would have obtained the following equation with decimal coefficients   0.80x + 0.50(280 - x) = 0.70(280)

When decimals occur in equations, we may:

(a)   solve the equation in the normal fashion; or

(b)   multiply both sides of the equation by the highest power of ten needed to change all decimals in the equation to whole numbers.

Example(a):

Solve $0.2x + 0.59(x - 2) = 3.678$

Solution:

$$0.2x + 0.59(x - 2) = 3.678$$

$$0.2x + 0.59x - 1.18 = 3.678$$

$$0.79x = 4.858$$

$$x = \frac{4.858}{0.79}$$

$$x = 6.15$$

Example (b):

Solve $0.2x + 0.59(x - 2) = 3.678$

Solution:

$$0.2x + 0.59(x - 2) = 3.678$$

$$1000[(0.2x) + 0.59(x - 2)] = 1000(3.678)$$

$$1000(0.2x) + 1000[0.59(x - 2)] = 1000(3.678)$$

$$200x + 590(x - 2) = 3678$$

$$790 x = 4858$$

$$x = 6.15$$

CHECK POINT 6

1. An alloy of nickel and silver weighing 400 g contains 80 g of silver. How many grams of nickel must be added to obtain a new alloy containing 10% silver?

2. A chemistry laboratory has a 5% solution and a 20% solution of hydrochloric acid. How many litres of each are required to make 4 litres of a 10% solution?

3. A company investing $500 000 has a guaranteed income of $74 500 each year. If part of the investment earns 14%, and the other 17%, find how much has been invested at 14% and how much at 17%.

DRILL EXERCISES

Translate each of the following English phrases into algebraic expressions.

1. The sum of a number and seventeen.

2. Six times a number.

3. Nine less than four times a number.

4. The sum of a number and 9 times its square.

5. Three more than eleven times a number.

Solve the following word problems. (Show all steps.)

6. A wire, 200 cm long, is cut into two pieces. One piece is 68 cm shorter than the other. Find the length of each piece.

DRILL EXERCISES (continued)

7. A man bought a house and lot worth $98 000. If the house costs $10 000 less than five times the cost of the lot, how much is the house alone worth?

8. The owner of a service station purchased a shipment of 32 car batteries. Type A battery costs $50 and type B battery costs $65. Find the number of each type of battery purchased if the entire shipment cost him $1780.

9. A coin collection worth $2.70 consists of nickels, dimes and quarters. If there are 17 coins and twice as many quarters as nickels, how many of each coin are there?

Number Problems

10. Six times a number is 9 more than 3 times that number. Find the number.

11. If we decrease the product of four times a number by eleven, the result is nine. What is the number?

12. If 6 is added to a certain number and the sum is divided by 5, the result is 3. Find the number.

13. If a number is increased by 12%, the result is 84. Find the number.

14. The sum of two numbers is 78. If the smaller number is 12 less than one-fifth of the larger number, find the two numbers.

Percent Problems

15. A man borrows $3000 for one year at 13%. He arranges to pay the loan back in one installment at the end of the year. How much interest does he have to pay?

16. An employee now makes $25 300 per year. One year ago, he got a new contract in which he received a raise of 15%. How much was he making before he got the new contract?

17. A service station buys shock absorbers for $18. What profit does the station make (in percents) if:
    (a) the shock absorbers are regularly sold at $24?
    (b) the shock absorbers are put on sale for $21?

18. A manufacturing company borrows $15 000 to replace worn machinery. The interest rate is 15% per year. After a year, they pay back $5000. What percent of this payment is interest?

19. A business invests $60 000. Part is invested at 18% and the rest at 12%. The total investment earns $9300 per year in interest. How much is invested at 12% and how much is invested at 18%?

Geometric Problems

20. A rectangle has a perimeter of 94 cm. The width is 2 cm more than $\frac{2}{3}$ the length. Find the length and the width.

21. A triangle has a perimeter of 19 m. The second side is 4 m more than $\frac{2}{5}$ of the first side and the third side is $\frac{8}{11}$ of the sum of the first and second side. Find the lengths of the sides.

DRILL EXERCISES (continued)

22. A beam 5 m long is cut into two pieces. One piece is $\frac{1}{2}$ m less than $\frac{4}{7}$ the other piece. Find the length of each piece.

Mixture Problems

23. We have 500 ml of a 15% solution of sulfuric acid. How many millilitres of water must be added to reduce this to a 6% sulfuric acid solution?

24. The IMCO company wants to produce 150 000 kg of a 35% nickel alloy. To produce this alloy, the company is going to use an alloy which contains 65% nickel and an alloy which contains 20% of nickel. Find the amount of 65% nickel alloy and the amount of 20% nickel alloy the company has to use in the production of the 35% nickel alloy.

25. A parts manufacturer has 40 kg of a copper and zinc alloy which contains 8 kg of zinc. Find the amount of copper which must be added if the manufacturer requires a new alloy containing 10% zinc.

ASSIGNMENT EXERCISES

Solve the following word problems. (Show all steps.)

1. A technician buys some transistors for $1.00 each and some others for $2.50 each. He buys 3 less of the expensive transistor than of the less expensive transistor, and spends a total of $24.00. How many of each type did he buy?

2. A water main which carries 1300 litres per minute separates into two branches. The first branch carries 1100 litres per minute less than twice the amount carried by the second branch. Find the amount carried per minute by each branch.

3. A collection of 14 coins worth $1.08 consists of dimes, nickels and pennies. If there are 3 times as many pennies as there are nickels, how many of each coin are in the collection?

4. A T.V. repairman is paid $55 for repairing colour sets and $30 for repairing black-and-white sets. One week he earned $615 for repairing 13 T.V. sets. How many of each type did he service?

5. An investor buys several shares of stocks X, Y and Z for a total of $976. If he spent $44 less on stock Z than he did on stock X, and $44 more on stock Y than twice the amount spent on X, how much did he spend on each stock?

6. A coin collector had a collection of silver coins worth $205. There were 5 times as many quarters as half dollars and 200 fewer dimes than quarters. How many of each type of coin did the collector have?

7. The first of 4 circuits carries 2 amperes more than twice the number carried by the third. The fourth carries 3 amperes less than the first and the second carries twice the number carried by the fourth. If the system carries 89 amperes in total, how many are carried by each circuit?

8. A water main separates into three branches. The first branch carries 900 litres per minute more than the third. The second branch carries 200 litres per minute less than twice the sum of the amount carried by the other two. Find the capacity of each branch if the main carries 2800 litres per minute.

ASSIGNMENT EXERCISES (continued)

9. The diameter of a rivet head is twice the diameter of the rivet. If the diameter of the rivet is increased by 12 mm, it will be 13 mm less than the diameter of the head. Find the diameter of the rivet head.

10. A developer bought 900 acres of land for $680 000. Some was bought at $600 per acre, some for $800 per acre and some for $1000 per acre. How much of each did he buy if he bought twice as many acres at $600 as he did at $1000?

Number Problems

11. Two numbers add to 39. One number is 9 less than 7 times the other. What are the numbers?

12. Twice a number less 11 is 3 more than the number. What is the number?

13. If a certain number is increased by 19, the result is 4 more than 6 times that number. Find the number.

14. The sum of three numbers is 27. The third number is 3 less than twice the second. The first number is twice the sum of the other two. Find the numbers.

15. If a number is increased by 40% of itself, the result is 49. Find the number.

16. If $\frac{2}{3}$ of a number is added to the number, the result is 130. Find the number.

17. If $\frac{3}{4}$ of a number is subtracted from the number, the result is 91. Find the number.

18. The sum of $\frac{2}{7}$ of a number and $\frac{3}{4}$ of the same number is 58. Find the number.

19. If a certain number is multiplied by three, the result is equal to 16 less than one-third of ten times the number. Find the number.

20. If a certain number is multiplied by two, the result is equal to one-half of three times the number, plus 17. Find the number.

Percent Problem

21. A retailer buys car batteries from the manufacturer at $60 each. He sells the batteries at $72 each. What is the mark-up in percents?

22. An item which is selling for $75 has been marked up 50% from the cost price. Find the cost price.

23. A service station marks down the price of shock absorbers from $25 to $21. Find the mark-down in percent.

24. A company gave an employee a raise of 15%. The employee's salary is now $18 400 a year. Find his salary before he got the raise.

25. An electronics technician borrows $5000 at 14% per year to start his own business. After a year, he pays back $1500. How much does he still owe?

ASSIGNMENT EXERCISES (continued)

26. A large corporation wants to invest $1 200 000. Part of this amount is to be invested at 12 1/2 % and part at 15%. The total earnings on the investment should be $155 000 per year. How much has to be invested at 12 1/2 % and how much at 15% to yield the desired yearly income?

27. A man has invested $5000. Part of this amount is invested in a term deposit at 17% per year and the rest is invested in a saving account at 15%. After one year, he has earned $810 in interest. Find how much he has invested in the term deposit and how much in the savings account.

Geometric Problems

28. The perimeter of a rectangle is 8.0 cm. The width is 3/5 of the length. Find the width and the length.

29. A rectangle has a perimeter of 300 cm. The length is 10 cm more than 4/3 the width. Find the length and the width.

30. The perimeter of a triangle is 40 cm. The second side is 1 cm less than twice the first side. The third side is 1 cm more than twice the first side. Find the lengths of the three sides.

31. A triangle has a perimeter of 112 cm. The second side is 1 cm more than 1/6 the first side. The third side is 3 cm more than 1/3 the first side. Find the lengths of the three sides.

32. A wooden molding 180 cm long is to be cut into 4 pieces to make a picture frame. The width of the frame is 6 cm more than 4/10 its length. Find the width and the length of the frame.

33. A rod 55 cm long is cut into two pieces. One piece is 3 cm more than 1/3 of the length of the other. Find the length of each piece.

34. A cedar board 100 cm long is cut into two pieces. One is 4 cm less than 1/3 the length of the other. Find the length of each piece.

Mixture Problems

35. Vinegar is 5% acetic acid. How much water must be added to 5 litres of vinegar to reduce it to a 3% solution of acetic acid?

36. How much water must be added to 5 litres of a 20% solution of salt water in order to change it to an 18% solution?

37. There are 7 litres of turpentine in a mixture of 21 litres of water and turpentine. How many litres of turpentine must be added to make a new mixture that is 75% turpentine?

38. How much water must be added to a litre of 90% pure alcohol to reduce it to 80% pure alcohol?

39. A company wants to produce 910 kg of an alloy that is 50% iron. The company wants to use a 90% iron alloy and a 20% iron alloy. How much of each of the 90% and the 20% iron alloys has to be used?

40. A satellite component which weighs 15 g has to contain 50% gold. If the company which produces this component has 65% and 15% gold alloys, how much of each does the company need to produce the component?

Fanshawe College

MODULE 2

**Operations involving
Algebraic Expressions**

Table of contents   **Page**

MULTIPLICATION OF ALGEBRAIC EXPRESSIONS, SPECIAL PRODUCTS

Objectives: After having worked through this unit, the student should be capable of:

1. classifying certain algebraic expressions as monomials, binomials, trinomial and multinomials;
2. multiplying (using the horizontal or vertical form) monomials and multinomials;
3. finding "special products" using shortcut methods.

1

INTRODUCTION

In the previous unit, we have solved problems which did not require the multiplication and/or division of algebraic expressions.
However, to solve problems involving, for example, areas and volumes, we may have to multiply or divide algebraic expressions.

2

Example 1:

The area of a rectangular plate is 98 cm^2.
The length is 4 cm longer than twice the width.
Find its dimensions.

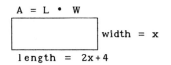

Since the area equals the length times the width,
we have: $98 = x(2x + 4)$.

3

Example 2:

The volume of a box is 1370 cm^3.
The length is 6 cm longer than its width
and the height is 3 cm less than its width.
Find the dimensions of the box.

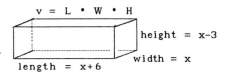

Since the volume equals the length times the width times the height, we
have: $1370 = x(x + 6)(x - 3)$.

Before we can solve the equations derived in examples 1 and 2 above, we
have to learn how to multiply algebraic expressions.

4

REVIEW: TERMS

A term is a product of numbers and variables.
A number or a variable by itself is also considered to be a term.

The expressions $3y^2$,   16,   x, $\dfrac{x^2 y}{6}$, $2a^7 b^5 c$ are terms.

5

The expression $3y^2 - 9y + 16$ has three terms.

The expression $x - 4$ has two terms.

The expression $5a^3 b^2 + 3a^2 b - 2a - 35$ has four terms.

6

1. Algebraic expressions consisting of only **one** term are called **monomials**.

    Example: The expression $3x^2 yz$ is a monomial.

2. Algebraic expressions consisting of **more than one** term are called
    multinomials.

    Example: The expression $5a^2 b^2 + 3a^2 b - 2a - 35$ is a multinomial.

    (a) Multinomials consisting of **two** terms are also called **binomials**.

        Example: The expression $2a + b$ is a binomial.

    (b) Multinomials consisting of **three** terms are also called **trinomials**.

        Example: The expression $9y^2 - 3y + 16$ is a trinomial.

7

CHECK POINT 1

Identify each of the following expressions as a monomial, binomial, or trinomial.

1. $5y^3 + 3y^2 - 10y$
2. 18
3. $6m^3 - n^2$
4. $x + 1$
5. $-7a^2b^3$
6. x

8

MULTIPLYING A MONOMIAL TIMES A MONOMIAL

To multiply two or more monomials, we make use of the commutative property of multiplication and the exponential law which states that

$$x^m \cdot x^n = x^{m+n}$$

Example:

$$(3x^2y^4)(2xy^3) = 3 \cdot x^2 \cdot y^4 \cdot 2 \cdot x \cdot y^3$$
$$= 3 \cdot 2 \cdot \underbrace{x^2 \cdot x} \cdot \underbrace{y^4 \cdot y^3}$$
$$= 6x^3y^7$$

9

SHORT CUT METHOD FOR MULTIPLYING MONOMIALS

Step 1.

Multiply all numerical coefficients.

Step 2.

Add exponents of identical variables.

10

Example 1: Multiply $(3x^2y^4)(2xy^3)$

Solution: $(3x^2y^4)(2xy^3) = 6x^3y^7$

11

Example 2: Multiply $(-2m^5n^4t^3)(8m^2n^{-4}t^{-2})$

Solution: $(-2m^5n^4t^3)(8m^2n^{-4}t^{-2}) = -16m^7n^0t^1$
$$= -16m^7t$$

12

The same method can be applied to multiplying more than two monomials

Examples:

1. $(-4z^2)(2xy^9z^6)(-3y^{-4}) = 24xy^5z^8$

2. $(4a^2bc^4)(-5a^3b^2c^{-2})(2a^{-5}b^3c) = -40a^0b^6c^3$
    $$= -40b^6c^3$$

13

CHECK POINT 2

Multiply the following.

1. $(2x^3y)(4x^2y^2)$
2. $(-a^3b^{-3})(4a^2b^3)$
3. $(5m^3n^2t)(-2mn^{-1}t^3)(-m^{-2}n^3t^{-2})$
4. $(p^4q)(-3pqt)(6p^{-5}q^{-2}t^2)$

14

MULTIPLYING A MONOMIAL TIMES A MULTINOMIAL
To multiply a monomial times a multinomial, we use the distributive property of multiplication over addition.

$$a(b + c) = ab + ac$$

Examples:
1. $7(x^2 + y^3) = 7x^2 + 7y^3$

2. $5x(3x^2 - 4y) = (5x)(3x^2) - (5x)(4y)$
$$= 15x^3 - 20xy$$

15

MORE EXAMPLES
1. $-a^2bc(a^3b + b^2c) = (-a^2bc)(a^3b) + (-a^2bc)(b^2c)$
$$= -a^5b^2c - a^2b^3c^2$$

2. $-x^3y^4(x - 3y + 4) = (-x^3y^4)(x) - (-x^3y^4)(3y) + (-x^3y^4)(4)$
$$= -x^4y^4 + 3x^3y^5 - 4x^3y^4$$

3. $m^2n(m^3 - 4n^2 + 3p - 5q - 10)$
$$= (m^2n)(m^3) - (m^2n)(4n^2) + (m^2n)(3p) - (m^2n)(5q) - (m^2n)(10)$$
$$= m^5n - 4m^2n^3 + 3m^2np - 5m^2nq - 10m^2n$$

16

CHECK POINT 3

Multiply each of the following expressions.

1. $5(a^2b - c^2)$

2. $-xy(3x^4z + 2y^2z^3)$

3. $2pq(5p - 3q + 9)$

4. $-3mn^2(2m^2 - 4n - t + 6)$

17

MULTIPLYING A MULTINOMIAL TIMES A MULTINOMIAL
To multiply a multinomial times a multinomial, we use again the distributive property of multiplication over addition.

Each term of one multinomial has to be multiplied by each term of the other multinomial.

Example:
Multiply $(a + b)[c + d + e]$.

$$(\ )[c + d + e] = (\ )c + (\ )d + (\ )e$$
$$(a + b)[c + d + e] = (a + b)c + (a + b)d + (a + b)e$$
$$= ac + bc + ad + bd + ae + be$$

MORE EXAMPLES

1. $(2x + y)(x - 3y + 5)$

   $= (2x + y)x - (2x + y)3y + (2x + y)5$

   $= (2x^2 + xy) - (6xy + 3y^2) + (10x + 5y)$

   $= 2x^2 + xy - 6xy - 3y^2 + 10x + 5y$

   $= 2x^2 - 3y^2 - 5xy + 10x + 5y$

2. $(x^2 + 2x - 4)(3x^2 - x + 5)$

   $= (x^2 + 2x - 4)3x^2 - (x^2 + 2x - 4)x + (x^2 + 2x - 4)5$

   $= (3x^4 + 6x^3 - 12x^2) - (x^3 + 2x^2 - 4x) + (5x^2 + 10x - 20)$

   $= 3x^4 + 6x^3 - 12x^2 - x^3 - 2x^2 + 4x + 5x^2 + 10x - 20$

   $= 3x^4 + 5x^3 - 9x^2 + 14x - 20$

Another way of multiplying multinomials is to distribute the second factor over the terms of the first factor.

Examples:

1. $(a + b)[c + d + e] = a[c + d + e] + b[c + d + e]$

   $= ac + ad + ae + bc + bd + be$

2. $(3x - 5)(2x + 4) = 3x(2x + 4) - 5(2x + 4)$

   $= (6x^2 + 12x) - (10x + 20)$

   $= 6x^2 + 12x - 10x - 20$

   $= 6x^2 + 2x - 20$

CHECK POINT 4

I Multiply each of the following expressions by distributing the second factor over the terms of the first factor.

   1. $(y + 6)(3y^2 + 4y + 1)$ 2. $(m - n)(m^2 - mn + n^2)$

II Multiply the following expressions by distributing the first factor over the terms of the second factor.

   3. $(3a + 2)(a^2 - 2a - 3)$ 4. $(2x^2 + 3x - 1)(5x^2 - 2x - 4)$

VERTICAL MULTIPLICATION

   When multiplying multinomials, it is sometimes convenient to use the following vertical form.

Example 1:
   Multiply $(a - 3b)(a^2 - ab + b^2)$

Solution:

Step 1.　Write the factors vertically.

$$a^2 - ab + b^2$$
$$\underline{\hspace{1.5em} a - 3b}$$

Step 2.　Multiply each term of　$a^2 - ab + b^2$　by each term of　$a - 3b$. Arrange like terms in columns and add.

$$
\begin{array}{l}
a^2 - ab + b^2 \\
\underline{\hspace{2em} a - 3b} \\
a^3 - a^2b + ab^2 \\
\underline{\hspace{1em} - 3a^2b + 3ab^2 - 3b^3} \\
a^3 - 4a^2b + 4ab^2 - 3b^3
\end{array}
$$

$\longleftarrow$ $a(a^2 - ab + b^2)$

$\longleftarrow$ $-3b(a^2 - ab + b^2)$

Hence, $(a - 3b)(a^2 - ab + b^2) = a^3 - 4a^2b + 4ab^2 - 3b^3$

22

Example 2:

Multiply　$(m^2 + 2m - 4)(3m^2 - m + 5)$

Solution:

Write the factors vertically and multiply each term of　$3m^2 - m + 5$　by each term of　$m^2 + 2m - 4$.

Arrange like terms in columns and add.

$$
\begin{array}{l}
3m^2 - m + 5 \\
\underline{m^2 + 2m - 4} \\
3m^4 - m^3 + 5m^2 \\
\quad + 6m^3 - 2m^2 + 10m \\
\underline{\quad\quad\quad - 12m^2 + 4m - 20} \\
3m^4 + 5m^3 - 9m^2 + 14m - 20
\end{array}
$$

$\longleftarrow$ $m^2(3m^2 - m + 5)$

$\longleftarrow$ $2m(3m^2 - m + 5)$

$\longleftarrow$ $-4(3m^2 - m + 5)$

Hence,　$(m^2 + 2m - 4)(3m^2 - m + 5) = 3m^4 + 5m^3 - 9m^2 + 14m - 20$.

23

Example 3:

Multiply　$(3y + z)(2y - 4z - 5)$.

Solution:

Write the factors vertically and multiply each term of　$2y - 4z - 5$　by each term of　$3y + z$. Arrange like terms in columns and add.

$$
\begin{array}{l}
2y - 4z - 5 \\
\underline{\hspace{2em} 3y + z} \\
6y^2 - 12yz - 15y \\
\underline{\quad\quad 2yz \quad\quad - 4z^2 - 5z} \\
6y^2 - 10yz - 15y - 4z^2 - 5z
\end{array}
$$

Hence,　$(3y + z)(2y - 4z - 5) = 6y^2 - 10yz - 15y - 4z^2 - 5z$.

24

CHECK POINT 5

Multiply vertically each of the following expressions.

1. $(3x + y)(2x^2 - xy + y^2)$

2. $(4p^2 - 2p + 3)(p^2 + 3p - 4)$

3. $(a - 2b)(3a - 4b + 6)$

25

When multiplying certain multinomials, we can use shortcuts. Since these so called "**special products**" occur frequently, they should be readily recognized. Hence, in what follows, we shall first list these special products and then discuss each one of them.

26

SPECIAL PRODUCTS

$$(ax + b)(cx + d) = acx^2 + (bc + ad)x + bd$$

$$(x + y)(x - y) = x^2 - y^2$$

$$(x + y)^2 = x^2 + 2xy + y^2$$

$$(x - y)^2 = x^2 - 2xy + y^2$$

$$(x + y)(x^2 - xy + y^2) = x^3 + y^3$$

$$(x - y)(x^2 + xy + y^2) = x^3 - y^3$$

27

PRODUCT OF A BINOMIAL TIMES A BINOMIAL

When multiplying two binomials, we can use a shortcut called the **FOIL** method.

Examples:

1. $(2x + 3)(x + 4) = 2x(x + 4) + 3(x + 4)$

    $$= (2x)(x) + (2x)(4) + (3)(x) + (3)(4)$$

Product of the	**First** terms	**Outer** terms	**Inner** terms	**Last** terms
	F	**O**	**I**	**L**

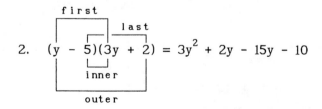

2. $(y - 5)(3y + 2) = 3y^2 + 2y - 15y - 10$

THE FOIL SHORTCUT

The word FOIL will help us to mentally multiply two binomials.

F stands for the product of the **first** terms of each binomial.

O stands for the product of the **outer** terms of each binomial.

I stands for the product of the **inner** terms of each binomial.

L stands for the product of the **last** terms of each binomial.

Examples:

1. $(x - 3)(x + 2) = \overset{\mathbf{F}}{x^2} + \overset{\mathbf{O}}{2x} - \overset{\mathbf{I}}{3x} - \overset{\mathbf{L}}{6}$

   $= x^2 - x - 6$

2. $(2y + 5)(3y + 4) = \overset{\mathbf{F}}{6y^2} + \overset{\mathbf{O}}{8y} + \overset{\mathbf{I}}{15y} + \overset{\mathbf{L}}{20}$

   $= 6y^2 + 23y + 20$

3. $(4a - b)(2a - 3b) = \overset{\mathbf{F}}{8a^2} - \overset{\mathbf{O}}{12ab} - \overset{\mathbf{I}}{2ab} + \overset{\mathbf{L}}{3b^2}$

   $= 8a^2 - 14ab + 3b^2$

When a product consists of more than two factors, multiply any two factors and then multiply the result by the third factor and so on.

Examples:

1. $2x(x + 6)(x - 3) = 2x(\overset{\mathbf{F}}{x^2} - \overset{\mathbf{O}}{3x} + \overset{\mathbf{I}}{6x} - \overset{\mathbf{L}}{18})$

   $= 2x(x^2 + 3x - 18)$

   $= 2x^3 + 6x^2 - 36x$

2. $(x + 1)(x - 4)(x + 3) = (\overset{\mathbf{F}}{x^2} - \overset{\mathbf{O}}{4x} + \overset{\mathbf{I}}{x} - \overset{\mathbf{L}}{4})(x + 3)$

   $= (x^2 - 3x - 4)(x + 3)$

   $= (x^2 - 3x - 4)(x) + (x^2 - 3x - 4)(3)$

   $= (x^3 - 3x^2 - 4x) + (3x^2 - 9x - 12)$

   $= x^3 - 3x^2 - 4x + 3x^2 - 9x - 12$

   $= x^3 - 13x - 12$

CHECK POINT 6

Multiply the following using the **FOIL** shortcut.

1. $(x - 2)(x + 3)$ 2. $(2a - b)(5a - 4b)$

3. $3x(x + 2)(x - 1)$ 4. $(x - 3)(x - 1)(x + 2)$

THE PRODUCT OF A SUM AND A DIFFERENCE

If we have to find the product of two binomials which are identical except for the sign between the two terms, we can write the product without multiplying.

Example:

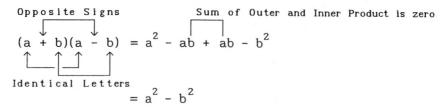

$$= a^2 - b^2$$

Hence, the product of the sum and the difference of the same two values is the square of the first term minus the square of the second term.

Examples:

1. $(x + y)(x - y) = x^2 - y^2$

2. $(3x + 4)(3x - 4) = (3x)^2 - (4)^2$
$$= 9x^2 - 16$$

3. $(5a + 2b)(5a - 2b) = (5a)^2 - (2b)^2$
$$= 25a^2 - 4b^2$$

4. $(4m^3 + 3n^4)(4m^3 - 3n^4) = (4m^3)^2 - (3n^4)^2$
$$= 16m^6 - 9n^8$$

CHECK POINT 7

Find the following products without multiplying.

1. $(y + 3)(y - 3)$

2. $(5x + 2y)(5x - 2y)$

3. $(3y^2 + 4z)(3y^2 - 4z)$

4. $(2p^4 + 3q^3)(2p^4 - 3q^3)$

THE SQUARE OF A BINOMIAL

Example 1:

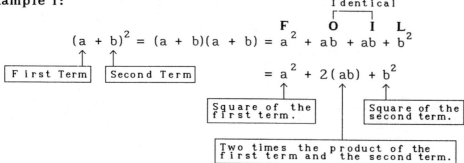

Example 2:

$$(x - 3)^2 = (x - 3)(x - 3) = x^2 - 3x - 3x + 9$$

$$= x^2 - 2(3x) + 9$$

$$= x^2 - 6x + 9 \qquad \boxed{\text{Two times the product of the first and second term.}}$$

Hence, when squaring a binomial,

(i) square the first term,
(ii) double the product of the first and second term,
(iii) square the last term.

35

Examples:

$$\boxed{\text{Double the product of the first and second term.}}$$

1. $(y + 5)^2 = y^2 + 2(5y) + 5^2$

$$= y^2 + 10y + 25$$

$$\boxed{\text{Second Term}}$$

2. $(m - 6)^2 = m^2 + 2(-6m) + (-6)^2$ $\qquad$ Note: $(m - 6) = (m + (-6))$

$$= m^2 - 12m + 36$$

3. $(3x - 2y)^2 = (3x)^2 + 2(-6xy) + (-2y)^2$

$$= 9x^2 - 12xy + 4y^2$$

36

CHECK POINT 8

Multiply the following squares by use of the shortcut.

1. $(x + 7)^2$ $\qquad$ 2. $(p - 3)^2$ $\qquad$ 3. $(2m + 3n)^2$ $\qquad$ 4. $(3y - 5z)^2$

37

THE PRODUCT OF A BINOMIAL AND A RELATED TRINOMIAL

When multiplying certain binomials and trinomials, we obtain a simple product. If we recognize the relationship between the terms of the binomial and trinomial, we shall be able to write their product without multiplying.

Examples:

1. $(x + y)(x^2 - xy + y^2) = (x + y)x^2 - (x + y)xy + (x + y)y^2$

$$= x^3 + yx^2 - x^2y - xy^2 + xy^2 + y^3$$
$$\qquad\qquad 0 \qquad\qquad 0$$

$$= x^3 + y^3 \quad \text{This expression is called}$$
$$\qquad\qquad \textbf{the sum of cubes.}$$

2. $(x - y)(x^2 + xy + y^2) = (x - y)x^2 + (x - y)xy + (x - y)y^2$

$$= x^3 - yx^2 + x^2y - xy^2 + xy^2 - y^3$$
$$\qquad\qquad 0 \qquad\qquad 0$$

$$= x^3 - y^3 \quad \text{This expression is called}$$
$$\qquad\qquad \textbf{the difference of cubes.}$$

Note:

In both of the examples in the previous frame, we see that

(a) the first term of the trinomial is the square of the first term of the binomial; the last term of the trinomial is the square of the last term of the binomial.

$$(x + y)(x^2 - xy + y^2) \qquad (x - y)(x^2 + xy + y^2)$$

(b) the middle term of the trinomial is the negative of the product of the two terms in the binomial.

$$(x + y)(x^2 - xy + y^2) \qquad (x - y)(x^2 + xy + y^2)$$

(c) the sign in the binomial factor is the same as the sign between the two cubes in the product.

$$(x + y)(x^2 - xy + y^2) = x^3 + y^3$$

$$(x - y)(x^2 + xy + y^2) = x^3 - y^3$$

When the terms of a binomial and a trinomial are related as in (a), (b) and (c), then we can use the following shortcut to find their product.

Examples:

1. $(a + 1)(a^2 - a + 1) = a^3 + 1^3$ or $a^3 + 1$

Terms of binomial squared
Product of binomial terms with opposite sign.

2. $(m - 2n)(m^2 + 2mn + 4n^2) = m^3 - (2n)^3$ or $m^3 - 8n^3$

Terms of binomial squared
Product of binomial terms with opposite sign.

3. $(5p + 3v)(25p^2 - 15pv + 9v^2) = (5p)^3 + (3v)^3$ or $125p^3 + 27v^3$

Terms of binomial squared
Product of binomial terms with opposite sign.

CHECK POINT 9

If possible, find the following products without multiplying.

1. $(r + 1)(r^2 - r + 1)$
2. $(m - 1)(m^2 + m - 1)$

3. $(2a + 3b)(4a^2 - 6ab + 9b^2)$
4. $(v - 5p)(v^2 + 5vp + 25p^2)$

DRILL EXERCISES

Identify each of the following expressions as a monomial, binomial or trinomial.

1. 33 2. $5m^2 - 2m + 6$ 3. y 4. $x - 1$ 4. $-9xy^2$

Multiply the following.

6. $(3m^2n^3)(2mn)$

7. $(2x^{-2}y^4)(-x^2y)$

8. $(-2x^2yz)(-x^2yz^2)(3x^{-4}yz^{-1})$

9. $3(p^3q + r^3)$

10. $-ab(5a^3c^2 + 4b^3c^4)$

11. $4mn(5m + 2n - 6)$

12. $-5x^2y(2x^3 + y - z - 4)$

13. $(x + 4)(2x^2 + 3x + 2)$

14. $(3m - 2n)(4m - n)$

15. $2x(x - 3)(x + 2)$

16. $(x - 2)(x + 3)(x - 1)$

17. $(a - b)(a^2 - 2ab + b^2)$

18. $(3m + 4)(m^2 - m - 1)$

19. $(x^2 + 2x - 3)(2x^2 - x - 5)$

Vertically multiply each of the following expressions.

20. $(x + 4y)(3x^2 + 2xy - y^2)$

21. $(2m - n)(m - 3n - 7)$

22. $(a^2 - 3a + 2)(5a^2 + 2a - 3)$

Multiply the following using a shortcut method.

23. $(y + 8)^2$

24. $(x + 4)(x + 4)$

25. $(a + 4)(a - 4)$

26. $(x - 4)^2$

27. $(3a + 3b)^2$

28. $(2m + 3n)(2m - 3n)$

29. $(6x^2 + 2y)(6x^2 - 2y)$

30. $(5y - 2z)^2$

31. $(v + 1)(v^2 - v + 1)$

32. $(2m - 1)(4m^2 + 2m + 1)$

33. $(3x - 5)(9x^2 + 15x + 25)$

34. $(x + 4y)(x^2 - 4xy + 16y^2)$

ASSIGNMENT EXERCISES

Multiply the following. If possible, use shortcuts.

1. $(-xy)(7x^2y^{-1})$

2. $(a - 5)(a + 2)$

3. $(p + 3)^2$

4. $(5a^4b)(2a^3b)$

5. $6(mn^2 - m)$

6. $(x + 5)(x - 5)$

7. $(a^2b^3)(4abc)(-a^{-3}b^{-2}c)$

8. $(m - 5)^2$

9. $(p - 1)(p^2 + p + 1)$

10. $-yz(3x^5y - 8x^2z^3)$

11. $-2r^2s^2(5r^2 - 3s + t - 2)$

12. $(5x - y)(3x - 3y)$

13. $5m(m - 2)(m + 1)$

14. $(m - 5)(m^2 + 5m + 25)$

15. $(y + 2)(4y^2 + 3y + 3)$

16. $(3r - 4s)^2$

17. $(4a + 2b)(4a - 2b)$

18. $(2a - 3)(2a^2 - a - 4)$

19. $(a + 2b)(a^2 - 2ab + 4b^2)$

20. $(x - y)(2x^2 - xy + 2y^2)$

21. $(x - 1)(x - 2)(x - 3)$

22. $(m^2 + 3n^2)(m^2 - 3n^2)$

23. $(3m^2 - 2m + 1)(2m^2 - m - 2)$

24. $(2x + 3y)^2$

25. $(2x + 3y)(4x^2 - 6xy + 9y^2)$

Vertically multiply each of the following expressions.

26. $(2m - n)(5m - 6n - 2)$

27. $(5x^3 - 3x + 2)(2x^4 + 3x^2 - 4)$

28. $(2a + 3b)(3a^2 - ab + b^2)$

29. $(x + 1)(x^2 - x + 1)$

UNIT **8**

DIVISION OF
ALGEBRAIC EXPRESSIONS

Objectives: After having worked through this unit, the student should be capable of:

1. dividing a monomial by a monomial;
2. dividing a multinomial by a monomial;
3. dividing a multinomial by a multinomial (long division)

1

DIVIDING A MONOMIAL BY A MONOMIAL

To divide a monomial by a monomial, we first divide the numerical coefficients and then we make use of the division law 3.2 of exponential expressions which states:

$$\frac{x^m}{x^n} = x^{m-n} \text{ for } m > n \quad \text{or} \quad \frac{x^m}{x^n} = \frac{1}{x^{n-m}} \text{ for } n > m, \ x \neq 0.$$

Examples:

1. $\dfrac{8x^5}{4x^3} = 2x^{5-3} = 2x^2$

2. $\dfrac{3a^3b^2}{a^2b} = 3a^{3-2}b^{2-1} = 3ab$

3. $\dfrac{6m^4n^7t^3}{3m^2n^4t^3} = 2m^{4-2}n^{7-4}t^{3-3} = 2m^2n^3t^0 = 2m^2n^3$

4. $\dfrac{-15x^3y^2z^2}{3xyz^4} = \dfrac{-5x^{3-1}y^{2-1}}{z^{4-2}} = \dfrac{-5x^2y}{z^2}$

2

SHORT CUT METHOD FOR DIVIDING MONOMIALS

1. Reduce numerical coefficients to lowest terms.
2. Subtract the smaller from the larger exponent of identical bases.

Examples:

1. $\dfrac{5x^3y^2}{xy} = 5x^2y$

2. $(6m^2n^5) \div (3mn^2) = 2mn^3$

3. $\dfrac{-p^3q^4t^7}{6p^2qt^3} = \dfrac{-pq^3t^4}{6}$

4. $(-16x^5y^3z^4) \div (8x^2y^3z^3) = -2x^3z$

5. $\dfrac{-24a^6b^5c^2}{-8a^6b^4c^5} = \dfrac{3b}{c^3}$

3

CHECK POINT 1

Divide the following monomials.

1. $\dfrac{3m^4n^7}{m^3n^4}$

2. $(-6x^2y^5) \div (3xy^2)$

3. $\dfrac{-a^4b^3c}{7a^4bc^2}$

4. $(-15p^5qt^4) \div (-5p^4qt^2)$

4

DIVIDING A MULTINOMIAL BY A MONOMIAL

Since $\dfrac{15}{5} + \dfrac{20}{5} = \dfrac{15 + 20}{5}$, we can also write $\dfrac{15 + 20}{5} = \dfrac{15}{5} + \dfrac{20}{5}$.

We divide the denominator into each term of the numerator.

Similarly, we can divide each term of the multinomial by the monomial.

Example:

$$\frac{4x^2 + 6xy}{2x} = \frac{4x^2}{2x} + \frac{6xy}{2x} = 2x + 3y$$

5

More Examples:

1. $\dfrac{6x^7 + 5x^4 - 12x^2}{3x^2} = \dfrac{6x^7}{3x^2} + \dfrac{5x^4}{3x^2} - \dfrac{12x^2}{3x^2} = 2x^5 + \dfrac{5x^2}{3} - 4$

2. $\dfrac{4a^3b^2 - 8a^2b^3 + 6ab^4}{-2ab^2} = \dfrac{4a^3b^2}{-2ab^2} - \dfrac{8a^2b^3}{-2ab^2} + \dfrac{6ab^4}{-2ab^2} = -2a^2 + 4ab - 3b^2$

3. $\dfrac{-15x^7y^3z^4 + 20x^3y^4z^5 - 10x^6y^5z^3}{-5x^3y^3z^3} = \dfrac{-15x^7y^3z^4}{-5x^3y^3z^3} + \dfrac{20x^3y^4z^5}{-5x^3y^3z^3} - \dfrac{10x^6y^5z^3}{-5x^3y^3z^3}$

$$= 3x^4z - 4yz^2 + 2x^3y^2$$

6

CHECK POINT 2

Divide the following.

1. $\dfrac{4a^3 - 8a^2b}{4a}$

2. $\dfrac{8y^6 - 3y^4 + 6y^2}{2y^2}$

3. $\dfrac{-16m^8n^2 + 8m^5n^3 - 4m^3n^5}{4m^3n^2}$

4. $\dfrac{36x^7y^5z^2 - 9x^5y^3z^3 + 18x^4y^2z^4}{-9x^4y^2z^2}$

7

DIVIDING A MULTINOMIAL BY A MULTINOMIAL

The division of a multinomial by a multinomial is similar to the process of long division in arithmetic.

Example 1:
Divide $(8x^2 + 14x - 15) \div (2x + 5)$

Solution:
Step 1. Write the division in long division form.

$$2x + 5 \overline{)8x^2 + 14x - 15}$$

Step 2. Divide the first term "$8x^2$" of the dividend ($8x^2 + 14x - 15$) by the first term "2x" of the divisor (2x + 5).
Write the result as the first term of the quotient.

$$\begin{array}{r} 4x \\ 2x + 5 \overline{\smash{\big)}\, 8x^2 + 14x - 15} \end{array} \qquad (8x^2) \div (2x) = 4x$$

Step 3. Multiply the divisor (2x + 5) by (4x).
Write the result under the dividend so that like terms are in the same column.

$$\begin{array}{r} 4x \\ 2x + 5 \overline{\smash{\big)}\, 8x^2 + 14x - 15} \\ \underline{8x^2 + 20x} \end{array} \qquad 4x(2x + 5) = 8x^2 + 20x$$

Step 4. Subtract and "bring down" the next term "-15".

$$\begin{array}{r} 4x \\ 2x + 5 \overline{\smash{\big)}\, 8x^2 + 14x - 15} \\ \text{Subtract: } \underline{8x^2 + 20x} \\ - 6x - 15 \end{array} \qquad \begin{array}{r} 8x^2 + 14x \\ \text{Add: } \underline{-8x^2 - 20x} \\ 0 \quad - 6x \end{array}$$

Step 5. Divide "-6x" by "2x" and place the result into the quotient as the second term.

$$\begin{array}{r} 4x - 3 \\ 2x + 5 \overline{\smash{\big)}\, 8x^2 + 14x - 15} \\ \underline{8x^2 + 20x} \\ - 6x - 15 \end{array} \qquad (-6x) \div (2x) = -3$$

Step 6. Multiply the divisor (2x + 5) by (-3).
Write the result below (-6x - 15).

$$\begin{array}{r} 4x - 3 \\ 2x + 5 \overline{\smash{\big)}\, 8x^2 + 14x - 15} \\ \underline{8x^2 + 20x} \\ - 6x - 15 \\ \underline{- 6x - 15} \end{array} \qquad -3(2x + 5) = -6x - 15$$

Step 7.   Subtract as in Step 4.

$$
\begin{array}{r}
4x - 3 \\
2x + 5 \overline{\smash{\big)}\ 8x^2 + 14x - 15} \\
8x^2 + 20x \\
\hline
-6x - 15
\end{array}
$$

Subtract: $- 6x - 15$ $\longleftarrow$ $-6x - 15$

Add: $+6x + 15$

$\overline{0}$ $\overline{0 + 0}$

Hence, $(8x^2 + 14x - 15) \div (2x + 5) = 4x - 3.$ Remainder = 0 or R0.

or $\dfrac{8x^2 + 14x - 15}{2x + 5} = 4x - 3 + \dfrac{0}{2x + 5} = 4x - 3$

Check: $(2x + 5)(4x - 3) + 0 = 8x^2 + 14x - 15$

Divisor   Quotient   Dividend

Remainder

Note:

To use this method of "long division", the terms in the divisor and dividend must be arranged so that the exponents appear in **descending order from left to right.**

8

Example 2:

Divide $(3m + 8 - 2m^2) \div (1 + m)$.

Solution:

Step 1.   Write the division in long division form, rearranging the terms so that the exponents appear in descending order from left to right.

$$
m + 1 \overline{\smash{\big)}\ -2m^2 + 3m + 8}
$$

Step 2. $m + 1 \overline{\smash{\big)}\ -2m^2 + 3m + 8}$ with $-2m$ above

$(-2m^2) \div m = -2m$

Step 3. $m + 1 \overline{\smash{\big)}\ -2m^2 + 3m + 8}$ with $-2m$ above and $-2m^2 - 2m$ below

$-2m(m + 1) = -2m^2 - 2m$

Step 4. $m + 1 \overline{\smash{\big)}\ -2m^2 + 3m + 8}$

Subtract: $-2m^2 - 2m$

$5m + 8$

$-2m^2 + 3m$

Add: $+2m^2 + 2m$

$\overline{0 + 5m}$

Step 5. $(5m) \div m = 5$

$m + 1 \overline{\smash{\big)}\ -2m^2 + 3m + 8}$ with $-2m + 5$ above

$-2m^2 - 2m$

$\overline{5m + 8}$

Step 6.
$$\begin{array}{r} -2m + 5 \\ m + 1 \overline{\smash{\big)}\ -2m^2 + 3m + 8} \\ \underline{-2m^2 - 2m} \\ 5m + 8 \end{array}$$

Subtract: $\dfrac{5m + 5}{3}$ ⟵——————— $5(m + 1)$

Since 3 cannot be evenly divided by "m", the first term of the divisor, we have a remainder of 3.

Hence, $(-2m^2 + 3m + 8) \div (m + 1) = -2m + 5$ R3

or $\dfrac{-2m^2 + 3m + 8}{m + 1} = -2m + 5 + \dfrac{3}{m + 1}$

Check: $\underbrace{(m + 1)}_{\text{Divisor}}\underbrace{(-2m + 5)}_{\text{Quotient}} + \underbrace{3}_{\text{Remainder}} = -2m^2 + 3m + 5 + 3 = \underbrace{-2m^2 + 3m + 8}_{\text{Dividend}}$

9

CHECK POINT 3

Use the long division process to perform the indicated division.

1. $(6a^2 - a - 2) \div (2a + 1)$ 2. $(-2x - 1 + 3x^2) \div (x - 2)$

10

MORE COMPLEX PROBLEMS

Example 1:
Divide $(3a^2 + 2a^3 - 24a - 16) \div (a + 4)$.

Solution:

Step 1. Write the division in long division form, rearranging the terms so that the exponents appear in descending order from left to right.

$$a + 4 \overline{\smash{\big)}\ 2a^3 + 3a^2 - 24a - 16}$$

Step 2.
$$\begin{array}{r} 2a^2 \\ \mathbf{a + 4} \overline{\smash{\big)}\ \mathbf{2a^3} + 3a^2 - 24a - 16} \end{array}$$
⟵————— $(2a^3) \div a = 2a^2$

Step 3.
$$\begin{array}{r} 2a^2 \\ a + 4 \overline{\smash{\big)}\ 2a^3 + 3a^2 - 24a - 16} \\ \underline{2a^3 + 8a^2} \end{array}$$
⟵————— $2a^2(a + 4) = 2a^3 + 8a^2$

Step 4.
$$\begin{array}{r} 2a^2 \\ a + 4 \overline{\smash{\big)}\ 2a^3 + 3a^2 - 24a - 16} \\ \text{Subtract: } \underline{2a^3 + 8a^2 \qquad} \\ -5a^2 - 24a \end{array}$$

Step 5.

$$\begin{array}{r} 2a^2 - 5a \\ a + 4 \overline{\smash{\big)}\, 2a^3 + 3a^2 - 24a - 16} \\ \underline{2a^3 + 8a^2} \\ -5a^2 - 24a \end{array}$$

$(-5a^2) \div a = -5a$

Step 6.

$$\begin{array}{r} 2a^2 - 5a \\ a + 4 \overline{\smash{\big)}\, 2a^3 + 3a^2 - 24a - 16} \\ \underline{2a^3 + 8a^2} \\ -5a^2 - 24a \\ \underline{-5a^2 - 20a} \end{array}$$

$\longleftarrow$ $-5a(a + 4) = -5a^2 - 20a$

Step 7.

$$\begin{array}{r} 2a^2 - 5a \\ a + 4 \overline{\smash{\big)}\, 2a^3 + 3a^2 - 24a - 16} \\ \underline{2a^3 + 8a^2} \\ -5a^2 - 24a \\ \underline{-5a^2 - 20a} \\ -4a - 16 \end{array}$$

Subtract:

Add:
$$\begin{array}{r} -5a^2 - 24a \\ +5a^2 + 20a \\ \underline{} \\ -4a \end{array}$$

Step 8.

$$\begin{array}{r} 2a^2 - 5a - 4 \\ a + 4 \overline{\smash{\big)}\, 2a^3 + 3a^2 - 24a - 16} \\ \underline{2a^3 + 8a^2} \\ -5a^2 - 24a \\ -5a^2 - 20a \\ \underline{ -4a - 16} \end{array}$$

$(-4a) \div a = -4$

Step 9.

$$\begin{array}{r} 2a^2 - 5a - 4 \\ a + 4 \overline{\smash{\big)}\, 2a^3 + 3a^2 - 24a - 16} \\ \underline{2a^3 + 8a^2} \\ -5a^2 - 24a \\ \underline{-5a^2 - 20a} \\ -4a - 16 \end{array}$$

Subtract:
$$\begin{array}{r} -4a - 16 \\ \underline{} \\ 0 \end{array}$$
$\longleftarrow$ $-4(a + 4) = -4a - 16$

Hence, $(2a^3 + 3a^2 - 24a - 16) \div (a + 4) = 2a^2 - 5a - 4$ R0.

Check: $(a + 4)(2a^2 - 5a - 4) = 2a^3 + 3a^2 - 24a - 16$

Example 2:
  Divide $(x^3 - 1) \div (x - 1)$.

Solution:
  Step 1.   We note that in $(x^3 - 1)$ there is neither an x^2 nor an x term. In this case, we add these terms by giving them zero coefficients.

$$x - 1 \overline{\smash{\big)}\ x^3 + 0x^2 + 0x - 1}$$

Step 2. $x - 1 \overline{\smash{\big)}\ x^3 + 0x^2 + 0x - 1}$ with quotient x^2 ← $\quad x^3 \div x = x^2$

Step 3. $x - 1 \overline{\smash{\big)}\ x^3 + 0x^2 + 0x - 1}$ with quotient x^2
$$\underline{x^3 - x^2}$$ ← $\quad x^2(x - 1) = x^3 - x^2$

Step 4. $x - 1 \overline{\smash{\big)}\ x^3 + 0x^2 + 0x - 1}$ with quotient x^2
Subtract: $\underline{x^3 - x^2}$
$$x^2 + 0x$$

Add:
$$\begin{aligned} x^3 + 0x^2 \\ \underline{-x^3 + x^2} \\ 0 + x^2 \end{aligned}$$

Step 5. $x - 1 \overline{\smash{\big)}\ x^3 + 0x^2 + 0x - 1}$ with quotient $x^2 + x$
$$\underline{x^3 - x^2}$$
$$x^2 + 0x$$ ← $\quad x^2 \div x = x$

Step 6. $x - 1 \overline{\smash{\big)}\ x^3 + 0x^2 + 0x - 1}$ with quotient $x^2 + x$
$$\underline{x^3 - x^2}$$
$$x^2 + 0x$$
$$\underline{x^2 - x}$$ ← $\quad x(x - 1) = x^2 - x$

Step 7.

$$\begin{array}{r} x^2 + x \\ x - 1 \overline{\smash{\big)}\, x^3 + 0x^2 + 0x - 1} \\ \underline{x^3 - x^2} \\ x^2 + 0x \end{array}$$

Subtract: $\underline{x^2 - x}$

$x - 1$

Add:
$$\begin{array}{r} x^2 + 0x \\ \underline{-x^2 + x} \\ 0 + x \end{array}$$

Step 8.

$$\begin{array}{r} x^2 + x + 1 \\ x - 1 \overline{\smash{\big)}\, x^3 + 0x^2 + 0x - 1} \\ \underline{x^3 - x^2} \\ x^2 + 0x \\ \underline{x^2 - x} \\ x - 1 \end{array}$$

$x \div x = 1$

Step 9.

$$\begin{array}{r} x^2 + x + 1 \\ x - 1 \overline{\smash{\big)}\, x^3 + 0x^2 + 0x - 1} \\ \underline{x^3 - x^2} \\ x^2 + 0x \\ \underline{x^2 - x} \\ x - 1 \end{array}$$

Subtract: $\underline{x - 1}$ $\longleftarrow$ $1(x - 1)$

0

Hence, $(x^3 - 1) \div (x - 1) = x^2 + x + 1$ R0.

Check: $(x - 1)(x^2 + x + 1) = x^3 - 1$

Note:

The preceding examples gave explanations and showed the detailed steps involved in the long division process. However, normally, we display a long division as illustrated by the following example.

12

Example 3:

Divide $(m^3 - 7m) \div (m^2 + 3m - 1)$.

Solution:

$$\begin{array}{r} m - 3 \\ m^2 + 3m - 1 \overline{\smash{\big)}\, m^3 + 0m^2 - 7m + 0} \\ \underline{m^3 + 3m^2 - m} \\ -3m^2 - 6m + 0 \\ \underline{-3m^2 - 9m + 3} \\ +3m - 3 \end{array}$$

Since the m in $3m$ is not evenly divisible by m^2, we have a remainder of $3m - 3$.

Hence, $(m^3 - 7m) \div (m^2 + 3m - 1) = m - 3 + \dfrac{3m - 3}{m^2 + 3m - 1}$.

Check: $(m^2 + 3m - 1)(m - 3) + (3m - 3) = (m^3 - 10m + 3) + (3m - 3)$

$= m^3 - 7m$

Example 4:

Divide $(8a^3 + 27b^3) \div (2a + 3b)$.

Solution:

$$
\begin{array}{r}
4a^2 - 6ab + 9b^2 \\
2a + 3b\,\overline{\smash{\big)}\,8a^3 + 27b^3} \\
\underline{8a^3 + 12a^2b } \\
-12a^2b \\
\underline{-12a^2b - 18ab^2 } \\
18ab^2 + 27b^3 \\
\underline{18ab^2 + 27b^3} \\
0
\end{array}
$$

Hence, $(8a^3 + 27b^3) \div (2a + 3b) = 4a^2 - 6ab + 9b^2$.

Check: $(2a + 3b)(4a^2 - 6ab + 9b^2) = 8a^3 + 27b^3$.

Note:

$(2a + 3b)(4a^2 - 6ab + 9b^2) = 8a^3 + 27b^3$ is a special product which we discussed in the previous unit. When we recognize in a division problem that the divisor is a factor of a special product, we simply write the second factor of the special product as the quotient.

Examples:

1. $(x^2 - y^2) \div (x + y) = x - y$

    since $(x + y)(x - y) = x^2 - y^2$

2. $(a^3 + b^3) \div (a + b) = a^2 - ab + b^2$

    since $(a + b)(a^2 - ab + b^2) = a^3 + b^3$

CHECK POINT 4

Perform the indicated divisions.

1. $(6x^3 - 11x^2 - 7x + 15) \div (2x - 3)$

2. $(x^3 + 1) \div (x + 1)$

3. $(4m^3 - 31m + 15) \div (2m^2 + m - 5)$

4. $(27m^3 + 125n^3) \div (3m + 5n)$

DRILL EXERCISES

Divide the following. If possible, use a shortcut.

1. $\dfrac{24x^3y^2z^2}{4xyz^2}$

2. $-16a^4b^6 \div (-8a^2b^2c)$

3. $(12y^2 - 9y) \div (3y)$

4. $\dfrac{-12x^4yz^3}{3x^2y^4z}$

5. $18r^3s^2t \div (-4r^5st^2)$

6. $\dfrac{-8vt + 20v^2t^2}{4vt}$

7. $\dfrac{4a^3b^2 + 16ab - a^2}{2a^2b}$

8. $(3x^3 + 16xy^2 - 12x^4yz^4) \div (-2x^2yz)$

9. $\dfrac{-4ab^3 - 3a^2bc - 12a^3b^2c^4}{-2ab^2c^3}$

10. $(v^2 - 1) \div (v + 1)$

11. $(x^2 - x - 2) \div (x + 1)$

12. $(m^3 - n^3) \div (m - n)$

13. $\dfrac{6a^3 + 7a^2 - 9a - 9}{2a + 3}$

14. $(8m^3 + 6m + 1) \div (2m - 1)$

15. $\dfrac{6a^2 - ab - b^2}{3a + b}$

16. $(9t^2 - 4) \div (3t + 2)$

17. $\dfrac{8x^3 - 27}{2x - 3}$

18. $(a^3 + 6a^2 + 8) \div (a + 2)$

19. $(3x^2 - 14x - 5) \div (x - 5)$

20. $(v^4 + 1) \div (v - 1)$

21. $(4a^3 - 6a^2 + 2) \div (2a + 1)$

22. $(27y^3 - 125) \div (9y^2 + 15y + 25)$

23. $(m^2 + 7m) \div (m - 2)$

24. $(x^4 + 6x^3 + 7x^2 - 5x + 3) \div (x + 3)$

25. $(4a^3 - a^2) \div (2a + 1)$

26. $(5m^2 - 9mn - 2n^2) \div (m - 2n)$

27. $(v^3 + 10) \div (v - 5)$

28. $(3t - 5t^2 + 1 + 2t^3) \div (2t - 1)$

ASSIGNMENT EXERCISES

Divide the following. If possible, use a shortcut.

1. $\dfrac{8r^5st^2}{-12r^3s^2t}$

2. $\dfrac{-8a^2b^2c}{-16a^4b^4}$

3. $-4xyz^2 \div [12(xyz)^2]$

4. $\dfrac{2a^3b^2 - 8ab - 3a^2}{-4ab^2}$

5. $(3x^2 - 16xy^3 - 9x^3yz) \div (-2x^2yz)$

6. $(-15m^3n + 3mn^2) \div (-3n)$

7. $\dfrac{25r^4s^3t^2 - 10rs^2t}{5s^2t}$

8. $(9a^2 - b^2) \div (3a - b)$

9. $(x^2 - 6x - 16) \div (x + 2)$

10. $(y^3 - 27) \div (y - 3)$

11. $(2m^2 + 7mn + 3n^2) \div (m + 3n)$

12. $(v^2 - 9) \div (v + 3)$

13. $(4t^2 + 15t - 4) \div (4t - 1)$

14. $(x - x^2 + 7) \div (x - 6)$

15. $(a^2 - 1 + a^3 + 5a) \div (a + 1)$

16. $(x^2 - 5) \div (x - 3)$

17. $(125v^3 - 1) \div (25v^2 + 5v + 1)$

18. $(8v^3 - 9) \div (2v - 1)$

19. $(4m^2 - m^3) \div (m + 2)$

20. $(6t^2 - 5tv - 6v^2) \div (3t + 2v)$

21. $\dfrac{8a^3 + 36a^2 + 36a + 26}{2a + 3}$

22. $\dfrac{a^4 - 3a^3 - a + 1}{a^2 - a + 1}$

23. $(2x^3 - 3x^2 - 2x + 3) \div (2x - 3)$

24. $(x^4 - 1) \div (x + 1)$

25. $(x^4 + 1) \div (x + 1)$

UNIT **9**

FACTORING
ALGEBRAIC EXPRESSIONS

Objectives: After having worked through this unit, the student should be capable of:

1.   factoring multinomials with common monomial factors;
2.   factoring certain trinomials into the product of two binomials;
3.   factoring a "difference of two squares" into the product of two binomials;
4.   factoring a "sum or difference of two cubes" into the product of a binomial times a trinomial.

1

FACTORING

Factoring is the reverse process of multiplication.

Examples:

1. Multiplying $2x(x^2 + 3)$, we obtain $2x^3 + 6x$.

   Factoring $2x^3 + 6x$, we obtain $2x(x^2 + 3)$.

2. Multiplying $(x + 1)(x - 3)$, we obtain $x^2 - 2x - 3$.

   Factoring $x^2 - 2x - 3$, we obtain $(x + 1)(x - 3)$.

2

To simplify algebraic fractions and to solve equations involving multinomial expressions, we need to know how to "break up" complex expressions into the product of simpler ones.

We should note that many expressions cannot be factored or are very difficult to factor. In this unit, we will deal with expressions which are simple to factor.

3

We shall discuss the following three approaches to factoring:

1. Use of the distributive property in reverse.

   $ab + ac = a(b + c)$

   This approach involves searching an expression for a **common factor** which is present in every term.

2. Use of the **FOIL** method of multiplying binomials in reverse.

   $x^2 - 2x - 3 = (x + 1)(x - 3)$

3. Use of the shortcut of multiplying the following special products in reverse.

   $x^2 - y^2 = (x + y)(x - y)$
   $x^3 + y^3 = (x + y)(x^2 - xy + y^2)$
   $x^3 - y^3 = (x - y)(x^2 + xy + y^2)$

   This approach involves the recognition of a **difference of squares** and the **sum and difference of cubes**.

4

FACTORING MULTINOMIALS WITH COMMON MONOMIAL FACTORS

Previously, we used the distributive property of multiplication over addition to multiply multinomials.

Example:

$6x(a + b - c) = 6xa + 6xb - 6xc$

Note:

6 and **x** are factors of each term in **6xa + 6xb - 6xc**.

COMMON MONOMIAL FACTORS

A factor which occurs in each term of a multinomial is called a common monomial factor.

Examples:
1. **5** is a common monomial factor of **5x + 5y.**

2. m^3 is a common monomial factor of $4m^3n^2 + 7m^3$.

3. The expression $3x + 5y$ has no common monomial factor.

4. 6, x^3 and y are common monomial factors of $12x^7y^3 - 6x^5y^2 + 18x^3y$.

$$12x^7y^3 - 6x^5y^2 + 18x^3y = \underline{6} \cdot 2 \cdot \underline{x^3} \cdot x^4 \cdot \underline{y} \cdot y^2 - \underline{6} \cdot \underline{x^3} \cdot x^2 \cdot \underline{y} \cdot y + \underline{6} \cdot 3 \cdot \underline{x^3} \cdot \underline{y} =$$
$$= \underline{6x^3y}(2x^4y^2) - \underline{6x^3y}(x^2y) + \underline{6x^3y}(3)$$

THE GREATEST COMMON MONOMIAL FACTOR

The product of all common monomial factors of a multinomial is called the greatest common monomial factor. In what follows, we shall call a common monomial factor simply a **common factor.**

Examples:
1. **5** is the greatest common factor of **5x + 5y.**

2. $6x^3$ is **not** the greatest common factor of $12x^7y^3 - 6x^5y^2 + 18x^3y$.

   $6x^3y$ is the greatest common factor of $12x^7y^3 - 6x^5y^2 + 18x^3y$.

3. $5m^3$ is the greatest common factor of $10m^5n^2 - 5m^3$.

CHECK POINT 1

For each of the following expressions find the greatest common factor.

1. $4a - 16b$

2. $6x^3y + 2x$

3. $2m^5 - 8m^3 + 10m^2$

4. $15p^5q^2 - 10p^4q^3 - 20p^3q^4$

5. $6x^2y^3z - 3x^2y + 8z$

6. $3x^4y^2z - 6x^3y^2z^2 + 15x^2yz^2$

Since $6x(a + b - c) = 6xa + 6xb - 6xc$, we can reverse the multiplication by using the symmetry property of equality.

Hence, $6xa + 6xb - 6xc = 6x(a + b - c)$.

In this approach to factoring, we are "factoring out" of the multinomial the greatest common factor.

COMMON FACTOR METHOD

Step 1. Find the greatest common factor of the multinomial to be factored.

Step 2. Divide the multinomial by its greatest common factor.

Step 3. The factored form of the multinomial is given by the greatest common factor times the quotient obtained in Step 2.

Step 4. Check by multiplication.

Example 1:
Factor $6x^2 + 3xy$.

Solution:

Step 1. The greatest common factor is $3x$.

Step 2. Divide $6x^2 + 3xy$ by $3x$.

$$\frac{6x^2 + 3xy}{3x} = \frac{6x^2}{3x} + \frac{3xy}{3x} = \underline{2x + y}$$
$$\text{Quotient}$$

Step 3. Factored form: $3x\underline{(2x + y)}$

Step 4. Check: $3x(2x + y) = 6x^2 + 3xy$.

Hence, $3x(2x + y)$ is the correct factored form of $6x^2 + 3xy$.

Example 2:
Factor $10a^3 - 15a^2 + 25a$.

Solution:

Step 1. The greatest common factor is $5a$.

Step 2. Divide $10a^3 - 15a^2 + 25a$ by $5a$.

$$\frac{10a^3 - 15a^2 + 25a}{5a} = \frac{10a^3}{5a} - \frac{15a^2}{5a} + \frac{25a}{5a} = \underline{2a^2 - 3a + 5}$$

Step 3. Factored form: $5a\underline{(2a^2 - 3a + 5)}$

Step 4. Check: $5a(2a^2 - 3a + 5) = 10a^3 - 15a^2 + 25a$

Example 3:
    Factor $3x^4y^2z - 6x^3y^2z^2 + 15x^2yz^3$.

Solution:
    Step 1.   The greatest common factor is $3x^2yz$.
              (Note: Write down all the common variables, then identify for each
              variable the lowest power.)

    Step 2.   Divide $3x^4y^2z - 6x^3y^2z^2 + 15x^2yz^3$ by $3x^2yz$.

$$\frac{3x^4y^2z - 6x^3y^2z^2 + 15x^2yz^3}{3x^2yz} = \frac{3x^4y^2z}{3x^2yz} - \frac{6x^3y^2z^2}{3x^2yz} + \frac{15x^2yz^3}{3x^2yz}$$

$$= x^2y - 2xyz + 5z^2$$

    Step 3.   Factored form: $(3x^2yz)(x^2y - 2xyz + 5z^2)$

    Step 4.   Check: $(3x^2yz)(x^2y - 2xyz + 5z^2)$

$$= (3x^2yz)(x^2y) - (3x^2yz)(2xyz) + (3x^2yz)(5z^2)$$

$$= 3x^4y^2z - 6x^3y^2z^2 + 15x^2yz^3$$

Note:
    When factoring multinomials, we look for common factors first and then we
    try other factoring methods.

CHECK POINT 2

Factor, if possible, the following expressions.

1. $7a - 7b$ 2. $3x + 2xy$ 3. $10xy - 3z$

4. $8m^2 - 4mn$ 5. $6x^3 + 12x^2 - 3x$ 6. $4x^3y - x^2y - 8xy^2$

7. $9m^3n - 3m^2t + 2nt^2$ 8. $5a^4b^2c + 2a^3b^3c^2 - 8a^2b^4c^3$

FACTORING TRINOMIALS

Certain trinomials can be factored into two binomials by the following trial and error method.

Type 1.

All terms of the trinomial are positive.

Previously, we used the **FOIL** shortcut method to multiply two binomials.

Example:

$$(x + 4)(x + 2) = x \cdot x + 2 \cdot x + 4 \cdot x + 4 \cdot 2$$
$$= x^2 + 6x + 8$$

1. $F = x^2$,   the product of the two first terms of the binomials.

2. $L = 8$,   the product of the two last terms of the binomials.

3. $O + I = 6x$,   the sum of the inner and outer products.

Note: $F \cdot L = 8 \cdot x^2$
       $O \cdot I = (2x)(4x) = 8x^2$ } Hence, $F \cdot L = O \cdot I$.

We will use the above facts to factor trinomials.

Example 1:
   Factor $x^2 + 7x + 12$ into two binomials.

Solution:
   Step 1.
       Find the product of $F = x^2$ and $L = 12$.
       $x^2 \cdot 12 = 12x^2$

   Step 2.
       Determine two factors of the product obtained in Step 1 such that their sum is equal to the middle term of the trinomial.

   Note:
       We normally list these factors in order as indicated below until we discover the pair whose sum is the middle term of the trinomial.

       $12x^2 = (x)(12x)$ and $x + 12x = 13x$

       $12x^2 = (2x)(6x)$ and $2x + 6x = 8x$

       $12x^2 = (3x)(4x)$ and $3x + 4x = 7x$ (middle term)

       Here, the factors 4x and 3x give us the sum of 7x which is the middle term (O + I) of the trinomial to be factored. We can choose either 3x or 4x to be the outer product. Let's choose 4x to be the outer product.

Step 3.

Write $x^2 + 7x + 12$ in **F.O.I.L.** form and group the first two and the last two terms.

$$\overset{\textbf{F}\quad\textbf{O}\quad\textbf{I}\quad\textbf{L}}{x^2 + 7x + 12 = x^2 + 4x + 3x + 12}$$

$$= (x^2 + 4x) + (3x + 12)$$

Step 4.

Factor the two groups.

$(x^2 + 4x) + (3x + 12) = x(x + 4) + 3(x + 4)$ **Note**: $x + 4$ is a common

$= (x + 4)[x + 3]$ factor in both terms.

Step 5.

Check by multiplying the binomials.

$$(x + 4)(x + 3) = x^2 + 4x + 3x + 12$$

$$= x^2 + 7x + 12$$

Therefore, $x^2 + 7x + 12 = (x + 4)(x + 3)$.

17

Example 2:

Factor $x^2 + 10x + 24$.

Solution:

Step 1.

$$F \cdot L = x^2 \cdot 24 = 24x^2$$

Step 2.

Determine two factors of $24x^2$ whose sum is $10x$.

$(x)(24x)$	and	$x + 24x = 25x$
$(2x)(12x)$	and	$2x + 12x = 14x$
$(3x)(8x)$	and	$3x + 8x = 11x$
$(4x)(6x)$	and	$4x + 6x = \mathbf{10x}$

Let's choose $4x$ to be the outer and $6x$ to be the inner product.

Step 3.

Write $x^2 + 10x + 24$ in **F.O.I.L.** form and group the first two and the last two terms.

$$\overset{\textbf{F}\quad\textbf{O}\quad\textbf{I}\quad\textbf{L}}{x^2 + 10x + 24 = x^2 + 4x + 6x + 24}$$

$$= (x^2 + 4x) + (6x + 24)$$

Step 4.

Factor the two groups.

$(x^2 + 4x) + (6x + 24) = x(x + 4) + 6(x + 4)$

$= (x + 4)[x + 6]$

Step 5.

Check: $(x + 4)(x + 6) = x^2 + 6x + 4x + 24$

$= x^2 + 10x + 24$

Therefore, $x^2 + 10x + 24 = (x + 4)(x + 6)$.

18

CHECK POINT 3

Factor the following trinomials.

1. $x^2 + 5x + 6$ 2. $x^2 + 11x + 18$

3. $x^2 + 8x + 15$ 4. $x^2 + 13x + 36$

19

Type 2.   Only the middle term of the trinomial is negative.

Example:

$$(x - 2)(x - 3) = x^2 \overset{F \quad\quad O \quad\; I \quad\; L}{- 3x - 2x + 6}$$

$$= x^2 \underset{O+I}{\underline{- 5x}} + 6$$

Note:

1. $-5x$, the sum of the outer and inner products is negative.
2. 6, the product of the last terms is positive.

20

Example:

Factor $x^2 - 11x + 24$ into two binomials.

Solution:

Step 1. $F \cdot L = 24x^2$

Step 2. Determine two factors of $24x^2$ whose sum is $-11x$.

Since the product of the two factors is positive and their sum is negative, the two factors must be negative.

$(-x)(-24x)$ and $(-x) + (-24x) = -25x$
$(-2x)(-12x)$ and $(-2x) + (-12x) = -14x$
$(-3x)(-8x)$ and $(-3x) + (-8x) = -11x$

Let's choose $-3x$ as the outer product, then $-8x$ is the inner product.

Step 3. Write $x^2 - 11x + 24$ in **F.O.I.L.** form and group the first two and the last two terms.

$$x^2 - 11x + 24 = x^2 \overset{F \quad\;\; O \quad\quad\; I \quad\quad L}{+ (-3x) + (-8x) + 24}$$

$$= (x^2 - 3x) + (-8x + 24)$$

Step 4. Factor the two groups.

$$(x^2 - 3x) + (-8x + 24) = x(x - 3) - 8(x - 3)$$

$$= (x - 3)[x - 8]$$

Step 5. Check: $(x - 3)(x - 8) = x^2 - 8x - 3x + 24$

$$= x^2 - 11x + 24$$

Therefore, $x^2 - 11x + 24 = (x - 3)(x - 8)$.

CHECK POINT 4

Factor the following trinomials.

1. $x^2 - 10x + 16$　　　　　　　　2. $x^2 - 11x + 30$

Type 3.　The last term of the trinomial is negative.

Examples:

1. $(x + 8)(x - 5) = x^2 - 5x + 8x - 40$

$$= x^2 + 3x - 40$$

2. $(x - 8)(x + 5) = x^2 + 5x - 8x - 40$

$$= x^2 - 3x - 40$$

Note:

−40, the product of the last terms in both examples is negative. This indicates that the last terms of the two binomials have **opposite** signs.

Example 1:

Factor　$x^2 + 7x - 18$

Solution:

Step 1.　$F \cdot L = -18x^2$

Step 2.　Determine two factors of　$-18x^2$　whose sum is 7x.

Since the product is negative, the two factors must have opposite signs.

Since the middle term is positive, the larger of the absolute values of the two factors must be positive. Hence, listing the factors, we place the negative sign with the smaller.

$(-x)(18x)$　　　and　　$(-x) + 18x = 17x$
$(-2x)(9x)$　　　and　　$(-2x) + 9x = 7x$

Let's choose　$-2x$　as the outer product, then　$9x$　is the inner product.

Step 3.　Write　$x^2 + 7x - 18$　in **F. O. I. L.** form, group and factor,

$$x^2 + 7x - 10 = \overset{F}{x^2} + \overset{O}{(-2x)} + \overset{I}{9x} - \overset{L}{18}$$

$$= (x^2 - 2x) + (9x - 18)$$

$$= x(x - 2) + 9(x - 2)$$

$$= (x - 2)[x + 9]$$

Check:　$(x - 2)(x + 9) = x^2 + 9x - 2x - 18$

$$= x^2 + 7x - 18$$

Therefore, $x^2 + 7x - 18 = (x - 2)(x + 9)$

Example 2:
Factor $m^2 - 3m - 28$

Solution:
Step 1. $F \cdot L = -28m^2$

Step 2.   Determine two factors of $-28m^2$ whose sum is $-3m$.

Since $-28m^2$ is negative, the two factors have opposite signs.

Since the middle term is negative, the larger of the absolute values of the two factors must be negative. Hence, listing the factors, we place the negative sign with the larger.

$(m)(-28m)$ and $m + (-28m) = -27m$
$(2m)(-14m)$ and $2m + (-14m) = -12m$
$(4m)(-7m)$ and $4m + (-7m) = \mathbf{-3m}$

Use $4m$ as the outer and $(-7m)$ as the inner product.

Step 3. $m^2 - 3m - 28 = m^2 + 4m + (-7m) - 28$

$$= (m^2 + 4m) + (-7m - 28)$$

$$= m(m + 4) + (-7)(m + 4)$$

$$= m(m + 4) - 7(m + 4)$$

$$= (m + 4)[m - 7]$$

Check: $(m + 4)(m - 7) = m^2 - 7m + 4m - 28$

$$= m^2 - 3m - 28$$

Therefore, $m^2 - 3m - 28 = (m + 4)(m - 7)$

CHECK POINT 5

Factor the following trinomials.

1. $x^2 - 3x - 18$

2. $y^2 + y - 30$

3. $a^2 + 6a - 16$

4. $x^2 - 5x - 24$

5. $p^2 + 9p - 36$

TRINOMIALS WITH COMMON FACTORS

As noted earlier, in factoring an expression we look first for common monomial factors and then try other factoring methods.

Example 1:
Factor $2x^2 + 14x + 20$

Solution:
Step 1.   Remove the common factor of 2: $2(x^2 + 7x + 10)$

Step 2.   Factor the trinomial factor $x^2 + 7x + 10$. $F \cdot L = 10x^2$

Step 3. (x)(10x)   and   x + 10x = 11x
         (2x)(5x)   and   2x + 5x = **7x**

Use 2x as the outer and 5x as the inner product.

Step 4. Write $x^2 + 7x + 10$ in **F. O. I. L.** form, group and factor.

$$x^2 + 7x + 10 = x^2 + 2x + 5x + 10$$
$$= (x^2 + 2x) + (5x + 10)$$
$$= x(x + 2) + 5(x + 2)$$
$$= (x + 2)[x + 5]$$

Hence, $2x^2 + 14x + 20 = 2(x^2 + 7x + 10)$
$$= 2(x + 2)(x + 5)$$

Check: $2(x + 2)(x + 5) = 2(x^2 + 7x + 10) = 2x^2 + 14x + 20$

27

Example 2:
    Factor $4y^3 + 8y^2 - 60y$

Solution:
   Step 1.   Remove the common factor of 4y.

$$4y^3 + 8y^2 - 60y = 4y(y^2 + 2y - 15)$$

   Step 2.   Factor $y^2 + 2y - 15$.

$$F \cdot L = -15y^2$$

   Step 3.   (-y)(15y)   and   -y + 15y = 14y
             (-3y)(5y)   and   -3y + 5y = **2y**

   Step 4. $y^2 + 2y - 15 = y^2 - 3y + 5y - 15$
$$= (y^2 - 3y) + (5y - 15)$$
$$= y(y - 3) + 5(y - 3)$$
$$= (y - 3)[y + 5]$$

Hence, $4y^3 + 8y^2 - 60y = 4y(y^2 + 2y - 15)$
$$= 4y(y - 3)(y + 5)$$

Check: $4y(y - 3)(y + 5) = 4y(y^2 + 2y - 15)$
$$= 4y^3 + 8y^2 - 60y$$

28

CHECK POINT 6

Factor the following trinomials.

1. $5x^2 + 40x + 60$ 2. $3m^3 - 12m^2 - 36m$

FACTORING TRINOMIALS WITH NO COMMON FACTORS AND WHERE THE COEFFICIENT OF THE FIRST TERM IS NOT ONE.

Examples: 1. $2x^2 + 5x + 3$ 2. $3m^2 - 7m + 4$ 3. $6y^2 - 17y + 5$

Example 1:
Factor $2x^2 + 5x + 3$.

Solution:
Step 1. $F \cdot L = 6x^2$

Step 2. $(x)(6x)$
$(2x)(3x)$ and $2x + 3x =$ **$5x$**

Step 3. $2x^2 + 5x + 3 = 2x^2 + 2x + 3x + 3$

$= (2x^2 + 2x) + (3x + 3)$

$= 2x(x + 1) + 3(x + 1)$

$= (x + 1)[2x + 3]$

Hence, $2x^2 + 5x + 3 = (x + 1)(2x + 3)$

Check by multiplying the factors.

Example 2:
Factor $6a^2 - 17a + 5$.

Solution:
Step 1. $F \cdot L = 30a^2$

Step 2.   **Note:**   Only the middle term is negative.
Hence, both factors are negative.

$(-a)(-30a)$
$(-2a)(-15a)$ and $(-2a) + (-15a) =$ **$-17a$**

Step 3. $6a^2 - 17a + 5 = 6a^2 + (-2a) + (-15a) + 5$

$= (6a^2 - 2a) - (15a - 5)$

$= 2a(3a - 1) - 5(3a - 1)$

$= (3a - 1)[2a - 5]$

Hence, $6a^2 - 17a + 5 = (3a - 1)(2a - 5)$

Check by multiplying the factors.

32

Example 3:
Factor $8x^2 + 14x - 15$

Solution:
Step 1. $F \cdot L = -120x^2$

Step 2.   **Note:**   The last term is negative and the middle term is positive. Hence, the larger of the absolute values of the two factors must be positive.

$(-x)(120x)$
$(-2x)(60x)$
$(-3x)(40x)$
$(-4x)(30x)$
$(-5x)(24x)$
$(-6x)(20x)$ and $(-6x) + 20x = 14x$

Step 3. $8x^2 + 14x - 15 = 8x^2 + (-6x) + 20x - 15$

$$= (8x^2 - 6x) + (20x - 15)$$

$$= 2x(4x - 3) + 5(4x - 3)$$

$$= (4x - 3)[2x + 5]$$

Hence, $8x^2 + 14x - 15 = (4x - 3)(2x + 5)$
Check by multiplying the factors.

33

CHECK POINT 7

Factor the following trinomials.

1. $2x^2 + 3x + 1$

2. $6m^2 + m - 5$

3. $5x^2 - 6x + 1$

4. $3a^2 - 7a + 4$

34

Note: The above process can become very laborious. However, with practice we can develop the skill of making the trials mentally and choosing for calculations only those factors which look promising for giving us the correct middle term.

You should note that many trinomials cannot be factored by the trial and error method shown in this unit. Later in this text we shall present a method called "completing the square" which can be used to factor any trinomial.

35

SHORT CUT
When the coefficient of the squared term is 1, we can use the following short cut to factor the trinomial.

$$(x + a)(x + b) = x^2 + bx + ax + ab$$
$$= x^2 + (b + a)x + a \cdot b$$

Note:
1.   The coefficient of the middle term of the trinomial is equal to the sum of the factors **a** and **b** which make up its last term.

2. The two factors **a** and **b** of the last term in the trinomial $x^2 + (b + a)x + ab$ are the two last terms in the binomials $(x + a)$ and $(x + b)$.

3. $(x + a)$ and $(x + b)$ are the factors of the trinomial

$$x^2 + (b + a)x + a \cdot b$$

We can use these facts to factor certain trinomials quickly.

36

Example 1: Factor $x^2 + 3x + 2$

Solution:

Step 1.   Write the trinomial as an indicated product of two binomials with **x** as the first terms.

$$x^2 + 3x + 2 = (x \quad)(x \quad)$$

Step 2.   Determine two factors of the last term of the trinomial such that their sum is the coefficient of the middle term.
$$1 \cdot 2 = 2 \quad \text{and} \quad 1 + 2 = 3$$

Step 3.   Use the factors **1** and **2** as the second terms in the two binomials.

$$\therefore \ x^2 + 3x + 2 = (x + 1)(x + 2)$$

Check answers by multiplying the binomial factors.

37

Example 2: Factor $m^2 - m - 6$

Solution:

Step 1. $m^2 - m - 6 = (m \quad)(m \quad)$

Step 2.   Since the middle term is negative and the last term is negative, the factors have opposite signs and the larger factor is negative.

$$(1)(-6) \quad \text{and} \quad 1 + (-6) = -5$$
$$(2)(-3) \quad \text{and} \quad 2 + (-3) = -1$$

Step 3. $(m^2 - m - 6) = (m + 2)(m - 3)$

Check answer by multiplying the binomial factors.

38

Example 3: Factor $t^2 - 7t + 12$

Solution:

Step 1. $t^2 - 7t + 12 = (t \quad)(t \quad)$

Step 2.   Since the middle term is negative and the last term is positive, both factors are negative.

$$(-1)(-12) \quad \text{and} \quad (-1) + (-12) = -13$$
$$(-2)(-6) \quad \text{and} \quad (-2) + (-6) \ = -8$$
$$(-3)(-4) \quad \text{and} \quad (-3) + (-4) \ = \textbf{-7}$$

Step 3. $t^2 - 7t + 12 = (t - 3)(t - 4)$

Check answer by multiplying the binomial factors.

39

CHECK POINT 8

Factor the following trinomials using the short cut method.

1. $x^2 + 6x + 5$ 2. $m^2 - 2m - 8$ 3. $t^2 - 9t + 18$

4. $a^2 + 4a - 21$ 5. $y^2 + 12y + 20$ 6. $n^2 - 5n - 24$

40

FACTORING THE DIFFERENCE OF TWO SQUARES

Binomials which can be written as the difference of two squares are easily factored by using the reverse of the special product.

$(x + y)(x - y) = x^2 - y^2$

Example: Factor $x^2 - 9$

Solution:

Step 1. Write both terms of the binomial as a square. $x^2 - 9 = x^2 - 3^2$

Use the special product $(x + y)(x - y) = x^2 - y^2$ in reverse.

Therefore, $x^2 - 3^2 = (x + 3)(x - 3)$.

41

Examples:

1. $a^2 - 25 = a^2 - 5^2$
    $= (a + 5)(a - 5)$

2. $4x^2 - 1 = (2x)^2 - 1^2$
    $= (2x + 1)(2x - 1)$

3. $9y^2 - 64z^2 = (3y)^2 - (8z)^2$
    $= (3y + 8z)(3y - 8z)$

4. $16 - 25m^2 = 4^2 - (5m)^2$
    $= (4 + 5m)(4 - 5m)$

5. $a^4 - b^4 = (a^2)^2 - (b^2)^2$ Reverse exponential Law 3.3
    $= (a^2 + b^2)(a^2 - b^2)$
    $= (a^2 + b^2)(a + b)(a - b)$

Note: The sum of two squares $a^2 + b^2$ cannot be factored.

42

DIFFERENCE OF SQUARES WITH COMMON FACTORS

For some differences of squares, we may not be able to write both terms as squares. However, that may be possible after factoring out a common factor.

Examples:

1. $5p^2 - 125 = 5(p^2 - 25)$
    $= 5(p^2 - 5^2)$
    $= 5(p + 5)(p - 5)$

2. $3x^3 - 48x = 3x(x^2 - 16)$
    $= 3x(x^2 - 4^2)$
    $= 3x(x + 4)(x - 4)$

3. $-6m + 54m^3 = -6m(1 - 9m^2)$
    $= -6m(1 - (3m)^2)$
    $= -6m(1 + 3m)(1 - 3m)$

CHECK POINT 9

Factor the following expressions.

1. $x^2 - y^2$ 2. $4a^2 - 9b^2$ 3. $36m^3 - m$

4. $100 - 9p^2$ 5. $p^4 - q^4$ 6. $-7x + 63x^3$

FACTORING THE SUM OR DIFFERENCE OF TWO CUBES

To factor expressions which are the sum or the difference of two cubes, we use the reverse of the special products.

$$(x + y)(x^2 - xy + y^2) = x^3 + y^3$$
$$(x - y)(x^2 + xy + y^2) = x^3 - y^3$$

Example:
Factor $8a^3 - 125b^3$.

Solution:

Step 1.   Write both terms of the binomial as a cube.

$$8a^3 - 125b^3 = (2a)^3 - (5b)^3$$

Step 2.   Use the reverse of the special product.

$$(x^3 - y^3) = (x - y)(x^2 + xy + y^2)$$

Therefore, $(2a)^3 - (5b)^3 = (2a - 5b)(4a^2 + 10ab + 25b^2)$.

More Examples:

1. $m^3 + 1 = m^3 + 1^3 = (m + 1)(m^2 - m + 1)$

2. $64 - a^3 = 4^3 - a^3 = (4 - a)(16 + 4a + a^2)$

3. $27x^3 - 8 = (3x)^3 - 2^3 = (3x - 2)(9x^2 + 6x + 4)$

4. $p^6 - 1 = (p^2)^3 - 1^3 = (p^2 - 1)(p^4 + p^2 + 1)$
$\quad\quad = (p + 1)(p - 1)(p^4 + p^2 + 1)$

Note:

Example 4 could also be factored first as a difference of squares and then as a difference and a sum of cubes as shown below.

$$p^6 - 1 = (p^3)^2 - 1^2 \ = (p^3 - 1)(p^3 + 1)$$
$$= (p - 1)(p^2 + p + 1)(p + 1)(p^2 - p + 1)$$

THE SUM OR DIFFERENCE OF CUBES WITH COMMON FACTORS

As was the case for the difference of squares, we may be able to factor an expression as a sum or difference of cubes after removing a common factor.

Examples:

1. $5xy^3 - 5x^4 = 5x(y^3 - x^3)$
$$= 5x(y - x)(y^2 + xy + x^2)$$

2. $54a^4 + 16ab^3 = 2a(27a^3 + 8b^3)$
$$= 2a[(3a)^3 + (2b)^3]$$
$$= 2a(3a + 2b)(9a^2 - 6ab + 4b^2)$$

3. $m^8 - m^2n^3 = m^2(m^6 - n^3)$
$$= m^2[(m^2)^3 - n^3]$$
$$= m^2(m^2 - n)(m^4 + m^2n + n^2)$$

CHECK POINT 10

Factor the following expressions.

1. $p^3 + 1$

2. $8 - a^3$

3. $125m^3 + 27$

4. $24x^3y - 3y^4$

5. $128pt^3 + 250p^4$

6. $3b^6 - 81$

DRILL EXERCISES

Factor, if possible, the greatest common monomial from each of the following expressions.

1. $9x + 3y$

2. $8ab^2 - 4b$

3. $10n^6 + 5n^5 - 15n^2$

4. $3x^3y^2z - 6x^2z + 5y$

5. $21p^4q^5 - 7p^3q^3 - 14pq^4$

6. $3x^5yz^2 + 9x^4y^3z - 12x^3yz^2$

Factor, if possible, the following trinomials.

7. $5x^3 + 10x^2 - 40x$

8. $m^2 - 11m + 18$

9. $2x^2 + 28x + 90$

10. $x^2 + 6x + 9$

11. $a^2 - 2a - 48$

12. $x^2 + x - 6$

13. $x^2 + 9x + 20$

14. $h^2 + 6h - 27$

15. $x^2 + 9x + 30$

16. $y^2 - 11y + 28$

17. $y^2 - 10y - 24$

18. $2x^2 - 13x + 15$

19. $10a^2 + 19a - 15$

20. $3y^2 + 7y + 4$

21. $8x^2 - 2x - 21$

22. $12m^2 + 11m - 15$

23. $14y^2 - 11y + 2$

24. $54a^2 + 87a + 28$

25. $11x^2 - 75x - 14$

26. $15t^2 - 34t + 15$

27. $8p^2 - 2p - 1$

DRILL EXERCISES (continued)

28. $21r^2 + r - 10$

29. $8y^2 + 26y + 15$

30. $4x^2 - 4x - 3$

31. $6m^2 + m - 12$

Factor, if possible, the following expressions.

32. $a^2 - b^2$

33. $9x^2 - 16y^2$

34. $t^3 + 1$

35. $64 - 25m^2$

36. $4x^4 - y^4$

37. $27 - k^3$

38. $9y^2 + 4$

39. $49a^3 - a$

40. $64d^3 + 27$

41. $-12x + 3x^3$

42. $2t^6 - 16$

43. $375n^4 + 24m^3n$

ASSIGNMENT EXERCISES

Factor, if possible, the following expressions.

1. $5m - 25n$

2. $a^2 + 6a + 8$

3. $a^2 + 3a - 54$

4. $8x^2y^4 + 16y$

5. $m^2 - 36$

6. $x^2 - 4x - 32$

7. $4x^2 + 9x + 5$

8. $x^2 - 5x + 6$

9. $m^4 - n^4$

10. $3y^2 - 5y + 2$

11. $1 - x^3$

12. $m^2 + 19m + 48$

13. $27a^3 + 64b^3$

14. $25 + 4x^2$

15. $3x^2 - 30x + 63$

16. $5a^2 - 3a - 14$

17. $16t^3 - 2p^6$

18. $25x^2 - 4y^2$

19. $-32a + 8a^3$

20. $y^2 - 15y + 56$

21. $8k^3 - 1$

22. $8x^2 + 14x - 49$

23. $125u^3 + v^6$

24. $5y^3 - 20y^2 - 60y$

25. $81 - 100x^2$

26. $81a^3b - 3b^4$

27. $10t^2 - 33t + 20$

28. $4m^2 + 7m - 2$

29. $3a^2 + 10a + 7$

30. $4y^2 + 4y - 3$

31. $9x^2 - 29x + 6$

32. $2r^2 - 13r + 20$

33. $9m^2 - 9m - 4$

34. $2x^2 - 13x - 24$

35. $12a^2 + 20a + 3$

36. $18p^2 + 27p + 7$

37. $4x^3y^4z^2 - 8x^2y^3z^5 - 2xy^3z$

38. $6a^5b^2 - 3a^4b + 9a^3b^3$

39. $7p^4q^3r^2 + 14p^2q + 2r$

UNIT **10**

SOLVING QUADRATIC EQUATIONS BY FACTORING, WORD PROBLEMS

Objectives: After having worked through this unit, the student should be capable of:

1.   solving certain quadratic equations by factoring;
2.   solving word problems involving the solution of a quadratic equation.

1

REVIEW

LINEAR EQUATIONS IN ONE UNKNOWN. STANDARD FORM.

Previously, we learned how to solve equations in one variable raised to the first power.

These equations we called linear equations in one unknown. Any of these equations can be written in the **standard form**

$bx + c = 0$, where b and c are constants and $b \neq 0$.

2

Example:

Write $3(2x + 5) - 5x = 2(3 - x)$ in the form $bx + c = 0$.

Solution:

Step 1.　Remove all parentheses.

$3(2x + 5) - 5x = 2(3 - x)$

$6x + 15 - 5x = 6 - 2x$

Step 2.　Transpose all terms from the right to the left side of the equation.

$6x + 15 - 5x - 6 + 2x = 0$

Step 3.　Collect like terms.

$3x + 9 = 0$

Here, $b = 3$ and $c = 9$.

3

Now that we know how to multiply and factor certain algebraic expressions, we shall learn how to solve problems of the following nature.

Example:

The area of a rectangular metal plate is 30 cm^2.
The length is 7 cm longer than the width.
Find the dimensions of the plate.

Since the area equals the width time the length, we have

$30 = x(x + 7)$　or　$x(x + 7) = 30$

$x^2 + 7x = 30$

$$\boxed{\text{Area } = 30 \text{ cm}^2} \quad x$$

$x + 7$

4

QUADRATIC EQUATIONS IN ONE UNKOWN. STANDARD FORM

To solve problems such as the one above, we have to learn how to solve equations which have the standard form

$ax^2 + bx + c = 0$　where a, b and c are constants and $a \neq 0$.

We call these equations quadratic because 2 is the highest exponent present when the equation is written in standard form.

The restriction $a \neq 0$ is necessary since if $a = 0$ and $b \neq 0$, we have

$0x^2 + bx + c = 0$ or
$bx + c = 0$

which is the general **linear** equation in one unknown written in standard form.

5

Examples:

1. $x^2 - 2x + 1 = 0$ is a quadratic equation in x written in standard form where $a = 1$, $b = -2$, $c = 1$.

2. $-7y^2 + 5y - 19 = 0$ is a quadratic equation in y written in standard form where $a = -7$, $b = 5$, $c = -19$.

3. $2x^3 - 4x + 3 = 0$ is **not** a quadratic equation. The highest power of x is **not** two.

4. $3x^2 + 6x = 0$ is a quadratic equation in x written in standard form where $a = 3$, $b = 6$, $c = 0$.

5. $16m^2 - 9 = 0$ is a quadratic equation in m written in standard form where $a = 16$, $b = 0$, $c = -9$.

6

COMPLETE QUADRATIC EQUATIONS

Equations in the form $ax^2 + bx + c = 0$ which have all three terms, are referred to as complete quadratic equations.

Examples:

1. $6x^2 - 5x + 1 = 0$
2. $m^2 + 2m + 4 = 0$

7

INCOMPLETE QUADRATIC EQUATIONS

If $c = 0$, we have $ax^2 + bx = 0$. These equations which do not have the constant term are referred to as incomplete quadratic equations.

Examples:

1. $4y^2 + 3y = 0$
2. $-9a^2 + 4a = 0$

8

PURE QUADRATIC EQUATIONS

If $b = 0$, we have $ax^2 + c = 0$. These equations which do not have a middle term are referred to as pure quadratic equations.

Examples:

1. $100x^2 - 25 = 0$
2. $p^2 + 17 = 0$

9

In practical problems, quadratic equations normally do not appear in the standard form.

However, in order to solve these equations by factoring, we first have to put them into the standard form.

Example:
Write $(x + 2)(x - 3) = 2(3x - 2)$ in the form $ax^2 + bx + c = 0$.

Solution:

Step 1.   Remove all parentheses.

$(x + 2)(x - 3) = 2(3x - 2)$

$x^2 - x - 6 = 6x - 4$

Step 2.   Transpose all terms from the right to the left side of the equation.

$x^2 - x - 6 - 6x + 4 = 0$

Step 3.   Collect like terms.

$x^2 - 7x - 2 = 0$

10

CHECK POINT 1

Change the following equations into the standard form.

1. $x(x + 6) = 30$

2. $(y - 5)(y + 1) = -9$

3. $m(3m + 2) = (m + 1)^2 + 7$

4. $6(x + 3) + 6(x - 2) = (x + 2)(x + 3)$

5. $4p^2 + p = 2(2p^2 + 3p + 1)$

11

In order to solve quadratic equations by factoring, we will use the following property:

If a and b are any two numbers and $a \cdot b = 0$,
then either $a = 0$, or $b = 0$, or both $a = 0$ and $b = 0$.

This can easily be shown to be true.
If $a = 0$, then the statement above is true.
If $a \neq 0$, then we can divide both sides by a: $\dfrac{a \cdot b}{a} = \dfrac{0}{a}$, $\therefore$ $b = 0$.

12

Examples:

1.   If $x(x - 2) = 0$, then either $x = 0$, or $x - 2 = 0$.

2.   If $(x + 3)(x - 4) = 0$, then either $x + 3 = 0$, or $x - 4 = 0$.

3.   If $7(x + 9) = 0$, then $x + 9 = 0$.

SOLVING INCOMPLETE QUADRATIC EQUATIONS

To solve an incomplete quadratic equation, we

1. remove the greatest common factor from both terms,

2. equate each factor to zero and solve the resulting linear equations,

3. check the solutions by substitution into the original equation.

Example 1:

Solve $x^2 + 7x = 0$.

Solution:

Step 1. Remove the greatest common factor from $x^2 + 7x$.

$$x^2 + 7x = 0$$

$$x(x + 7) = 0$$

Step 2. Equate each factor to zero and solve the resulting equations.

$$x = 0 \ \text{ or } \ x + 7 = 0$$

$$\therefore \ x = 0 \ \text{ or } \ \ \ \ \ x = -7$$

Step 3. Check: 1. If $x = 0$, $x^2 + 7x = 0^2 + 7(0)$
$$= 0$$

2. If $x = -7$, $x^2 + 7x = (-7)^2 + 7(-7)$
$$= 49 - 49$$
$$= 0$$

Hence, the solutions to $x^2 + 7x = 0$ are $x = 0$ and $x = -7$.

Note:

Do not solve incomplete quadratic equations by dividing both sides by the greatest common factor.

Example:

$$x^2 + 7x = 0$$

$$x(x + 7) = 0$$

$$\frac{x(x + 7)}{x} = \frac{0}{x}$$

$$\therefore \ x + 7 = 0$$

$$x = -7$$

If we use this method of solving incomplete quadratic equations, **we lose the solution $x = 0$.**

16

Example 2:

Solve $-5m^2 + 20m = 0$

Solution:

Step 1. $-5m^2 + 20m = 0$

$-5m(m - 4) = 0$

Step 2. Either $-5m = 0$ or $m - 4 = 0$

$\therefore$ $m = 0$ or $m = 4$

Step 3. Check: 1. If $m = 0$, $-5m^2 + 20m = -5(0^2) + 20(0)$

$= 0$

2. If $m = 4$, $-5m^2 + 20m = -5(4^2) + 20(4)$

$= -5(16) + 80$

$= 0$

Hence, the solutions to $-5m^2 + 20m = 0$ are $m = 0$ and $m = 4$.

17

CHECK POINT 2

Solve the following equations by factoring:

1. $3y^2 - 12y = 0$ 2. $-4p^2 + 28p = 0$

18

SOLVING PURE QUADRATIC EQUATIONS

For us to solve pure quadratic equations, we have to put them into the form of a difference of squares.

At this point, we are not able to solve equations such as

$x^2 + 16 = 0$ or $5x^2 - 7 = 0$

19

To solve pure quadratic equations which are in the form of a difference of squares, we:

1. factor the difference of squares,

2. equate each factor to zero and solve the resulting linear equations,

3. check the solutions by substitution into the original equation.

20

Example 1:

Solve $x^2 - 9 = 0$

Solution:

Step 1. Factor $x^2 - 9$.

$x^2 - 9 = 0$

$\therefore$ $(x + 3)(x - 3) = 0$

Step 2.   Equate each factor to zero and solve the equations.

$$x + 3 = 0 \quad \text{or} \quad x - 3 = 0$$

$$\therefore \quad x = -3 \quad \text{or} \quad x = 3$$

Step 3.   Check: 1.   If $x = -3$, $x^2 - 9 = (-3)^2 - 9$

$$= 9 - 9$$

$$= 0$$

2.   If $x = 3$, $x^2 - 9 = 3^2 - 9$

$$= 0$$

Hence, $x = -3$ and $x = 3$ are the solutions to $x^2 - 9 = 0$.

21

Example 2:

Solve $-a^2 + 64 = 0$.

Solution:
Step 1.   Change $-a^2 + 64 = 0$ into the form of a difference of squares and factor.

$$(-1)(-a^2 + 64) = 0(-1) \qquad \text{Multiply both sides by } -1.$$

$$a^2 - 64 = 0$$

$$a^2 - 8^2 = 0$$

$$\therefore \quad (a + 8)(a - 8) = 0$$

Step 2.   Either $a + 8 = 0$ or $a - 8 = 0$.

$$\therefore \quad a = -8 \quad \text{or} \quad a = 8$$

Step 3.   Check: 1.   If $a = -8$, $-a^2 + 64 = -(-8)^2 + 64$
$$= 0$$

2.   If $a = 8$, $-a^2 + 64 = -8^2 + 64$
$$= 0$$

Hence, $a = -8$ and $a = 8$ are the solutions to $-a^2 + 64 = 0$.

22

CHECK POINT 3

Solve the following equations by factoring:

1. $m^2 - 49 = 0$ 　　　　　　　　　　　　2. $-y^2 + 100 = 0$

23

SOLVING COMPLETE QUADRATIC EQUATIONS
To solve complete quadratic equations , we:

1.   factor the trinomials by trial and error,

2.   equate each factor to zero and solve the resulting linear equations,

3.   check the solutions by substitution into the original equation.

24

Example 1:

Solve $x^2 + 3x + 2 = 0$

Solution:

Step 1.   Factor $x^2 + 3x + 2$.

$$x^2 + 3x + 2 = 0$$

$$\therefore \ (x + 2)(x + 1) = 0$$

Step 2.   Equate each factor to zero and solve the equations

$$x + 2 = 0 \quad \text{or} \quad x + 1 = 0$$

$$\therefore \qquad x = -2 \quad \text{or} \qquad x = -1$$

Step 3.   Check: 1.   If $x = -2$, $x^2 + 3x + 2 = (-2)^2 + 3(-2) + 2$

$$= 4 - 6 + 2$$

$$= 0$$

2.   If $x = -1$, $x^2 + 3x + 2 = (-1)^2 + 3(-1) + 2$

$$= 1 - 3 + 2$$

$$= 0$$

Hence, $x = -2$ and $x = -1$ are the solutions to $x^2 + 3x + 2 = 0$.

25

Example 2:

Solve $-p^2 + 2p + 24 = 0$.

Solution:

Step 1.   Change $-p^2 + 2p + 24 = 0$ so that the coefficient of p^2 is positive and factor.

$$(-1)(-p^2 + 2p + 24) = 0(-1) \qquad \text{Multiply both sides by } -1.$$

$$p^2 - 2p - 24 = 0$$

$$(p + 4)(p - 6) = 0$$

Step 2.   Either $p + 4 = 0$ or $p - 6 = 0$.

$$\therefore \qquad p = -4 \ \text{or} \qquad p = 6$$

Step 3.   Check: 1.   If $p = -4$, $-p^2 + 2p + 24 = -(-4)^2 + 2(-4) + 24$

$$= -16 - 8 + 24$$

$$= 0$$

2.   If $p = 6$, $-p^2 + 2p + 24 = -(6)^2 + 2(6) + 24$

$$= -36 + 12 + 24$$

$$= 0$$

Remember: $-6^2 = -(6 \times 6) = -36$ **but** $(-6)^2 = (-6)(-6) = 36$

CHECK POINT 4

Solve the following equations by factoring.

1. $y^2 + 5y + 6 = 0$ 2. $-m^2 + 3m + 18 = 0$

SUMMARY: SOLVING QUADRATIC EQUATIONS BY FACTORING

Method:

Step 1. Put the given equations into the appropriate standard forms.

$$ax^2 + bx + c = 0$$
$$ax^2 + bx = 0$$
$$ax^2 + c = 0$$

Step 2. Factor the equation by the appropriate method depending on its standard form.

Step 3. Equate the factors to zero and solve the resulting linear equations.

Step 4. Check the solutions by substitution into the original equation

Example 1:

Solve $x^2 + 18x + 2 = 12x - 6$

Solution:

Step 1. Put the equation into its standard form.

$$x^2 + 18x + 2 = 12x - 6$$
$$x^2 + 18x + 2 - 12x + 6 = 0$$
$$x^2 + 6x + 8 = 0$$

Step 2. Factor $x^2 + 6x + 8$

$$(x + 2)(x + 4) = 0$$

Step 3. Equate both factors to zero and solve the equations.

$$x + 2 = 0 \quad \text{or} \quad x + 4 = 0$$
$$\therefore \quad x = -2 \quad \text{or} \quad x = -4$$

Step 4. Check: 1. Substitute $x = -2$ into $x^2 + 18x + 2 = 12x - 6$

$$(-2)^2 + 18(-2) + 2 = 12(-2) - 6$$
$$4 - 36 + 2 = -24 - 6$$
$$-30 = -30$$

2. Substitute $x = -4$ into $x^2 + 18x + 2 = 12x - 6$

$$(-4)^2 + 18(-4) + 2 = 12(-4) - 6$$
$$16 - 72 + 2 = -48 - 6$$
$$-54 = -54$$

Hence, $x = -2$ and $x = -4$ are the solutions to $x^2 + 18x + 2 = 12x - 6$.

Example 2:

Solve $(m - 5)(m + 1) = -9$

Solution:

Step 1.   Put equation into the standard form.

$$(m - 5)(m + 1) = -9$$
$$m^2 - 4m - 5 = -9$$
$$m^2 - 4m + 4 = 0$$

Step 2.   Factor $m^2 - 4m + 4$

$$(m - 2)(m - 2) = 0$$

Step 3.   Equate both factors to zero and solve for m.

$$m - 2 = 0 \quad \text{or} \quad m - 2 = 0$$
$$m = 2 \quad \text{or} \quad m = 2$$

Step 4.   Check: Substitute $m = 2$ into $(m - 5)(m + 1) = -9$

$$(2 - 5)(2 + 1) = -9$$
$$-9 = -9$$

Hence, the equation $(m - 5)(m + 1) = -9$ has the solution $m = 2$.

Example 3:

Solve $7p^2 + 15(p + 2) = 2(p^2 + 13) + 2(p^2 + 4p)$.

Solution:

Step 1.   Put equation into the standard form.

$$7p^2 + 15p + 30 = 2p^2 + 26 + 2p^2 + 8p$$
$$3p^2 + 7p + 4 = 0$$

Step 2.   Factor $3p^2 + 7p + 4$. (If necessary, review unit 9)

$$(3p + 4)(p + 1) = 0$$

Step 3.   Equate both factors to zero and solve for p.

$$3p + 4 = 0 \quad \text{or} \quad p + 1 = 0$$
$$\therefore \quad p = -\frac{4}{3} \qquad\qquad p = -1$$

Step 4.   Check: 1.   If $p = -\frac{4}{3}$, we have

$$7(-\tfrac{4}{3})^2 + 15[(-\tfrac{4}{3}) + 2] = 2[(-\tfrac{4}{3})^2 + 13] + 2[(-\tfrac{4}{3})^2 + 4(-\tfrac{4}{3})]$$
$$7(\tfrac{16}{9}) + 15(\tfrac{2}{3}) = 2(\tfrac{133}{9}) + 2(-\tfrac{32}{9})$$
$$\frac{202}{9} = \frac{202}{9}$$

2.   If $p = -1$, we have

$$7(-1)^2 + 15[(-1) + 2] = 2[(-1)^2 + 13] + 2[(-1)^2 + 4(-1)]$$

$$7 + 15 = 28 - 6$$

$$22 = 22$$

Hence, $p = -\dfrac{4}{3}$ and $p = -1$ are the solutions to

$$7p^2 + 15(p + 2) = 2(p^2 + 13) + 2(p^2 + 4p).$$

Example 4:

     Solve $3y(y - 3) - 10 = 5 - y(9y + 20)$.

Solution:

     Step 1.   Put equation into the standard form.

$$3y^2 - 9y - 10 = 5 - 9y^2 - 20y$$

$$12y^2 + 11y - 15 = 0$$

     Step 2.   Factor $12y^2 + 11y - 15$

$$(3y + 5)(4y - 3) = 0$$

     Step 3.   Equate both factors to zero and solve for y.

$$3y + 5 = 0 \quad\text{or}\quad 4y - 3 = 0$$

$$y = -\frac{5}{3} \quad\text{or}\quad y = \frac{3}{4}$$

     Step 4.   Check:

          1.   If $y = -\dfrac{5}{3}$, $3(-\dfrac{5}{3})[-\dfrac{5}{3} - 3] - 10 = 5 - (-\dfrac{5}{3})[9(-\dfrac{5}{3}) + 20]$

$$3(-\frac{5}{3})(-\frac{14}{3}) - 10 = 5 - (-\frac{5}{3})[-15 + 20]$$

$$\frac{40}{3} = \frac{40}{3}$$

          2.   If $y = \dfrac{3}{4}$, $3(\dfrac{3}{4})[\dfrac{3}{4} - 3] - 10 = 5 - (\dfrac{3}{4})[9(\dfrac{3}{4}) + 20]$

$$(\frac{9}{4})[-\frac{9}{4}] - 10 = 5 - (\frac{3}{4})[\frac{107}{4}]$$

$$-\frac{241}{16} = -\frac{241}{16}$$

Hence, $y = -\dfrac{5}{3}$ and $y = \dfrac{3}{4}$ are the solutions to

$$3y(y - 3) - 10 = 5 - y(9y + 20)$$

CHECK POINT 5

Solve the following equations by factoring.

1. $x(x + 7) = 30$ 2. $6(m + 3) + 6(m - 2) = (m + 2)(m + 3)$

3. $a(3a + 2) = (a + 1)^2 + 7$ 4. $7y(y - 1) + 8 = 3y(y - 4) - 4y + 3$

WORD PROBLEMS

We are now ready to solve word problems which involve the solutions of quadratic equations.

REWIEW:

Method:

Step 1.   Represent the unknown which describes the others by a variable and, if possible, present the problem pictorially.

Step 2.   Express the other unknowns in terms of the chosen variable.

Step 3.   Analyze the word statements and translate them into an equation using the expressions obtained in Steps 1 and 2.

Step 4.   Solve the equation.

Step 5.   Evaluate the unknowns expressed in Step 2.

Step 6.   Interpret, check, and express the solutions in terms of the original statement.

Rewiew:

If W is the width and L the length of a rectangle, then:

Area = W • L   Perimeter = 2W + 2L

If s is one side of a square, then:

Area = s^2 Perimeter = 4s

Example:

The area of a rectangular metal plate is 30 cm^2.
The length is 7 cm longer than the width.
Find the dimensions of the plate.

Solution:

Step 1.   Let w = the width of the metal plate.

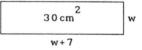

Step 2.   w + 7 = the length of the metal plate.

Step 3.   Since the area of a rectangle equals the width times the length, we have:
30 = w(w + 7)

Step 4.   Solve the equation: 30 = w(w + 7)

$$30 = w^2 + 7w$$

$$w^2 + 7w - 30 = 0$$

$$(w + 10)(w - 3) = 0$$

$$\therefore \quad w + 10 = 0 \quad \text{or} \quad w - 3 = 0$$

$$w = -10 \quad \text{or} \quad w = 3$$

Step 5.   If w = -10, w + 7 = -3 and if w = 3, w + 7 = 10.

Step 6.   Since the width and the length of the metal plate cannot be negative, we discard the solution w = -10.

Hence, the width is 3 cm, the length is 10 cm and the area is 30 cm^2.

Example 2:

The numerical value of the area of a square is three times the numerical value of its perimeter. Find the length of one side.

Solution:

Step 1.   Let x = the length of one side.

Step 2.   Hence, 4x = the perimeter of the square.

x^2 = the area of the square.

Step 3.   "The area of a square is three times its perimeter."

$x^2 = 3(4x)$

Step 4.   Solve $x^2 = 3(4x)$

$x^2 = 12x$

$x^2 - 12x = 0$

$x(x - 12) = 0$

∴ x = 0   or   x - 12 = 0

x = 12

Step 5.   If x = 0, 4x = 0 and $x^2 = 0$.

If x = 12, 4x = 48 and $x^2 = 144$.

Step 6.   If x = 0, we have no area. Hence, we discard this solution.

If x = 12, the perimeter is 48 units and three times 48 is 144.

Hence, the length of one side of the square is 12 units.

Example 3:

The sum of two numbers is 2 and their product is -24. Find the numbers.

Solution:

Step 1.   Let n = one number.

Step 2.   Hence, 2 - n = the other number.

Step 3.   The product of the two numbers is -24.   n(2 - n) = -24

Step 4.   Solve the equation: n(2 - n) = -24

$2n - n^2 = -24$

$-n^2 + 2n + 24 = 0$

$(-1)(-n^2 + 2n + 24) = (-1)0$

$n^2 - 2n - 24 = 0$

$(n + 4)(n - 6) = 0$

∴ n + 4 = 0   or   n - 6 = 0

n = -4   n = 6

Step 5.   If n = -4, 2 - n = 2 - (-4) = 6.

If n = 6, 2 - n = 2 - 6 = -4

Step 6.   The sum of 6 and -4 is 2 and their product is -24.

Hence, the two numbers are 6 and -4.

37

Example 4:

A rectangular platform is to be constructed so that the length is 3 units more than its width. The numerical value of the area of the platform must be 14 more than the numerical value of the perimeter. Find the dimensions of the platform.

Solution:

Step 1.   Let w = the width of the platform.

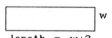

length = w+3

Step 2.   Hence, w + 3 = the length of the platform.

Step 3.   "The area of the platform must be 14 more than the perimeter."

Area = w(w + 3)   and   Perimeter = 2w + 2(w + 3)

$\therefore$ w(w + 3) = 2w + 2(w + 3) + 14

Step 4.   Solve the equation.

$$w(w + 3) = 2w + 2(w + 3) + 14$$
$$w^2 + 3w = 2w + 2w + 6 + 14$$
$$w^2 - w - 20 = 0$$
$$(w + 4)(w - 5) = 0$$

$\therefore$ w + 4 = 0   or   w - 5 = 0
         w = -4   or   w = 5

Step 5.   We discard w = -4 since the width cannot be negative.
If w = 5, w + 3 = 5 + 3 = 8.

Step 6.   The area less 14 is (5)(8) - 14 = 26   and
the perimeter is 2(5) + 2(8) = 26.

Hence, the width must be 5 units and the length 8 units.

38

Example 5:

A variable voltage in a given electrical circuit is given by the formula

$V = t^2 - 14t + 58.$

Find the values of t when the voltage is 10.

Solution:

Step 1.   Substitute 10 for V into the given equation.

Hence, $10 = t^2 - 14t + 58.$

Step 2.   Solve the equation.

$$t^2 - 14t + 58 = 10$$

144

$$t^2 - 14t + 48 = 0$$

$$(t - 8)(t - 6) = 0$$

$$\therefore \quad t - 8 = 0 \quad \text{or} \quad t - 6 = 0$$

$$t = 8 \quad \text{or} \quad t = 6$$

Step 3.   Check: $10 = 8^2 - 14(8) + 58$ and $10 = 6^2 - 14(6) + 58$

$$10 = 64 - 112 + 58 \qquad\qquad 10 = 36 - 84 + 58$$

$$10 = 10 \qquad\qquad\qquad\qquad 10 = 10$$

Hence, $t = 8$ or $t = 6$ when the voltage is 10.

39

Example 6:

To construct a container from a square piece of sheet metal, we cut a 5 cm square from each corner and turn the sides up. If the volume of the container must be 2000 cubic centimetres, find the size of the square piece of sheet metal required to construct the container.

Solution:

Step 1.   Let x = the length in centimetres of one side of the square piece of sheet metal

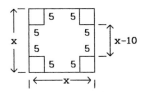

Step 2.   Since we are cutting 5 cm from each end of the side, the length of one side of the container will be $x - 2(5) = x - 10$ and the height 5 cm.

Step 3.   Volume of container = (width)(length)(height).
Hence, $2000 = (x - 10)(x - 10)(5)$.

Step 4.   Solve the equation.

$$5(x - 10)(x - 10) = 2000$$

$$(x - 10)(x - 10) = 400$$

$$x^2 - 20x + 100 = 400$$

$$x^2 - 20x - 300 = 0$$

$$(x + 10)(x - 30) = 0$$

$$\therefore \quad x + 10 = 0 \quad \text{or} \quad x - 30 = 0$$

$$x = -10 \quad \text{or} \quad x = 30$$

Step 5.   Since the length of the piece cannot be negative, we discard $x = -10$.

If $x = 30$, $x - 10 = 20$.

Step 6.   Check: The volume of the container is $(20)(20)(5) = 2000$ cm^3.

Hence, the size of the piece of sheet metal required is 30 cm by 30 cm.

CHECK POINT 6

Solve the following problems.

1. The area of a rectangular piece of land should be 5000 m^2. If the width is to be 50 m less than the length, find the required dimensions of this piece of land.

2. The numerical value of the area of a square metal plate must be ten times the numerical value of its perimeter. Find the length of one side.

3. The sum of two numbers is 19 and their product is 48. Find the two numbers.

4. A rectangular window is to be installed so that the length is 7 units more than its width. The numerical value of the area of the window must be 40 more than the numerical value of the perimeter. Find the dimensions of the window.

5. To construct a container from a square piece of sheet metal, we cut a 10 cm square from each corner and turn the sides up.

   The volume of the container has to be 9000 cm^3. Find the dimensions of the piece of sheet metal required to make the container.

DRILL EXERCISES

I   Change the following equations into the standard form.

   1. $y(y - 1) = 12$

   2. $(x + 2)(x - 3) = -26$

   3. $3(a^2 - a - 2) = 3a^2 + 6a$

   4. $5(x + 4) - 8(x + 1) = (x - 3)(x - 4)$

   5. $(x - 1)^2 + 20 = x(2x - 7) - 3$

II   Solve the following equations by factoring.

   6. $4x^2 - 16x = 0$ 7. $4x^2 - 19x + 12 = 0$

   8. $-9p^2 + 4 = 0$ 9. $-y^2 + 2y + 15 = 0$

DRILL EXERCISES (continued)

10. $m^2 - 81 = 0$ 11. $-5y^2 + 30y = 0$

12. $x(x - 2) = 3$ 13. $(a + 2)^2 + 5 = 4a(a + 1) - 18$

14. $(y + 2)(y - 5) = 2(3y - 8) + 3(y + 2)$

15. $4m(m - 4) + 5 = 5m(m - 4) + 9$

III Solve the following problems.

16. The sum of two numbers is 21 and their product is 80. Find the two numbers.

17. A rectangular swimming pool is to be constructed. If the area of the pool is to be 170 m^2 and the length is to be 7 m more than the width, find the required dimensions of the pool.

18. A rectangular deck is to be constructed so that the length is 2 units more than its width. The numerical value of the area of the deck must be 20 units more than the numerical value of the perimeter. Find the dimensions of the deck.

19. Two consecutive positive numbers have a product of 132. What are the numbers?

20. When a positive number is subtracted from the square of the number, the result is 30. Find the number.

21. A rectangular sheet of metal is twice as long as it is wide. From each corner a 2 cm square is cut out and the ends are turned up so as to form a box. If the volume of the box is 672 cm^3, find the dimensions of the box.

22. The length of a rectangle is three times its width. If the width is decreased by 1 cm and the length is increased by 3 cm, the area will be 72 cm^2. Find the dimensions of the original rectangle.

23. A garden plot, rectangular in shape, is 80 units long and 20 units wide. A walk of uniform width is built around the plot. If the area of the walk is 1344 square units, find the width of the walk.

ASSIGNMENT EXERCISES

I   Solve the following equations by factoring.

1. $x^2 - 3x - 10 = 0$

2. $2p^2 + 12p = 0$

3. $y(y + 4) = 12$

4. $4a^2 - 121 = 0$

5. $2a(a - 3) - 10 = 3a(a - 4) - 1$

6. $-3x^2 - 21x = 0$

7. $-4m^2 + 4m + 3 = 0$

8. $5(x + 4) - (x - 1) = (x - 7)(x - 3)$

9. $-y^2 + 16 = 0$

10. $m(3m - 4) - 10 = (m - 2)^2 + 18$

II Solve the following problems.

11. The area of a rectangular garden should be 52 m^2. If the width is 9 m shorter than the length, find the dimensions of the garden.

12. The sum of two numbers is 11 and their product is –42. Find the two numbers.

13. Two consecutive even positive numbers have a product of 168. What are the numbers?

14. The numerical value of the area of a square metal plate is eight times the numerical value of its perimeter. Find the length of one side.

15. The side of a square equals the width of a rectangle. The length of the rectangle is 6 cm longer than its width. The sum of the areas is 176 cm^2. Find the length of the side of the square.

16. A picture 8 cm by 12 cm is placed in a frame which has uniform width. If the area of the frame equals the area of the picture, find the width of the frame.

17. Determine the dimensions of a rectangle having perimeter 50 cm and area 150 cm^2.

18. A rectangular swimming pool is surrounded by a patio. The area of the pool is 288 m^2 and the outside dimensions of the patio are 20 m by 22 m. If the patio is of uniform width, how wide is the patio?

Fanshawe College

MODULE 3

Operations involving
Fractional Algebraic Expressions

Table of contents   **Page**

ALGEBRAIC FRACTIONS, EQUIVALENT FRACTIONS

Objectives: After having worked through this unit, the student should be capable of:

1. evaluating algebraic fractions for specific values of the variables;
2. indicating restrictions on the variables in algebraic fractions;
3. expanding algebraic fractions to higher terms;
4. reducing algebraic fractions to lowest terms;
5. understanding the three signs of an algebraic fraction.

1

MEANINGS OF A FRACTION

1. When we worked with numerical fractions, we looked at a fraction as representing one or more equal parts of a unit.

2. Another interpretation of a fraction is that it represents a ratio of two numbers which gives us a comparison of two numbers by division.

Example: The ratio of 7 to 21 may be written as $\frac{7}{21}$.

3. Finally, a fraction $\frac{c}{d}$ can be interpreted as indicating that c should be divided by d.

2

ALGEBRAIC FRACTIONS

Fractions which contain letters are referred to as algebraic fractions.

Examples: 1. $\frac{2x}{5y}$ 2. $\frac{m + 1}{n - 5}$ 3. $\frac{y^2 - 3y + 5}{y - 3}$

3

When working with algebraic fractions, it is more meaningful to consider fractions as just another way of writing a division problem.

Examples: 1. $\frac{2x}{5y}$ means $(2x) \div (5y)$

      2. $\frac{m + 1}{n - 5}$ means $(m + 1) \div (n - 5)$

      3. $\frac{y^2 - 3y + 5}{y - 3}$ means $(y^2 - 3y + 5) \div (y - 3)$

4

INTERPRETATION OF THE FRACTION SYMBOLS

When interpreting a fraction as an indicated division,

(i)   the numerator is the dividend,

(ii)   the denominator is the divisor,

(iii)   the fraction is the quotient.

Numerator $\rightarrow$ $x + 3$ Dividend   Divisor

Fraction Bar$\rightarrow$ ———— $= (x + 3) \div (y - 7)$

Denominator $\rightarrow$ $y - 7$ Quotient

5

FUNCTION OF THE FRACTION BAR

The fraction bar indicates division and acts as a grouping symbol.

Parentheses **must** be placed around the numerators and denominators containing several terms if we replace the fraction bar by a division sign.

Examples: Replace the fraction bar by a division sign.

1. $\dfrac{2x - 5}{7} = (2x - 5) \div 7$ correct

2. $\dfrac{m - 2n}{m + n} = (m - 2n) \div (m + n)$ correct

3. $\dfrac{y^2 - 3y + 5}{y - 5} = y^2 - 3y + 5 \div y - 5$ **incorrect**

4. $\dfrac{y^2 - 3y + 5}{y - 5} = (y^2 - 3y + 5) \div (y - 5)$ correct

6

Examples: Replace the division sign by a fraction bar

1. $(x - 2y) \div (y^2 - 1) = \dfrac{x - 2y}{y^2 - 1}$ correct

2. $m + 5m^2n \div (2m) - 3n = m + \dfrac{5m^2n}{2m} - 3n$ correct

3. $m + 5m^2n \div 2m - 3n = \dfrac{m + 5m^2n}{2m - 3n}$ **incorrect**

4. $(a^2 - 2a + 1) \div a - 1 = \dfrac{a^2 - 2a + 1}{a} - 1$ correct

7

CHECK POINT 1

I   Replace the fraction bar by a division sign in the following.

1. $\dfrac{3a}{5}$ 2. $\dfrac{5x - 1}{6}$ 3. $\dfrac{m^2 - 1}{m + 1}$

4. $\dfrac{3p^2 + 2p - 1}{p^2 - p + 6}$ 5. $\dfrac{8x}{2x - 1}$ 6. $\dfrac{1}{a + 1}$

II   Replace the division sign by a fraction bar in the following.

7. $(x + 1) \div (2x - 5)$ 8. $p - 3p^2 \div (6r) + 1$

9. $(y^2 - 2y + 1) \div y - 6$ 10. $m^2 - 3mn^2 \div (m - 1)$

8

EVALUATION OF ALGEBRAIC FRACTIONS

Example 1:

If $x = 10$, evaluate $\dfrac{4x + 20}{5}$.

Solution:

$$\frac{4x + 20}{5} = \frac{4(10) + 20}{5} = \frac{40 + 20}{5} = \frac{60}{5} = 12$$

Example 2:

If $m = 3$, evaluate $\dfrac{2m - 5}{7m}$.

Solution:

$$\frac{2m - 5}{7m} = \frac{2(3) - 5}{7(3)} = \frac{6 - 5}{21} = \frac{1}{21}$$

Example 3:

If $x = 2$, $y = 3$, evaluate $\dfrac{3x - 2y}{5xy}$.

Solution:

$$\frac{3x - 2y}{5xy} = \frac{3(2) - 2(3)}{5(2)(3)} = \frac{6 - 6}{30} = \frac{0}{30} = 0$$

Example 4:

If $y = 4$, evaluate $\dfrac{6y}{6y}$.

Solution:

$$\frac{6y}{6y} = \frac{6(4)}{6(4)} = \frac{24}{24} = 1$$

Example 5:

If $p = 2$, evaluate $\dfrac{p^2 - 3p + 5}{3p - 6}$.

Solution:

$$\frac{p^2 - 3p + 5}{3p - 6} = \frac{2^2 - 3(2) + 5}{3(2) - 6} = \frac{4 - 6 + 5}{6 - 6} = \frac{3}{0} \quad \text{Undefined.}$$

9

CHECK POINT 2

Evaluate the following algebraic fractions as indicated.

1. $\dfrac{5a - 2}{5}$, $a = 3$

2. $\dfrac{6y - 4}{9y}$, $y = 5$

3. $\dfrac{4x^3}{4x^3}$, $x = 2$

4. $\dfrac{3m - n^3}{m - n}$, $m = 9$, $n = 3$

5. $\dfrac{2(p - 1)}{p - 3}$, $p = 3$

5. $\dfrac{5x^2 + 3y}{2x(y - 1)}$, $x = 0$, $y = 1$

10

RESTRICTIONS ON VARIABLES

Since we cannot divide by zero, an algebraic fraction is not defined for values which will make the **denominator zero**.

Examples:

1. $\dfrac{5}{x}$ is not defined when $x = 0$.

2. $\dfrac{2m - 9}{m - 3}$ is not defined when $m - 3 = 0$; that is, $m = 3$.

3. $\dfrac{5y^2 + 2y - 6}{x^2 - 16}$ is not defined when $x^2 - 16 = 0$; that is, $x = 4$ or $x = -4$.

The following example shows how we normally indicate restrictions on variables.

4. $\dfrac{7x - 2y}{3x(y + 5)}$, $x \neq 0$, $y \neq -5$

11

CHECK POINT 3

Indicate the restrictions on the variables in the following expressions.

1. $\dfrac{2x}{5y}$ 2. $\dfrac{7a - 1}{a + 2}$ 3. $\dfrac{m^2 - 5n + 3}{9(m - 1)(n + 2)}$

12

CHANGING THE FORM OF A FRACTION

When working with fractions, we often have to change the form of the fraction.

Examples:

1. $\dfrac{1}{2}$ may be changed to $\dfrac{3}{6}$.

2. $\dfrac{3}{12}$ may be changed to $\dfrac{1}{4}$.

In changing the form of a fraction, we must make sure that the value of the fraction is not changed.

Hence, to change the form of a fraction, we use the following property of numbers.

$$\dfrac{a}{b} \cdot 1 = \dfrac{a}{b} \text{ for any numbers } a \text{ and } b \text{ where } b \neq 0.$$

Any number multiplied by 1 remains the same.

13

The number 1 can be written in many different forms.

Examples:

1. $\dfrac{5}{5} = 1$ 2. $\dfrac{xy}{xy} = 1$, $xy \neq 0$

3. $\dfrac{-1}{-1} = 1$ 4. $\dfrac{m^2 - 3}{m^2 - 3} = 1$, $m^2 - 3 \neq 0$

14

Multiplication of algebraic fractions is similar to multiplication of numerical fractions.

Rule:

When multiplying two fractions, multiply the numerators and then multiply the denominators.

Examples:

1. $\dfrac{2}{3} \cdot \dfrac{4}{5} = \dfrac{2 \cdot 4}{3 \cdot 5} = \dfrac{8}{15}$

2. $\dfrac{7}{11} \cdot \dfrac{x}{2y} = \dfrac{7 \cdot x}{11 \cdot 2y} = \dfrac{7x}{22y}$

15

MULTIPLYING FRACTIONS BY A FORM OF 1

Examples:

1. $\dfrac{3}{4} = \dfrac{5}{5} \cdot \dfrac{3}{4} = \dfrac{5 \cdot 3}{5 \cdot 4} = \dfrac{15}{20}$

2. $\dfrac{7}{8} = \dfrac{7}{8} \cdot \dfrac{xy}{xy} = \dfrac{7 \cdot xy}{8 \cdot xy} = \dfrac{7xy}{8xy}, \quad xy \neq 0$

16

REVIEW:

FUNDAMENTAL PRINCIPLE OF FRACTIONS

The value of a fraction is not changed when the numerator and denominator are either both multiplied or both divided by the same non-zero quantity.

1. If a, b and c are numbers such that $b \neq 0$ and $c \neq 0$, then

$$\dfrac{a}{b} = \dfrac{a \cdot c}{b \cdot c} \quad \text{or} \quad \dfrac{a}{b} = \dfrac{a \div c}{b \div c}$$

17

EQUIVALENT FRACTIONS

A fraction which can be obtained from a given fraction by use of the Fundamental Principle of Fractions is said to be equivalent to the given fraction.

18

Examples:

1. $\dfrac{18}{24} = \dfrac{18 \div 6}{24 \div 6} = \dfrac{3}{4}$

2. $\dfrac{2a^4}{7a^2} = \dfrac{(2a^4)(3a)}{(7a^2)(3a)} = \dfrac{6a^5}{21a^3}$

3. $\dfrac{2a^4}{7a^2} = \dfrac{(2a^4) \div a^2}{(7a^2) \div a^2} = \dfrac{2a^2}{7}$

$\left.\begin{array}{c} \\ \\ \end{array}\right\} \quad \therefore \quad \dfrac{2a^4}{7a^2} = \dfrac{6a^5}{21a^3} = \dfrac{2a^2}{7}$

Hence, $\dfrac{2a^4}{7a^2}, \dfrac{6a^5}{21a^3}$ and $\dfrac{2a^2}{7}$ are equivalent fractions.

Note:

We have changed the form in both examples 2 and 3 above, but not the value of the fraction $\dfrac{2a^4}{7a^2}$.

EXPANDING FRACTIONS TO HIGHER TERMS

If we change a fraction to an equivalent fraction having a greater numerator and denominator, we say that we expand the fraction to higher terms.

Examples:

The following fractions are being expanded to higher terms.

1. $\dfrac{2x}{5y} = \dfrac{2x(3y)}{5y(3y)} = \dfrac{6xy}{15y^2}$

2. $\dfrac{3x + 2}{x - 1} = \dfrac{(3x + 2)(x + 1)}{(x - 1)(x + 1)} = \dfrac{3x^2 + 5x + 2}{x^2 - 1}$

When working with unlike fractions, we frequently expand them to fractions having a common denominator.

This normally involves the following problem.

Example 1:

Find the missing numerator $\dfrac{x}{x + 1} = \dfrac{?}{x^2 - 1}$.

Solution:

Step 1. Factor $x^2 - 1$, to identify the factor by which the denominator $x + 1$ has to be multiplied to obtain $x^2 - 1$.

$x^2 - 1 = (x - 1)(x + 1)$, hence, the factor is $x - 1$.

Step 2. Multiply x, the numerator of $\dfrac{x}{x + 1}$ by $x - 1$ to obtain the missing numerator.

$\therefore \quad \dfrac{x}{x + 1} = \dfrac{x(x - 1)}{x^2 - 1} \quad$ or $\quad \dfrac{x^2 - x}{x^2 - 1}$

Example 2:

Expand $\dfrac{x - 4}{3x + 2}$ to a fraction having $3x^2 + 5x + 2$ as a denominator.

Solution:

Step 1. Factor $3x^2 + 5x + 2$, to identify the factor by which the denominator $3x + 2$ has to be multiplied to obtain $3x^2 + 5x + 2$.

$3x^2 + 5x + 2 = (3x + 2)(x + 1)$, hence, the factor is $x + 1$.

Step 2. Multiply the numerator **and** denominator of $\dfrac{x - 4}{3x + 2}$ by $x + 1$.

$\therefore \quad \dfrac{x - 4}{3x + 2} = \dfrac{(x - 4)(x + 1)}{(3x + 2)(x + 1)} \quad$ or $\quad \dfrac{x^2 - 3x - 4}{3x^2 + 5x + 2}$

22

Note:

In Step 1. of the previous examples, we can also determine the desired factor by division.

Examples:

1.
$$
\begin{array}{r}
x - 1 \\
x + 1 \overline{\smash{\big)}\ x^2 + 0x - 1} \\
\underline{x^2 + x} \\
-x - 1 \\
\underline{-x - 1} \\
0
\end{array}
$$

2.
$$
\begin{array}{r}
x + 1 \quad \leftarrow \text{Quotient} \\
3x + 2 \overline{\smash{\big)}\ 3x^2 + 5x + 2} \\
\underline{3x^2 + 2x} \\
3x + 2 \\
\underline{3x + 2} \\
0
\end{array}
$$

In each case the quotient is the desired factor.

23

CHECK POINT 4

Expand the following fractions so that they have denominators as indicated.

1. $\dfrac{2a}{b} = \dfrac{?}{3a^2b}$

2. $\dfrac{2}{n} = \dfrac{?}{n^2 + n}$

3. $\dfrac{m - 2}{m + 3} = \dfrac{?}{m^2 + 5m + 6}$

4. $\dfrac{m - 5}{3n + 2}$, new denominator $12n + 8$.

5. $\dfrac{x - 1}{y}$, new denominator x^2y^2.

6. $\dfrac{y + 4}{y - 2}$, new denominator $y^2 - 4$.

24

REDUCING A FRACTION TO LOWER TERMS

If we change a fraction to an equivalent fraction having a smaller numerator and denominator, we say the fraction has been reduced to lower terms.

Example:

$$\frac{6x^2}{8xy} = \frac{(6x^2) \div (2x)}{(8xy) \div (2x)} = \frac{3x}{4y}$$

25

THE LOWEST TERMS OF A FRACTION

When the numerator and denominator of a fraction have no common factor, the fraction is said to be in its lowest terms or simplest form.

Examples:

1. The fraction $\dfrac{3x}{4y}$ is in its lowest terms.

2. The fraction $\dfrac{6x^2}{8xy}$ is **not** in its lowest terms since the numerator and

denominator have the common factors 2 and x.

Note:

Final answers in fractional form should always be reduced to lowest terms.

REDUCING A FRACTION TO ITS LOWEST TERMS BY DIVISION

Method:

Step 1.   Find the greatest common factor of the numerator and denominator.

Step 2.   Divide both numerator and denominator by the greatest common factor.

Example 1:

Reduce $\dfrac{x^2 y^3}{3x^4 y^4}$ to its lowest terms.

Solution:

Step 1.   The greatest common factor is $x^2 y^3$.

Step 2. $\dfrac{x^2 y^3}{3x^4 y^4} = \dfrac{(x^2 y^3) \div (x^2 y^3)}{(3x^4 y^4) \div (x^2 y^3)} = \dfrac{1}{3x^2 y}$

Example 2:

Reduce $\dfrac{4a - 6b}{8c - 10d}$ to its lowest terms.

Solution:

Step 1.   The greatest common factor is 2.

Step 2. $\dfrac{4a - 6b}{8c - 10d} = \dfrac{2(2a - 3b) \div 2}{2(4c - 5d) \div 2} = \dfrac{2a - 3b}{4c - 5d}$

SHORTCUT FOR REDUCING FRACTIONS BY DIVISION

Example:

Reduce $\dfrac{7mn - 14n}{21np - 28nq}$ to lowest terms.

Solution:

Step 1.   The greatest common factor is 7n.

Step 2. $\dfrac{7mn - 14n}{21np - 28nq} = \dfrac{7n(m - 2) \div (7n)}{7n(3p - 4q) \div (7n)} = \dfrac{m - 2}{3p - 4q}$

SHORTCUT BY MENTAL DIVISION

Step 1.   The greatest common factor is 7n.

Step 2. $\dfrac{7mn - 14n}{21np - 28nq} = \dfrac{\overset{1}{\cancel{7n}}(m - 2)}{\underset{1}{\cancel{7n}}(3p - 4q)} = \dfrac{m - 2}{3p - 4q}$

The notation $\dfrac{\overset{1}{\cancel{7n}}(m - 2)}{\underset{1}{\cancel{7n}}(3p - 4q)}$ is used to indicate that we divided both the numerator and denominator by the same factor. In the above example, $\overset{1}{\cancel{7n}}$ indicates $\dfrac{7n}{7n} = 1$.

The process of mentally dividing both numerator and denominator by the same factor is commonly referred to as cancellation.

30

REDUCING FRACTIONS TO LOWEST TERMS BY CANCELLATION

Method:

Step 1.   Factor both numerator and denominator.

Step 2.   Cancel the greatest common factor of the numerator and denominator.

31

Example 1:

Reduce $\dfrac{9x^2 + 18xy}{3x^2 - 12xy}$ to lowest terms.

Solution:

Step 1.   Factor both numerator and denominator.

$$\frac{9x^2 + 18xy}{3x^2 - 12xy} = \frac{9x(x + 2y)}{3x(x - 4y)}$$

Step 2.   Cancel the greatest common factor of the numerator and denominator.

$$\frac{9x(x + 2y)}{3x(x - 4y)} = \frac{3 \cdot \overset{1}{\cancel{3x}}(x + 2y)}{\underset{1}{\cancel{3x}}(x - 4y)}$$

$$= \frac{3x + 6y}{x - 4y}$$

32

Example 2:

Reduce $\dfrac{6m^2 - 12m}{m - 2}$ to lowest terms.

Solution:

Step 1.   Factor both numerator and denominator.

$$\frac{6m^2 - 12m}{m - 2} = \frac{6m(m - 2)}{m - 2}$$

Step 2.   Cancel the greatest common factor of the numerator and denominator.

$$\frac{6m(m - 2)}{m - 2} = \frac{6m(\overset{1}{\cancel{m - 2}})}{\underset{1}{\cancel{m - 2}}}$$

$$= 6m$$

33

Note:

Normally we combine Step 1 and Step 2 and reduce fractions as shown in the following examples.

Example 3:

Reduce $\dfrac{5a - 10}{2a - 4}$ to lowest terms.

Solution:

$$\frac{5a - 10}{2a - 4} = \frac{5(\overset{1}{\cancel{a - 2}})}{2(\underset{1}{\cancel{a - 2}})} = \frac{5}{2}$$

Example 4:

Reduce $\dfrac{x}{7x - x^2}$ to lowest terms.

Solution:

$$\frac{x}{7x - x^2} = \frac{\overset{1}{\cancel{x}}}{\underset{1}{\cancel{x}}(7 - x)} = \frac{1}{7 - x}$$

34

Example 5:

Reduce $\dfrac{10m^2 - 5m}{25m^3 - 15m^2}$ to lowest terms.

Solution:

$$\frac{10m^2 - 5m}{25m^3 - 15m^2} = \frac{5m(2m - 1)}{5m^2(5m - 3)} = \frac{\cancel{5m}^1(2m - 1)}{\cancel{5m}m(5m - 3)_1}$$

$$= \frac{2m - 1}{m(5m - 3)} \quad \text{or} \quad \frac{2m - 1}{5m^2 - 3m}$$

Example 6:

Reduce $\dfrac{6x^2 - 13x - 5}{4x^2 - 20x + 25}$ to lowest terms.

Solution:

$$\frac{6x^2 - 13x - 5}{4x^2 - 20x + 25} = \frac{(3x + 1)\cancel{(2x - 5)}^1}{(2x - 5)\cancel{(2x - 5)}_1} = \frac{3x + 1}{2x - 5}$$

35

CHECK POINT 5

Reduce the following fractions to lowest terms by cancellation.

1. $\dfrac{4x^2 + x}{5x}$

2. $\dfrac{3a^2 - 6a}{a - 2}$

3. $\dfrac{3m}{6m^2 + 9m}$

4. $\dfrac{2a^2 - 7a + 3}{4a^2 - 1}$

5. $\dfrac{x^2 + 6x + 9}{4x^2 + 7x - 15}$

6. $\dfrac{8y^3 + 6y^2}{2y^4 - 4y^5}$

36

WORD OF CAUTION

Many students perform cancellation **incorrectly**.

Example: $\dfrac{x + 3}{x + 7} = \dfrac{\cancel{x} + 3}{\cancel{x} + 7} \neq \dfrac{3}{7}$

term $\downarrow$
term $\uparrow$

The cancelled expression must be a common factor of the numerator and the denominator, **not a term.**

37

CHECKING CANCELLATION

To check whether you cancelled correctly or not, **multiply** both numerator and denominator of the reduced fraction by the cancelled expression. The result must be the original fraction.

38

Example 1:

cancelled expression

Incorrect Cancellation. $\dfrac{\cancel{x} + 3}{\cancel{x} + 7} = \dfrac{3}{7}$ Check: $\dfrac{x}{x} \cdot \dfrac{3}{7} \neq \dfrac{x + 3}{x + 7}$

reduced fraction

Hence, $\dfrac{x + 3}{x + 7} \neq \dfrac{3}{7}$, x is incorrectly cancelled.

Example 2:

Incorrect
Cancellation.

cancelled expression

term↘

$$\frac{\cancel{8y} + 5}{\cancel{8y}} = 5 \qquad \text{Check: } \frac{8y}{8y} \cdot 5 \neq \frac{8y + 5}{8y}$$

factor↗ reduced expression

Hence, $\dfrac{8y + 5}{8y} \neq 5$; 8y is incorrectly cancelled.

Example 3:

Incorrect
Cancellation.

factor↘

$$\frac{\cancel{2x}(5x - y)}{\cancel{2x} + 4} = \frac{5x - y}{4} \qquad \text{Check: } \frac{2x}{2x} \cdot \frac{(5x - y)}{4}$$

term↗

$$= \frac{2x(5x - y)}{2x \cdot 4} \neq \frac{2x(5x - y)}{2x + 4}$$

Hence, $\dfrac{2x(5x - y)}{2x + 4} \neq \dfrac{5x - y}{4}$; 2x is incorrectly cancelled.

Remember: Do not cancel any terms.

CHECK POINT 6

Check whether or not the following cancellations are correct.

1. $\dfrac{\cancel{4xy} - 2y}{x^2 + \cancel{4xy}} \overset{?}{=} \dfrac{-2y}{x^2}$

2. $\dfrac{\cancel{4m^2}}{\cancel{4m^2}(m - 1)} \overset{?}{=} \dfrac{1}{m - 1}$

3. $\dfrac{6(\cancel{p} - 2)}{\cancel{p}} \overset{?}{=} -12$

4. $\dfrac{3x(\cancel{a + 1})}{\cancel{a + 1}} \overset{?}{=} 3x$

THE THREE SIGNS OF A FRACTION

There are three signs associated with a fraction:

(i)   the sign of the numerator,
(ii) the sign of the denominator,
(iii) the sign of the fraction itself.

If a fraction has **one** negative sign, it can be placed in front of either:

the numerator $\dfrac{-a}{b}$,

the denominator $\dfrac{a}{-b}$, or

the fraction bar $-\dfrac{a}{b}$.

To show that $\boxed{\dfrac{-a}{b} = \dfrac{a}{-b} = -\dfrac{a}{b}}$

we proceed as follows:

$$\boxed{\frac{-a}{b}} = \frac{(-1)(-a)}{(-1)(b)} = \boxed{\frac{a}{-b}} = \frac{(1)a}{(-1)b} = \frac{1}{-1} \cdot \frac{a}{b} = -1 \cdot \frac{a}{b} = \boxed{-\frac{a}{b}}$$

Examples:

1. $\dfrac{-5}{7} = -\dfrac{5}{7} = \dfrac{5}{-7}$

2. $\dfrac{m + 1}{-3} = \overset{\text{parentheses}\ \text{required}}{\dfrac{-(m + 1)}{3}} = -\dfrac{m + 1}{3}$

3. $-\dfrac{8x(x - 3)}{3 - x} = \dfrac{-8x(x - 3)}{3 - x} = \underset{\text{parentheses}\ \text{required}}{\dfrac{8x(x - 3)}{-(3 - x)}}$

Note:

If the numerator or denominator consists of several terms, you must place parentheses around the expression before placing the minus sign from the fraction bar in front of it.

Any two of the three signs of a fraction can be interchanged without changing the value of the fraction.

(a) $+\dfrac{-a}{-b} = -\dfrac{+a}{-b} = -\dfrac{-a}{+b} = +\dfrac{+a}{+b}$ Positive fraction.
(even number of negative signs)

(b) $-\dfrac{-a}{-b} = -\dfrac{+a}{+b} = +\dfrac{-a}{+b} = +\dfrac{+a}{-b}$ Negative fraction.
(odd number of negative signs)

Examples:

As usual, we will not write the plus sign.

1. $\dfrac{-15}{-21} = \dfrac{(\overset{1}{\cancel{-1}})(15)}{(\underset{1}{\cancel{-1}})(21)} = \dfrac{15}{21}$

2. $\dfrac{-(m + 1)}{-(m - 1)} = \dfrac{(m + 1)}{(m - 1)}$

3. $-\dfrac{-x^2}{-y^2} = -\dfrac{x^2}{y^2}$

4. $\dfrac{-2a(a - 1)}{(1 - a^2)} = \dfrac{2a(a - 1)}{-1(1 - a^2)}$

5. $-\dfrac{y - x}{2xy} \neq \dfrac{-y - x}{2xy}$

6. $\dfrac{-m^2}{m - 1} \neq \dfrac{m^2}{-m - 1}$

but $-\dfrac{y - x}{2xy} = \dfrac{-(y - x)}{2xy}$

but $\dfrac{-m^2}{m - 1} = \dfrac{m^2}{-(m - 1)}$

CHECK POINT 7

Indicate which of the following statements are true and correct the ones which are false.

1. $-\dfrac{7}{11} = \dfrac{7}{-11}$

2. $\dfrac{-3xy}{-x^2} = -\dfrac{3xy}{x^2}$

3. $\dfrac{-(2x - y)}{-(y - 2x)} = -\dfrac{2x - y}{-(y - 2x)}$

4. $\dfrac{-6a^2}{a^2 - 1} = \dfrac{6a^2}{-a^2 - 1}$

5. $\dfrac{-m^2}{-n^3} = -\dfrac{m^2}{n^3}$

6. $-\dfrac{-(x + 1)}{-(x - 2)} = -\dfrac{x + 1}{x - 2}$

REVIEW: OPPOSITES

The opposite of a number b is $-b$.

Examples:

1. The opposite of m^2n is $-(m^2n) = -m^2n$.

2. The opposite of $-3a^2$ is $-(-3a^2) = 3a^2$.

3. The opposite of $x - 2y$ is $-(x - 2y) = -x + 2y$.

4. The opposite of $-3a^2 + 2a - 1$ is $-(-3a^2 + 2a - 1) = 3a^2 - 2a + 1$

We see in examples 3 and 4 above that two expressions which have identical terms with opposite signs are opposites of each other.

Hence, we can write one as the negative of the other.

Example:

$5x - y$ and $y - 5x$ have identical terms with opposite signs.

Hence, $5x - y = -(y - 5x)$ or $y - 5x = -(5x - y)$.

We shall use this fact to reduce fractions when factors of the numerator and denominator are opposites of each other.

Example 1:

Reduce $\dfrac{x - y}{y - x}$ to lowest terms.

Solution:

Step 1. $y - x = -(x - y)$

Step 2. $\dfrac{x - y}{y - x} = \dfrac{x - y}{-(x - y)} = -\dfrac{\overset{1}{\cancel{x - y}}}{\underset{1}{\cancel{x - y}}} = -1$

Example 2:

Reduce $\dfrac{2m(3 - m)}{(m - 3)(m + 5)}$ to lowest terms.

Solution:

Step 1. $3 - m = -(m - 3)$

Step 2. $\dfrac{2m(3 - m)}{(m - 3)(m + 5)} = \dfrac{-2m\overset{1}{\cancel{(m - 3)}}}{\underset{1}{\cancel{(m - 3)}}(m + 5)} = -\dfrac{2m}{m + 5}$

CHECK POINT 8

Reduce the following fractions to lowest terms.

1. $\dfrac{4 - a}{a - 4}$

2. $\dfrac{5(x - 1)(x + 2)}{(x - 2)(1 - x)}$

DRILL EXERCISES

I (a) Replace the fraction bar by a division sign in the following.

1. $\dfrac{4b}{3a + 2}$ 2. $\dfrac{2x^2 + x + 1}{x^2 - 1}$ 3. $\dfrac{7}{n - 1}$

(b) Replace the division sign by a fraction bar in the following.

4. $(m^2 - 3) \div (2m + 7)$ 5. $y^2 - 3x^3 y \div (2x) + 3y$

6. $p^2 + 6p \div (2p - 1)$

II Evaluate the following algebraic fractions as indicated.

7. $\dfrac{4y - 3}{7y}, \ y = 2$ 8. $\dfrac{2x + y^2}{3(x - y)}, \ x = 5, \ y = 3$

9. $\dfrac{5m^2 + 4n}{(m - 2)(n - 3)}, \ m = 2, \ n = 3$

III Indicate the restrictions on the variables in the following expressions.

10. $\dfrac{4}{3x}$ 11. $\dfrac{5y + 2}{y + 3}$ 12. $\dfrac{3a - 1}{b(a - 2)}$

13. $\dfrac{2a^2 + 3b - 4}{3(a + 2)(b - 3)}$ 14. $\dfrac{2m + 3}{m^2 - 4}$

IV Expand the following fractions so they have denominators as indicated.

15. $\dfrac{7}{8} = \dfrac{?}{32}$ 16. $\dfrac{3m}{5n} = \dfrac{?}{20n^3}$ 17. $\dfrac{a + 2}{b} = \dfrac{?}{ab^2}$

18. $\dfrac{5}{x - 2} = \dfrac{?}{x^2 - x - 2}$ 19. $\dfrac{y + 1}{y - 4} = \dfrac{?}{y^2 - 16}$

20. $\dfrac{2m + 5}{3m - 2}$, new denominator $9m^2 - 12m + 4$.

21. $\dfrac{1}{a + 2}$, new denominator $a^2 - a - 6$.

V Reduce the following fractions to lowest terms.

22. $\dfrac{21}{49}$ 23. $\dfrac{27x^2 y^4}{36x^4 y^7}$ 24. $\dfrac{6a^2 - 8b}{2a^2 + 14b}$

25. $\dfrac{3m + 6n}{4m + 8n}$ 26. $\dfrac{7x}{7x^3 - 14x^2}$ 27. $\dfrac{3b^4 - 9b^3}{6b^5 + 15b^6}$

DRILL EXERCISES (continued)

28. $\dfrac{x^2 - 9}{x^2 + 2x - 15}$ 29. $\dfrac{m^2 - 2m - 8}{m^2 - 4m}$ 30. $\dfrac{a^2 + 9a + 20}{3a^2 + 13a - 10}$

VI Indicate which of the following statements are true and correct the ones which are false.

31. $\dfrac{-5}{13} = \dfrac{5}{-13}$ 32. $\dfrac{-a^2 b}{-b^2} = -\dfrac{a^2 b}{b^2}$

33. $\dfrac{x^2 - 1}{-3x^2} = \dfrac{-x^2 - 1}{3x^2}$ 34. $-\dfrac{-(2m + n)}{-(m + 2n)} = -\dfrac{2m + n}{m + 2n}$

VII Reduce the following fractions to lowest terms.

35. $\dfrac{m - 2}{2 - m}$ 36. $\dfrac{3(y + 3)(y - 4)}{9(4 - y)(y - 3)}$

ASSIGNMENT EXERCISES

I (a) Replace the fraction bar by a division sign in the following.

   1. $\dfrac{x^2 + 1}{x - 1}$ 2. $\dfrac{5a^2 - 2a - 3}{a^2 + a + 1}$

   (b) Replace the division sign by a fraction bar in the following.

   3. $(m^2 + 3m - 1) \div m + 2$ 4. $y^3 - 4xy^2 \div (y - 1)$

II Evaluate the following algebraic fractions as indicated.

   5. $\dfrac{3(x + 1)}{x - 2}$, $x = 2$ 6. $\dfrac{7a - b^2}{b - 2a}$, $a = 4$, $b = 5$

   7. $\dfrac{y^2 - 2y + 4}{9 - 3y}$, $y = 3$

III Indicate the restrictions on the variables in the following expressions.

   8. $\dfrac{3a}{7b}$ 9. $\dfrac{7x^2 - 3x + 2}{x^2 - 25}$ 10. $\dfrac{2m - 3n}{5m(n + 4)}$

IV Expand the following fractions so they have denominators as indicated.

   11. $\dfrac{4}{7} = \dfrac{?}{21}$ 12. $\dfrac{3y}{x} = \dfrac{?}{4x^3 y^3}$ 13. $\dfrac{5}{x^2 + 1} = \dfrac{?}{x(x^2 + 1)}$

   14. $\dfrac{n + 1}{n + 7} = \dfrac{?}{n^2 + 4n - 21}$

165

ASSIGNMENT EXERCISES (continued)

15. $\dfrac{a - 3}{3a - 4}$, new denominator $9a^2 - 16$.

16. $\dfrac{3m}{m + 2}$, new denominator $4m^2 + 7m - 2$.

17. $\dfrac{1}{x - y}$, new denominator $x^2 - 2xy + y^2$.

V   Reduce the following fractions to lowest terms.

18. $\dfrac{26}{39}$

19. $\dfrac{11a^2b^3}{22a^3b^2}$

20. $\dfrac{m}{m^2 + mn}$

21. $\dfrac{6x^2 - 4x}{4x^3 + 2x^2}$

22. $\dfrac{5y^2 - y}{7y}$

23. $\dfrac{6x^2 + 12x}{x + 2}$

24. $\dfrac{4m^2 - m - 5}{16m^2 - 25}$

25. $\dfrac{8a^2 + 16a}{4a^3 + 4a^2 - 4a}$

26. $\dfrac{x^2 - 3x - 18}{x^2 + 8x + 15}$

VI   Indicate which of the following statements are true and correct the ones which are false.

27. $-\dfrac{2}{3} = \dfrac{-2}{3}$

28. $\dfrac{-(a - 2b)}{-(a + b)} = -\dfrac{-(a - 2b)}{(a + b)}$

29. $-\dfrac{2m - n}{mn} = \dfrac{-2m - n}{mn}$

30. $\dfrac{-3x(x + 1)}{1 - x} = \dfrac{3x(x + 1)}{x - 1}$

VII Reduce the following to lowest terms.

31. $\dfrac{b - c}{c - b}$

32. $\dfrac{4(5 - x)}{x - 5}$

33. $\dfrac{2(a - 2)(a + 2)}{4(a + 3)(2 - a)}$

34. $\dfrac{3(y - 1)(2 - y)}{(y - 2)(1 - y)}$

MULTIPLICATION AND DIVISION OF ALGEBRAIC FRACTIONS

Objectives: After having worked through this unit, the student should be capable of:

1.   multiplying algebraic fractions;
2.   finding the reciprocals of algebraic fractions;
3.   dividing algebraic fractions.

1

MULTIPLICATION OF ALGEBRAIC FRACTIONS

Multiplication of fractions in algebra is similar to multiplication of fractions in arithmetic.

If a, b, c and d are numbers, then

$$\frac{a}{b} \cdot \frac{c}{d} = \frac{ac}{bd}, \text{ where } b \neq 0 \text{ and } d \neq 0.$$

Example: Multiply $\frac{2}{3}$ by $\frac{5}{7}$.

Solution:

$$\frac{2}{3} \cdot \frac{5}{7} = \frac{2 \cdot 5}{3 \cdot 7} = \frac{10}{21}$$

2

The above procedure for multiplying fractions in arithmetic can be equally well used to multiply fractions in algebra.

Examples: Multiply each of the following as indicated.

1. $\frac{3}{4} \cdot \frac{x}{y} = \frac{3 \cdot x}{4 \cdot y} = \frac{3x}{4y}$ or $(\frac{3}{4})(\frac{x}{y}) = \frac{(3)(x)}{(4)(y)} = \frac{3x}{4y}$

2. $\frac{4V}{5P} \cdot \frac{2V}{3Z} = \frac{4V \cdot 2V}{5P \cdot 3Z} = \frac{8V^2}{15PZ}$ or $(\frac{4V}{5P})(\frac{2V}{3Z}) = \frac{(4V)(2V)}{(5P)(3Z)} = \frac{8V^2}{15PZ}$

3. $\frac{2a + 3}{a - 1} \cdot \frac{a - 2}{a + 4} = \frac{(2a + 3)(a - 2)}{(a - 1)(a + 4)} = \frac{2a^2 - a - 6}{a^2 + 3a - 4}$

Hence, to multiply two algebraic fractions, we simply multiply numerator times numerator and denominator times denominator.

3

CHECK POINT 1

Multiply each of the following as indicated.

1. $\frac{1}{m^2} \cdot \frac{1}{7m}$

2. $\left(\frac{3x}{y}\right)\left(\frac{x}{4y}\right)$

3. $\left(\frac{-5}{R}\right)\left(\frac{2z}{R^2}\right)$

4. $\frac{3a}{4a - 2} \cdot \frac{2a + 3}{4a + 2}$

5. $\frac{m - 3}{m + 3} \cdot \frac{m - 3}{8m}$

6. $\frac{2x + 1}{x - 5} \cdot \frac{3x - 2}{4x + 5}$

4

RAISING ALGEBRAIC FRACTIONS TO A POWER

To multiply algebraic fractions raised to a power, we can use the following exponential law.

$$(\frac{x}{y})^n = \frac{x^n}{y^n}, \quad y \neq 0$$

Examples:

1. $\left(\dfrac{2}{5}\right)^4 = \dfrac{2^4}{5^4} = \dfrac{16}{625}$

2. $\left(\dfrac{3ab^3}{c^2d}\right)^2 = \dfrac{(3ab^3)^2}{(c^2d)^2} = \dfrac{9a^2b^6}{c^4d^2}$

3. $\left(\dfrac{x+2}{x-5}\right)^2 = \dfrac{(x+2)^2}{(x-5)^2} = \dfrac{(x+2)(x+2)}{(x-5)(x-5)} = \dfrac{x^2 + 4x + 4}{x^2 - 10x + 25}$

5

CHECK POINT 2

Multiply the following fractions raised to a power.

1. $\left(\dfrac{1}{3}\right)^4$

2. $\left(\dfrac{5x^2y}{2z^3}\right)^3$

3. $\left(\dfrac{2a-1}{a+4}\right)^2$

6

MULTIPLICATION OF ALGEBRAIC FRACTIONS BY NON-FRACTIONAL EXPRESSIONS

Just as any whole number can be expressed as a fraction by dividing it by 1, we can express an algebraic expression in fractional form by dividing it by 1.

Examples:

1. $5 = \dfrac{5}{1}$

2. $x = \dfrac{x}{1}$

3. $2m^3n = \dfrac{2m^3n}{1}$

4. $5y^2 + 2y - 1 = \dfrac{5y^2 + 2y - 1}{1}$

7

Examples:

1. $5 \cdot \dfrac{x}{y} = \dfrac{5}{1} \cdot \dfrac{x}{y} = \dfrac{5x}{y}$

2. $7m^2 \cdot \dfrac{3m}{5n^3} = \dfrac{7m^2}{1} \cdot \dfrac{3m}{5n^3} = \dfrac{21m^3}{5n^3}$

3. $\dfrac{10V}{3P} \cdot 2V^3 = \dfrac{10V}{3P} \cdot \dfrac{2V^3}{1} = \dfrac{20V^4}{3P}$

4. $2a^3 \cdot \dfrac{a-1}{a+1} = \dfrac{2a^3}{1} \cdot \dfrac{a-1}{a+1} = \dfrac{2a^3(a-1)}{a+1}$ or $\dfrac{2a^4 - 2a^3}{a+1}$

5. $(2x-1) \cdot \dfrac{x+2}{x-4} = \dfrac{2x-1}{1} \cdot \dfrac{x+2}{x-4} = \dfrac{(2x-1)(x+2)}{(x-4)}$ or $\dfrac{2x^2 + 3x - 2}{x-4}$

Hence, to multiply an algebraic fraction by a non-fractional expression, multiply the numerator by the non-fractional expression and leave the denominator as is.

8

CHECK POINT 3

Multiply each of the following as indicated.

1. $7 \cdot \dfrac{a}{3b}$

2. $(4x)\left(\dfrac{3x}{2y}\right)$

3. $\dfrac{6V^2}{5P} \cdot 3V$

4. $5m^2 \cdot \dfrac{m+1}{2m-3}$

5. $\dfrac{a-6}{3a+2} \cdot (a+6)$

9

WRITING A FRACTION AS THE PRODUCT OF A NON-FRACTIONAL EXPRESSION TIMES A FRACTION

1. $\dfrac{5}{7} = 5 \cdot \dfrac{1}{7}$

2. $\dfrac{-3x}{2y} = -3x \cdot \dfrac{1}{2y}$

3. $\dfrac{4a^2}{a^2 - 1} = 4a^2 \cdot \dfrac{1}{a^2 - 1}$

4. $\dfrac{x - 1}{x + 1} = (x - 1) \cdot \dfrac{1}{x + 1}$

Note:

If the numerator has several terms, we must place parentheses around the numerator as in example 4 above. Remember the fraction bar acts as a grouping symbol.

Examples:

1. $\dfrac{a^2 + 2a - 4}{a - 2} \neq a^2 + 2a - 4 \cdot \dfrac{1}{a - 2}$ (parentheses are missing)

2. $\dfrac{a^2 + 2a - 4}{a - 2} = (a^2 + 2a - 4) \cdot \dfrac{1}{a - 2}$

10

CHECK POINT 4

Write the following fractions as a product of a non-fractional expression times a fraction.

1. $\dfrac{12}{13}$

2. $\dfrac{-5m^2}{2n^3}$

3. $\dfrac{v^2 - 3v}{2P^2}$

4. $\dfrac{5a^2 - 2a + 3}{4a^2 - 9}$

11

SIMPLIFYING ALGEBRAIC FRACTIONS

When working with fractions, the final result should always be written in lowest terms.

Method:

Step 1.   Factor numerators and denominators.

Step 2.   Cancel all common factors and multiply the remaining fractions.

Example 1: Multiply $\dfrac{3x + 6}{5x}$ by $\dfrac{15x^2y}{2x + 4}$.

Solution:

Step 1.   Factor numerators and denominators.

$$\dfrac{3x + 6}{5x} \cdot \dfrac{15x^2y}{2x + 4} = \dfrac{3(x + 2)}{5x} \cdot \dfrac{3 \cdot 5x \cdot xy}{2(x + 2)}$$

Step 2.   Cancel all common factors and then multiply.

$$\dfrac{3(\cancel{x + 2})}{\cancel{5x}} \cdot \dfrac{3 \cdot \cancel{5x} \cdot xy}{2(\cancel{x + 2})} = \dfrac{9xy}{2}$$

Therefore, $\dfrac{3x + 6}{5x} \cdot \dfrac{15x^2y}{2x + 4} = \dfrac{9xy}{2}$

Example 2: $\dfrac{x^2 - 1}{2x^2 - 6x} \cdot \dfrac{x^2 - x - 6}{x^2 + 2x - 3}$

Solution:

Step 1. Factor numerators and denominators.

$$\frac{x^2 - 1}{2x^2 - 6x} \cdot \frac{x^2 - x - 6}{x^2 + 2x - 3} = \frac{(x - 1)(x + 1)}{2x(x - 3)} \cdot \frac{(x + 2)(x - 3)}{(x - 1)(x + 3)}$$

Step 2. Cancel all common factors and then multiply.

$$\frac{\cancel{(x - 1)}(x + 1)}{2x\cancel{(x - 3)}} \cdot \frac{(x + 2)\cancel{(x - 3)}}{\cancel{(x - 1)}(x + 3)} = \frac{(x + 1)(x + 2)}{2x(x + 3)} \quad \text{or} \quad \frac{x^2 + 3x + 2}{2x^2 + 6x}$$

Hence, $\dfrac{x^2 - 1}{2x^2 - 6x} \cdot \dfrac{x^2 - x - 6}{x^2 + 2x - 3} = \dfrac{x^2 + 3x + 2}{2x^2 + 6x}$

Example 3: $\dfrac{y^2 - 2y - 15}{y^2 + 6y + 8} \cdot \dfrac{y^2 - 16}{20 - 4y}$

Solution:

Step 1. Factor numerators and denominators.

$$\frac{y^2 - 2y - 15}{y^2 + 6y + 8} \cdot \frac{y^2 - 16}{20 - 4y} = \frac{(y + 3)(y - 5)}{(y + 2)(y + 4)} \cdot \frac{(y + 4)(y - 4)}{4(5 - y)} \longleftarrow$$

$$= \frac{(y + 3)(y - 5)}{(y + 2)(y + 4)} \cdot \frac{(y + 4)(y - 4)}{-4(y - 5)} \longleftarrow$$

$5 - y$ is the negative of $y - 5$

Step 2. Cancel all common factors and multiply.

$$\frac{(y + 3)\cancel{(y - 5)}}{(y + 2)\cancel{(y + 4)}} \cdot \frac{\cancel{(y + 4)}(y - 4)}{-4\cancel{(y - 5)}} = \frac{y + 3}{y + 2} \cdot \frac{y - 4}{-4}$$

$$= \frac{(y + 3)(y - 4)}{-4(y + 2)}$$

$$= -\frac{(y + 3)(y - 4)}{4(y + 2)} \quad \text{or} \quad -\frac{y^2 - y - 12}{4y + 8}$$

Note:

Normally, we combine Step 1 and Step 2 as shown in the following examples.

Example 4: $\dfrac{4m^3 - 2m}{3m^3 - 6m^2 - 24m} \cdot \dfrac{m^2 + 5m + 6}{2m^2 - 1}$

Solution:

$$\frac{4m^3 - 2m}{3m^3 - 6m^2 - 24m} \cdot \frac{m^2 + 5m + 6}{2m^2 - 1} = \frac{2m\cancel{(2m^2 - 1)}}{3\cancel{m}(m - 4)\cancel{(m + 2)}} \cdot \frac{(m + 3)\cancel{(m + 2)}}{\cancel{2m^2 - 1}}$$

$$= \frac{2(m + 3)}{3(m - 4)} \quad \text{or} \quad \frac{2m + 6}{3m - 12}$$

Therefore, $\dfrac{4m^3 - 2m}{3m^3 - 6m^2 - 24m} \cdot \dfrac{m^2 + 5m + 6}{2m^2 - 1} = \dfrac{2(m + 3)}{3(m - 4)}$

Example 5: $\dfrac{6a^2 - 3a - 9}{a^2 - 5a + 6} \cdot \dfrac{a^2 - 8a + 15}{2a^2 - 13a + 15}$

Solution:

$$\dfrac{6a^2 - 3a - 9}{a^2 - 5a + 6} \cdot \dfrac{a^2 - 8a + 15}{2a^2 - 13a + 15} = \dfrac{3(2a - 3)(a + 1)}{(a - 2)(a - 3)} \cdot \dfrac{(a - 3)(a - 5)}{(2a - 3)(a - 5)}$$

$$= \dfrac{3(a + 1)}{a - 2} \quad \text{or} \quad \dfrac{3a + 3}{a - 2}$$

Hence, $\dfrac{6a^2 - 3a - 9}{a^2 - 5a + 6} \cdot \dfrac{a^2 - 8a + 15}{2a^2 - 13a + 15} = \dfrac{3(a + 1)}{a - 2}$

CHECK POINT 5

Multiply as indicated.

1. $\dfrac{-2V^3}{p^2} \cdot \dfrac{Vp}{6V^2}$

   2. $\dfrac{-2}{5x - 5} \cdot \dfrac{x - 1}{6(x + 1)}$

3. $\dfrac{-7m}{m^2 - 3m + 2} \cdot \dfrac{4m^2 - 4m}{14m^3}$

   4. $\dfrac{9a^2 - 25}{a^2 + 2a - 8} \cdot \dfrac{a^2 + 5a + 4}{5 - 3a}$

5. $\dfrac{4x^3 - 10x^2 + 6x}{x^2 - 4x - 12} \cdot \dfrac{x^2 - 7x + 6}{2x^2 + 5x - 12}$

RECIPROCALS

Two quantities are a pair of reciprocals if their product is 1.

Examples:

1. $\dfrac{5}{7}$ and $\dfrac{7}{5}$ are a pair of reciprocals, since $\dfrac{5}{7} \cdot \dfrac{7}{5} = 1$.

2. $\dfrac{2x}{3}$ and $\dfrac{3}{2x}$ are a pair of reciprocals, since $\dfrac{2x}{3} \cdot \dfrac{3}{2x} = 1$.

3. $5a^2 - 1$ and $\dfrac{1}{5a^2 - 1}$ are a pair of reciprocals,

   since $(5a^2 - 1) \cdot \dfrac{1}{5a^2 - 1} = 1$.

4. $\dfrac{x^2 - 1}{x + 3}$ and $\dfrac{x + 3}{x^2 - 1}$ are a pair of reciprocals,

   since $\dfrac{x^2 - 1}{x + 3} \cdot \dfrac{x + 3}{x^2 - 1} = 1$.

The process of finding a reciprocal is also referred to as inverting the number.

18

CHECK POINT 6

Find the reciprocals of the following fractions.

1. $\dfrac{3m^2}{-5}$

2. $\dfrac{x - 5}{x + 1}$

3. $\dfrac{6V^2 + 5V}{RV^2}$

4. $\dfrac{a^2 - 2a + 1}{9a^2 - 16}$

5. 9

6. $3x^2 + 1$

19

DIVIDING A FRACTION BY A FRACTION

Example: Divide $\dfrac{2}{5}$ by $\dfrac{3}{4}$.

Solution:

$$\dfrac{2}{5} \div \dfrac{3}{4} = \dfrac{\dfrac{2}{5}}{\dfrac{3}{4}}$$

$$= \dfrac{\dfrac{2}{5} \cdot \dfrac{4}{3}}{\dfrac{3}{4} \cdot \dfrac{4}{3}}$$ Multiply numerator and denominator by the reciprocal of the denominator.

$$= \dfrac{\dfrac{2}{5} \cdot \dfrac{4}{3}}{1}$$

$$= \dfrac{2}{5} \cdot \dfrac{4}{3} = \dfrac{8}{15}$$

Note: In the above solution, we have shown that: $\dfrac{2}{5} \div \dfrac{3}{4} = \dfrac{2}{5} \cdot \dfrac{4}{3}$.

20

In general, if a, b, c and d are numbers, then

$$\dfrac{a}{b} \div \dfrac{c}{d} = \dfrac{a}{b} \cdot \dfrac{d}{c} \quad \text{where } b \neq 0, \ c \neq 0, \ d \neq 0.$$

change

invert ⟶ find reciprocal

21

DIVISION OF ALGEBRAIC FRACTIONS

Division of fractions in algebra is similar to division of fractions in arithmetic.

To divide one fraction by another, we proceed as follows:

1. Invert the divisor (the fraction in the denominator).

2. Multiply the dividend (the fraction in the numerator) by the inverted fraction.

22

Example 1: Divide $\dfrac{a - 2}{3b} \div \dfrac{b + 1}{2a}$.

Solution:

Step 1. The reciprocal of $\dfrac{b + 1}{2a}$ is $\dfrac{2a}{b + 1}$.

Step 2. $\dfrac{a - 2}{3b} \div \dfrac{b + 1}{2a} = \dfrac{a - 2}{3b} \cdot \dfrac{2a}{b + 1}$

$\qquad\qquad = \dfrac{2a(a - 2)}{3b(b + 1)} \quad$ or $\quad \dfrac{2a^2 - 4a}{3b^2 + 3b}$

23

Example 2: Divide $\dfrac{\dfrac{2m^2}{5n^3}}{\dfrac{n - 1}{m + 1}}$

Solution: $\dfrac{\dfrac{2m^2}{5n^3}}{\dfrac{n - 1}{m + 1}} = \dfrac{2m^2}{5n^3} \div \dfrac{n - 1}{m + 1}$

Step 1. The reciprocal of $\dfrac{n - 1}{m + 1}$ is $\dfrac{m + 1}{n - 1}$.

Step 2. Multiply the numerator by the reciprocal of the denominator.

$\dfrac{\dfrac{2m^2}{5n^3}}{\dfrac{n - 1}{m + 1}} = \dfrac{2m^2}{5n^3} \cdot \dfrac{m + 1}{n - 1}$

$\qquad\qquad = \dfrac{2m^2(m + 1)}{5n^3(n - 1)} \quad$ or $\quad \dfrac{2m^3 + 2m^2}{5n^4 - 5n^3}$

24

CHECK POINT 7

Divide the following fractions as indicated.

1. $\dfrac{2V}{3P^2} \div \dfrac{P}{V^2}$
2. $\dfrac{m + 2}{3n} \div \dfrac{7n - 5}{m^3}$
3. $\dfrac{\dfrac{10x^2}{7y}}{\dfrac{3y^2}{x - 1}}$

25

DIVIDING AN ALGEBRAIC FRACTION BY A NON-FRACTIONAL EXPRESSION

We have seen that any non-fractional expression can be written as a fraction.

Example: $3x + 2 = \dfrac{3x + 2}{1}$

When dividing an algebraic fraction by a non-fractional expression, change the expression to fractional form, then invert and multiply as before.

Example: Divide $\dfrac{V^2 + 1}{R} \div (R + 3)$

Solution:

Step 1. The reciprocal of $\dfrac{R + 3}{1}$ is $\dfrac{1}{R + 3}$.

Step 2. $\dfrac{V^2 + 1}{R} \div (R + 3) = \dfrac{V^2 + 1}{R} \cdot \dfrac{1}{R + 3}$

$$= \dfrac{V^2 + 1}{R(R + 3)} \quad \text{or} \quad \dfrac{V^2 + 1}{R^2 + 3R}$$

26

CHECK POINT 8

Divide the following fractions as indicated.

1. $\dfrac{5x}{y} \div z^2$

2. $\dfrac{m - 1}{n^2} \div (n + 1)$

3. $\dfrac{2V^2}{R + 1} \div (R + 1)$

27

EXAMPLES INVOLVING CANCELLATION

Example 1: Divide $\dfrac{3V^2 - 3V}{R^2} \div \dfrac{V - 1}{R}$

Solution:

Step 1. The reciprocal of $\dfrac{V - 1}{R}$ is $\dfrac{R}{V - 1}$.

Step 2. $\dfrac{3V^2 - 3V}{R^2} \div \dfrac{V - 1}{R} = \dfrac{3V^2 - 3V}{R^2} \cdot \dfrac{R}{V - 1}$

Step 3. Factor and cancel (if possible).

$$\dfrac{3V^2 - 3V}{R^2} \cdot \dfrac{R}{V - 1} = \dfrac{3V\,\overset{1}{(\cancel{V - 1})}}{R \cdot \underset{1}{\cancel{R}}} \cdot \dfrac{\overset{1}{\cancel{R}}}{\underset{1}{\cancel{V - 1}}} = \dfrac{3V}{R} \cdot 1 = \dfrac{3V}{R}$$

28

Example 2: Divide $\dfrac{p + 3}{4p^3} \div \dfrac{7p + 21}{2p^2}$

Solution:

Step 1. The reciprocal of $\dfrac{7p + 21}{2p^2}$ is $\dfrac{2p^2}{7p + 21}$.

Step 2. $\dfrac{p + 3}{4p^3} \div \dfrac{7p + 21}{2p^2} = \dfrac{p + 3}{4p^3} \cdot \dfrac{2p^2}{7p + 21}$

Step 3.   Factor and cancel (if possible).

$$\frac{p + 3}{4p^3} \cdot \frac{2p^2}{7p + 21} = \frac{\overset{1}{\cancel{p + 3}}}{2p \cdot \underset{1}{\cancel{2p^2}}} \cdot \frac{\overset{1}{\cancel{2p^2}}}{7(\underset{1}{\cancel{p + 3}})}$$

$$= \frac{1}{2p} \cdot \frac{1}{7} = \frac{1}{14p}$$

29

WORD OF CAUTION
When dividing two fractions, we **must first invert the divisor** and indicate multiplication before we can cancel any common factors.

Example 3: Divide $\dfrac{3(x - 1)}{2xy} \div \dfrac{5y}{x^2(x - 1)}$

Incorrect cancellation: $\dfrac{3(x \overset{1}{\cancel{- 1}})}{2x\cancel{y}} \div \dfrac{5\overset{1}{\cancel{y}}}{x^2\underset{1}{(x - 1)}} \neq \dfrac{3}{2x} \div \dfrac{5}{x^2}$

Correct cancellation:

Step 1.   The reciprocal of $\dfrac{5y}{x^2(x - 1)}$ is $\dfrac{x^2(x - 1)}{5y}$.

Step 2. $\dfrac{3(x - 1)}{2xy} \div \dfrac{5y}{x^2(x - 1)} = \dfrac{3(x - 1)}{2xy} \cdot \dfrac{x^2(x - 1)}{5y}$

$$= \frac{3(x - 1)}{2\underset{1}{\cancel{x}}y} \cdot \frac{\overset{1}{\cancel{x}} \cdot x (x - 1)}{5y}$$

$$= \frac{3x(x - 1)^2}{10y^2}$$

30

Note:
Normally we combine the steps as shown in the following examples.

Example 4: Divide $\dfrac{6x^2 - 3x}{8x} \div \dfrac{2x - 1}{4x}$

Solution:

$$\frac{6x^2 - 3x}{8x} \div \frac{2x - 1}{4x} = \frac{6x^2 - 3x}{8x} \cdot \frac{4x}{2x - 1} = \frac{3x(\overset{1}{\cancel{2x - 1}})}{\underset{2}{\cancel{8x}}} \cdot \frac{\overset{1}{\cancel{4x}}}{\underset{1}{\cancel{2x - 1}}} = \frac{3x}{2}$$

Hence, $\dfrac{6x^2 - 3x}{8x} \div \dfrac{2x - 1}{4x} = \dfrac{3x}{2}$

176

31

CHECK POINT 9

Divide the following fractions as indicated.

1. $\dfrac{R + 1}{3V^2} \div \dfrac{5R + 5}{V}$

2. $\dfrac{4(a - 3)}{ab} \div \dfrac{8b^2}{2a(a - 3)}$

32

MORE EXAMPLES:

Example 1: Divide $\dfrac{m^2 - 4}{2 - n} \div \dfrac{m + 2}{2n - 4}$

Solution:

Step 1.   The reciprocal of $\dfrac{m + 2}{2n - 4}$ is $\dfrac{2n - 4}{m + 2}$.

Step 2. $\dfrac{m^2 - 4}{2 - n} \div \dfrac{m + 2}{2n - 4} = \dfrac{m^2 - 4}{2 - n} \cdot \dfrac{2n - 4}{m + 2}$

Step 3. $= \dfrac{(m + 2)(m - 2)}{-(n - 2)} \cdot \dfrac{2(n - 2)}{m + 2}$

$= -(m - 2) \cdot 2$

$= -2(m - 2) \quad \text{or} \quad 4 - 2m$

33

Example 2: Divide $\dfrac{x^2 + x - 6}{2x^2 + 3x} \div \dfrac{x^2 - 9}{6x - 2}$

Solution:

Step 1.   The reciprocal of $\dfrac{x^2 - 9}{6x - 2}$ is $\dfrac{6x - 2}{x^2 - 9}$.

Step 2. $\dfrac{x^2 + x - 6}{2x^2 + 3x} \div \dfrac{x^2 - 9}{6x - 2} = \dfrac{x^2 + x - 6}{2x^2 + 3x} \cdot \dfrac{6x - 2}{x^2 - 9}$

$= \dfrac{(x - 2)(x + 3)}{x(2x + 3)} \cdot \dfrac{2(3x - 1)}{(x - 3)(x + 3)}$

$= \dfrac{2(x - 2)(3x - 1)}{x(2x + 3)(x - 3)} \quad \text{or} \quad \dfrac{6x^2 - 14x + 4}{2x^3 - 3x^2 - 9x}$

34

CHECK POINT 10

Divide the following fractions.

1. $\dfrac{5 - x}{4y^2} \div \dfrac{x^2 - 25}{2xy}$

2. $\dfrac{m^2 + 5m - 14}{3m - 6} \div \dfrac{2m + 14}{m^2 - 4}$

3. $\dfrac{a^2 + 9a + 18}{a^2 + 3a + 2} \div \dfrac{a^2 + 5a - 6}{a^2 - 2a - 8}$

COMPLEX FRACTIONS

Fractions which have fractions in their numerator and/or denominator are called complex fractions.

SIMPLIFYING COMPLEX FRACTIONS

To simplify a complex fraction is to write it as a common fraction in lowest terms.

METHOD

Step 1. Find the lowest common denominator (LCD) of all the fractions in the numerator and denominator.

Step 2. Multiply all terms in the numerator and in the denominator of the **complex fraction** by the LCD and simplify.

Step 3. Reduce the resulting common fraction to lowest terms. This may involve collecting like terms and factoring.

Example 1: Simplify: $\dfrac{2 + \dfrac{1}{x}}{5 - \dfrac{1}{x}}$

Solution:

Step 1. Find the LCD of the fractions in the numerator and the denominator. The LCD = x.

Step 2. Multiply all terms in the numerator and in the denominator of the complex fraction by the LCD and simplify.

$$\frac{2 + \dfrac{1}{x}}{5 - \dfrac{1}{x}} = \frac{\left(2 + \dfrac{1}{x}\right)(x)}{\left(5 - \dfrac{1}{x}\right)(x)} = \frac{2x + \dfrac{x}{x}}{5x - \dfrac{x}{x}} = \frac{2x + 1}{5x - 1}$$

Step 3. Numerator and denominator have no factors in common.

Hence, $\dfrac{2x + 1}{5x - 1}$ is in lowest terms.

Example 2: Simplify $\dfrac{\dfrac{1}{y} - \dfrac{1}{x}}{\dfrac{3}{x} + \dfrac{1}{y}}$

Solution:

Step 1. LCD = xy

Step 2. $\dfrac{\dfrac{1}{y} - \dfrac{1}{x}}{\dfrac{3}{x} + \dfrac{1}{y}} = \dfrac{\left(\dfrac{1}{y} - \dfrac{1}{x}\right)(xy)}{\left(\dfrac{3}{x} + \dfrac{1}{y}\right)(xy)} = \dfrac{\dfrac{x\cancel{y}}{\cancel{y}1} - \dfrac{\cancel{x}y}{\cancel{x}1}}{\dfrac{3\cancel{x}1y}{\cancel{x}} + \dfrac{x\cancel{y}1}{\cancel{y}}} = \dfrac{x - y}{3y + x}$

Step 3. Numerator and denominator have no factors in common.

Hence, $\dfrac{x - y}{3y + x}$ is in lowest terms.

Example 3: Simplify $\dfrac{2z - \dfrac{3}{z-1}}{\dfrac{5}{z+1} - \dfrac{1}{z-1}}$

Solution:

Step 1.   Multiply numerator and denominator of the complex fraction by the LCD, $(z + 1)(z - 1)$.

Step 2. $\dfrac{(2z - \dfrac{3}{z-1})(z+1)(z-1)}{(\dfrac{5}{z+1} - \dfrac{1}{z-1})(z+1)(z-1)}$

$= \dfrac{2z(z+1)(z-1) - \dfrac{3(z+1)(z-1)}{z-1}}{\dfrac{5(z+1)(z-1)}{z+1} - \dfrac{(z+1)(z-1)}{z-1}}$

$= \dfrac{2z(z+1)(z-1) - 3(z+1)}{5(z-1) - (z+1)}$

Step 3.   Factor and simplify common fraction.

$\dfrac{(z+1)[2z(z-1) - 3]}{5z - 5 - z - 1} = \dfrac{(z+1)(2z^2 - 2z - 3)}{2(2z - 3)}$

CHECK POINT 11

Simplify

1. $\dfrac{\dfrac{x}{y}}{\dfrac{x}{y} + 1}$

2. $\dfrac{\dfrac{1}{a} - b}{\dfrac{1}{b} + a}$

3. $\dfrac{\dfrac{x^2}{y^2} + \dfrac{2x}{y} - 8}{\dfrac{x}{y^2} - \dfrac{1}{y} - \dfrac{2}{x}}$

DRILL EXERCISES

I   Multiply each of the following as indicated.

1. $\dfrac{5b^2}{6a} \cdot \dfrac{1}{a^4}$

2. $\left(\dfrac{7m^2}{2n^4}\right)\left(\dfrac{3m}{4n^2}\right)$

3. $\left(\dfrac{3}{y^5}\right)\left(\dfrac{-2x}{y}\right)$

4. $\left(\dfrac{2a^3b}{3c^2}\right)^3$

5. $(2y)\left(\dfrac{5xy}{3z}\right)$

6. $\dfrac{n-2}{2n+1} \cdot 3n^3$

II   Write the following fractions as a product of a non-fractional expression times a fraction.

7. $\dfrac{2x}{-3y^2}$

8. $\dfrac{a^3 - 5a}{3b^2}$

9. $\dfrac{m^2 - 3m + 5}{m + 1}$

III   Multiply the following fractions.

10. $\dfrac{4a^3}{3b^2} \cdot \dfrac{15b}{16a}$

11. $\dfrac{3x}{5x - 5} \cdot \dfrac{2x - 2}{9x^3}$

DRILL EXERCISES (continued)

12. $\dfrac{-x^2y^2}{14z^2} \cdot \dfrac{-7x^3z^3}{4y^4}$

13. $(x^2 + 5x + 6) \cdot \dfrac{x - 1}{x^2 + 3x}$

14. $\dfrac{3b^2 + 3b}{15b^2} \cdot \dfrac{-5b}{b^2 - 6b - 7}$

15. $\dfrac{a^2 - 4}{a^2 + 7a + 12} \cdot \dfrac{7a^2 + 28a}{a^2 - 8a + 12}$

IV Find the reciprocals of the following fractions.

16. $\dfrac{-2x^3}{7}$

17. $\dfrac{a^2b^3}{3a^2 + 2b^2}$

18. $m^2n + mn + 2$

V Divide or multiply the following fractions as indicated.

19. $\dfrac{x - 3}{5y} \div \dfrac{y + 7}{6x}$

20. $\dfrac{\dfrac{m^3}{3n^2}}{\dfrac{n + 1}{m - 1}}$

21. $\dfrac{4a}{3b} \div c^3$

22. $\dfrac{x^2 - 2x}{y^3} \div (y + 5)$

23. $\dfrac{3m + 6}{m^3} \div \dfrac{m + 2}{m^2}$

24. $\dfrac{2(x + 3)}{x^2y} \div \dfrac{y}{6(x + 3)}$

25. $\dfrac{6x^2 + 3x}{6x} \cdot \dfrac{10x^2}{12x + 6}$

26. $\dfrac{16a^5}{3a - 15} \cdot \dfrac{a^2 - 25}{4a^3}$

27. $\dfrac{4y^2 - 20y}{3y + 6} \cdot \dfrac{y^2 - y - 6}{y - 5}$

28. $\dfrac{x^2 - 9}{3 - 3x} \cdot \dfrac{x^2 + 3x - 4}{x^2 + 5x + 6}$

29. $\dfrac{3x^3 - 18x^2 + 24x}{10x^3 + 5x} \cdot \dfrac{2x^2 + 1}{x^2 + 3x - 28}$

30. $\dfrac{9b^2 - 18b}{b^2 - 3b - 4} \cdot \dfrac{2b^2 + 16b + 14}{6 - 3b}$

31. $\dfrac{4y^2 - 9}{2y^3 + 2y^2 - 4y} \cdot \dfrac{3y^2 - 6y + 3}{2y^2 - y - 3}$

32. $\dfrac{x^2 + 10x + 21}{2x^2 + 7x - 15} \cdot \dfrac{x^2 + 10x + 25}{x^2 + 8x + 15}$

DRILL EXERCISES (continued)

VI   Divide the following fractions.

33. $\dfrac{3n - 9}{m + 4} \div \dfrac{3 - n}{m^2 - 16}$

34. $\dfrac{x^2 + 4x + 3}{x^2 - 4} \div \dfrac{3x^2 - x}{5x - 10}$

35. $\dfrac{a^2 + 3a}{a^2 - 3a - 4} \div \dfrac{a^2 + 2a - 3}{a^2 - 5a + 4}$

36. $\dfrac{2x^2 + 5x - 12}{x^2 - 10x + 25} \div \dfrac{2x^2 + 7x - 15}{x^2 - 25}$

37. $\dfrac{2y^2 - 7y + 5}{3y^2 + 16y - 12} \div \dfrac{6 - 5y - y^2}{3y^2 + y - 2}$

38. $\dfrac{9x^2 - 1}{4x^2} \div (6x + 2)$

VII   Simplify the following complex fractions.

39. $\dfrac{\dfrac{x^2 - 1}{x + 2}}{\dfrac{x + 1}{x - 1}}$

40. $\left(\dfrac{m}{K} + 3\right) \div \left(\dfrac{1}{3K} + \dfrac{1}{K}\right)$

41. $\dfrac{\dfrac{1}{x} - \dfrac{1}{y}}{\dfrac{1}{x} + \dfrac{1}{y}}$

42. $\dfrac{5 + \dfrac{4}{t - 1}}{\dfrac{7}{t + 5} - \dfrac{3}{t - 1}}$

43. $\dfrac{\dfrac{p^2}{a^2} - \dfrac{3p}{a} - 10}{\dfrac{p}{a^2} + \dfrac{1}{a} - \dfrac{2}{p}}$

44. $\dfrac{\dfrac{1}{a + h} - \dfrac{1}{h}}{h}$

ASSIGNMENT EXERCISES

I   Multiply the following fractions.

1. $\dfrac{1}{4y} \cdot \dfrac{5z}{x^2}$

2. $\left(\dfrac{-7}{x^2}\right)\left(\dfrac{-y}{x}\right)$

3. $\dfrac{8y^2}{x^2} \cdot \dfrac{-xy}{2y^3}$

4. $\dfrac{6a^3}{7b} \cdot 3a$

5. $\dfrac{3 - 3x}{2x} \cdot \dfrac{-x^2}{6x - 6}$

6. $4x^2 \cdot \dfrac{x - 7}{3x - 1}$

7. $\dfrac{2y + 12}{24y^2} \cdot \dfrac{6y^3}{y^2 - 36}$

8. $\dfrac{9x^2 - 3x}{15x^2} \cdot \dfrac{5x}{x - 1}$

9. $\dfrac{5y + 15}{y^2 - y - 12} \cdot \dfrac{y^3 - 4y^2}{5y^2}$

10. $\dfrac{2a^2 + 5a}{a^2 - 49} \cdot \dfrac{7 - a}{5a^3}$

11. $\dfrac{m^2 - 2m - 15}{9m^2 + 18m} \cdot \dfrac{m^2 + 3m + 2}{m^2 - 9}$

12. $\dfrac{5a^2 - 10a}{a^2 + 4a + 3} \cdot \dfrac{8a + 24}{20 - 10a}$

13. $\dfrac{2x^2 - 5x - 3}{x^2 - 3x - 4} \cdot \dfrac{x^2 - 2x - 8}{2x^2 + 11x + 5}$

14. $\dfrac{3b^2 + 2}{b^2 + b - 30} \cdot \dfrac{b^2 - 8b + 15}{12b^3 + 8b}$

Unit 12

ASSIGNMENT EXERCISES (continued)

15. $\dfrac{2a^2 - 32}{2a^2 - 15a + 18} \cdot \dfrac{2a^2 + 7a - 15}{4 + 3a - a^2}$

16. $\dfrac{6x^3 + 6x^2}{x^2 - 4} \cdot \dfrac{x^2 - x - 2}{x^2 + x}$

17. $\dfrac{x^2 + 9x + 14}{x^2 + 4x - 21} \cdot \dfrac{x^2 + 2x - 35}{2x^2 - 6x - 20}$

II   Divide the following fractions.

18. $\dfrac{5x^2}{2y} \div \dfrac{y^3}{x}$

19. $\dfrac{\frac{7x^2}{11y}}{\frac{5y^3}{x + 2}}$

20. $\dfrac{x^2}{2z} \div y^3$

21. $\dfrac{a - 2}{3a^4} \div \dfrac{5a - 10}{27a}$

22. $\dfrac{3m}{n - 1} \div (n - 1)$

23. $\dfrac{a^2 - 9}{6b^2} \div \dfrac{3 - a}{3a^2 b}$

24. $\dfrac{3a^2}{b + 3} \div \dfrac{2b^3}{a - 1}$

25. $\dfrac{m^2 - 9}{2m - 12} \div \dfrac{5m - 15}{m^2 - 2m - 24}$

26. $\dfrac{5x - 25}{x^2 - 9} \div \dfrac{x^2 - x - 20}{x^2 + 5x + 6}$

27. $\dfrac{9y^2 - 16}{3y^2 + 2y - 8} \div \dfrac{3y^2 - 5y - 12}{y^2 - y - 6}$

28. $\dfrac{4x^2 - 1}{9x - 3x^2} \div \dfrac{2x^2 - 7x - 4}{x^2 - 7x + 12}$

29. $\dfrac{2m^2 + 3m - 2}{2m^2 + 9m - 5} \div \dfrac{m^2 - m - 6}{m^2 - 8m + 15}$

30. $(4x - 1) \div \dfrac{16x^2 - 1}{24x^3 + 6x^2}$

III   Simplify the following complex fractions.

31. $\dfrac{\frac{1}{y + h + 1} + \frac{1}{y + 1}}{h}$

32. $\dfrac{\frac{2}{x + 1} + \frac{8}{x}}{\frac{6}{x + 1} - \frac{5}{x^2}}$

33. $\dfrac{\frac{4}{x^2 - 9} + \frac{1}{x - 3}}{\frac{2}{x + 3} + 3}$

34. $\dfrac{2 - \frac{5}{x} - \frac{4}{x^2}}{4x - \frac{1}{x}}$

ADDITION AND SUBTRACTION OF ALGEBRAIC FRACTIONS

Objectives: After having worked through this unit, the student should be capable of:

1.   finding the lowest common denominator (LCD) of a group of algebraic fractions by:
(a) inspection
(b) prime factorization;

2.   adding and/or subtracting algebraic fractions.

REVIEW

To add or subtract fractions in algebra, we use the same procedure as when we add or subtract fractions in arithmetic.

In general, if a, b and c are numbers and $c \neq 0$, then

1. $\dfrac{a}{c} + \dfrac{b}{c} = \dfrac{a + b}{c}$

2. $\dfrac{a}{c} - \dfrac{b}{c} = \dfrac{a - b}{c}$

Examples:

Add or subtract the following fractions as indicated.

1. $\dfrac{2x}{5} + \dfrac{3}{5} = \dfrac{2x + 3}{5}$

2. $\dfrac{V^2}{R} - \dfrac{2}{R} = \dfrac{V^2 - 2}{R}$

3. $\dfrac{1}{a} + \dfrac{1}{a} = \dfrac{1 + 1}{a} = \dfrac{2}{a}$

4. $\dfrac{1}{M + 1} - \dfrac{3N}{M + 1} = \dfrac{1 - 3N}{M + 1}$

5. $\dfrac{7y^2}{x^2 z} + \dfrac{5y^2}{x^2 z} = \dfrac{7y^2 + 5y^2}{x^2 z} = \dfrac{12y^2}{x^2 z}$

6. $\dfrac{-a^3 b}{a + b} + \dfrac{ab^3}{a + b} = \dfrac{-a^3 b + ab^3}{a + b}$

CHECK POINT 1

Add or subtract the following fractions as indicated.

1. $\dfrac{2}{T - 1} - \dfrac{9V}{T - 1}$

2. $\dfrac{3ab}{vw^2} + \dfrac{4ab}{vw^2}$

3. $\dfrac{-x^2 y^3}{2x + 3} + \dfrac{x^3 y}{2x + 3}$

ADDING FRACTIONS WITH MULTINOMIAL NUMERATORS

When adding fractions with multinomial numerators,

add the numerators,
collect like terms and
reduce the resulting fraction if possible.

Example:

Add $\dfrac{3x + 7}{15x^2} + \dfrac{2x + 3}{15x^2}$.

Solution:

$$\dfrac{3x + 7}{15x^2} + \dfrac{2x + 3}{15x^2} = \dfrac{3x + 7 + 2x + 3}{15x^2} \qquad \text{(Add numerators.)}$$

$$= \dfrac{5x + 10}{15x^2} \qquad \text{(Collect like terms.)}$$

$$= \dfrac{\overset{1}{\cancel{5}}(x + 2)}{\underset{1}{\cancel{5}}(3x^2)} \qquad \text{(Reduce fraction.)}$$

Hence, $\dfrac{3x + 7}{15x^2} + \dfrac{2x + 3}{15x^2} = \dfrac{x + 2}{3x^2}$.

SUBTRACTING FRACTIONS WITH MULTINOMIAL NUMERATORS

Example:

(a) $-\dfrac{5x - 7}{3y} = -(5x - 7) \div (3y)$ and

(b) $-\dfrac{5x - 7}{3y} = \dfrac{-(5x - 7)}{3y}$ (Remember: $-\dfrac{a}{b} = \dfrac{-a}{b} = \dfrac{a}{-b}$.)

$= \dfrac{-5x + 7}{3y}$

Note:

When the negative sign of a fraction is placed with a multinomial numerator, the **sign of every term in the numerator changes**.

Hence, when subtracting fractions with multinomial numerators, first enclose the numerator to be subtracted in parentheses; then proceed to combine the numerators and place over the common denominator.

Example 1:

Subtract $\dfrac{7x + 2}{3y} - \dfrac{5x - 7}{3y}$.

Solution:

$\dfrac{7x + 2}{3y} - \dfrac{5x - 7}{3y} = \dfrac{7x + 2}{3y} - \dfrac{(5x - 7)}{3y}$ (Place parentheses.)

$= \dfrac{7x + 2 - (5x - 7)}{3y}$ (Combine numerators over common denominator.)

$= \dfrac{7x + 2 - 5x + 7}{3y}$ (Remove parentheses.)

$= \dfrac{2x + 9}{3y}$ (Collect like terms.)

Hence, $\dfrac{7x + 2}{3y} - \dfrac{5x - 7}{3y} = \dfrac{2x + 9}{3y}$.

COMMON ERROR

error

$\dfrac{7x + 2}{3y} - \dfrac{5x - 7}{3y} = \dfrac{7x + 2 - 5x - 7}{3y}$

Note:

1.   The signs of all terms in the numerator of the fraction to be subtracted must be changed.

2.   To avoid the common error of changing only the sign of the first term, make use of parentheses as shown in example 1.

8

CHECK POINT 2

Add or subtract the following fractions as indicated.

1. $\dfrac{2a + 3}{6b} + \dfrac{4a - 1}{6b}$

2. $\dfrac{5m^2 - 3m}{2n} - \dfrac{m^2 - m}{2n}$

9

ADDING AND SUBTRACTING FRACTIONS WITH UNLIKE DENOMINATORS

Review: The lowest Common Denominator (LCD)

The lowest common denominator of a group of denominators is the smallest number into which all the denominators divide evenly.

Examples:

1. The LCD of $\dfrac{3}{4}$ and $\dfrac{5}{6}$ is 12.

2. The LCD of $\dfrac{x}{2y}$ and $\dfrac{7}{5y}$ is 10y.

3. The LCD of $\dfrac{1}{x - 1}$ and $\dfrac{1}{x + 1}$ is $x^2 - 1$.

10

FINDING THE LCD BY INSPECTION

In many practical problems, we can find the LCD by simply inspecting the given denominators.

This is particularly the case in the following two situations:

1. The denominators are prime or have no common factors.
2. One denominator is a factor of the other denominator.

11

If the denominators are prime or have **no** common factors, the LCD is simply the product of the two denominators.

Examples:

1. The LCD of $\dfrac{3}{R - 1}$ and $\dfrac{5P}{R + 3}$ is $(R - 1)(R + 3)$.

2. The LCD of $\dfrac{9}{x^3}$ and $\dfrac{4x}{x^2 + 1}$ is $x^3(x^2 + 1)$.

12

If one denominator is a factor of the other denominator, the larger denominator is the LCD.

Examples:

1. The LCD of $\dfrac{1}{(x - 1)^2}$ and $\dfrac{x^3}{x - 1}$ is $(x - 1)^2$,

   since $x - 1$ is a factor of $(x - 1)^2$.

2. The LCD of $\dfrac{a}{a^2 - a - 12}$ and $\dfrac{7}{a - 4}$ is $a^2 - a - 12$,

   since $a - 4$ is a factor of $a^2 - a - 12 = (a + 3)(a - 4)$.

13

CHECK POINT 3

Find the LCD of the following groups of fractions by inspection.

1. $\dfrac{m^2}{n-3}, \dfrac{n}{m+1}$ 2. $\dfrac{6}{R+3}, \dfrac{5R}{(R+3)^2}$ 3. $\dfrac{1}{4x^2-9}, \dfrac{1}{2x+3}$

14

PRIME FACTORIZATION

In algebra, as in arithmetic, an expression is prime if it can be factored only into 1 and itself.

Example 1:

The following expressions are prime: x,   2V – 1,   a,   7x + 9.
Each has no factors other than 1 and itself.

Example 2:

The following expressions are considered to be in prime factored exponential form:

x^5, $(2V - 1)^2$, $3^4 a^2 b^4$, $7x + 9$.

15

MORE EXAMPLES:

Example 1:

The prime factored exponential form of

$54x^2 - 108x = 54x(x - 2) = 2 \cdot 3^3 x(x - 2)$

Example 2:

The prime factored exponential form of

$4y^2 + 4y + 1 = (2y + 1)^2$

Example 3:

The prime factored exponential form of

$32a^2 - 50b^2 = 2(16a^2 - 25b^2) = 2(4a - 5b)(4a + 5b)$

Example 4:

The prime factored exponential form of

$8m^3 + 44m^2 - 24m = 4m(2m^2 + 11m - 6) = 2^2 m(m + 6)(2m - 1)$

16

CHECK POINT 4:

Write each of the following in prime factored exponential form.

Note:

You may want to review Unit 9, "Factoring Algebraic Expressions".

1. $8a^2 + 24a$ 2. $12m^2 - 48n^2$ 3. $6a^3 - 15a^2 - 9a$

FINDING THE LCD USING PRIME FACTORIZATION

Example 1:

Find the LCD of $\dfrac{16x^3}{4y^2 - 12y}$ and $\dfrac{3z}{34x^2y^3}$.

Solution:

Step 1.   Prime factor each denominator and write it in exponential form.

$$4y^2 - 12y = 4y(y - 3) = 2^2y(y - 3)$$
$$34x^2y^3 = 2 \cdot 17 \cdot x^2 \cdot y^3$$

Step 2.   Write a product using only once each different base.

$$2 \cdot 17 \cdot x \cdot y \cdot (y - 3)$$

Step 3.   Place the highest exponent appearing in Step 1 on the appropriate bases in Step 2.

$$\text{LCD} = 2^2 \cdot 17 \cdot x^2 \cdot y^3 \cdot (y - 3)$$

Step 4.   Multiply the numerical factors.

$$\text{LCD} = 68x^2y^3(y - 3)$$

Note:

Do not multiply out the binomial factors in the LCD. Having the algebraic factors in the LCD will make it easier for us to determine by what we have to multiply each denominator in order to expand it to the LCD.

Example 2:

Find the LCD of $\dfrac{5}{3m + 3}$ and $\dfrac{2m}{m^2 - 1}$.

Solution:

Step 1.   Prime factor each denominator.
$$3m + 3 = 3(m + 1)$$
$$m^2 - 1 = (m + 1)(m - 1)$$

Step 2.   Write a product using only once each different base.

$$3(m + 1)(m - 1)$$

Step 3.   Since each factor in Step 1 is to be the first power, we have
$\text{LCD} = 3(m + 1)(m - 1)$.   Do not multiply out the binomial factors.

Example 3:

Find the LCD of $\dfrac{R - 3}{R^2 + 4R + 4}$, $\dfrac{R^2 + 6}{R^2 - 3R - 10}$ and $\dfrac{7}{R^3}$.

Solution:

Step 1. Prime factor each denominator.

$$R^2 + 4R + 4 = (R + 2)(R + 2) = (R + 2)^2$$

$$R^2 - 3R - 10 = (R + 2)(R - 5)$$

$$R^3 = R^3$$

Step 2. Write a product using only once each different base.

$$R(R + 2)(R - 5)$$

Step 3. Place the highest exponent appearing in Step 1 on the appropriate bases in Step 2.

$$LCD = R^3(R + 2)^2(R - 5)$$

CHECK POINT 5

For each of the following groups of fractions, find the LCD.

1. $\dfrac{13V^2}{32R^3z}$ and $\dfrac{10V}{18R^3 + 36R^2}$

2. $\dfrac{a + 1}{5a - 10}$ and $\dfrac{17}{a^2 + 2a - 3}$

3. $\dfrac{4x + 1}{7x^3}$, $\dfrac{5x^3}{x^2 - 4}$ and $\dfrac{x - 1}{x^2 + 5x + 6}$

ADDITION AND SUBTRACTION OF FRACTIONS WITH UNLIKE DENOMINATORS

Method

Step 1. Find the Lowest Common Denominator (LCD).

Step 2. Expand each fraction to one which has the LCD.

Step 3. Add or subtract the numerators and reduce answer to lowest terms.

Example 1:

Subtract $\dfrac{8V^2}{5R} - \dfrac{-3V}{2P}$.

Solution:

Step 1. The LCD = (2P)(5R) = 10PR (by inspection).

Step 2. Expand each fraction to one which has the LCD.

$$\frac{8V^2}{5R} - \frac{-3V}{2P} = \frac{(8V^2)(2P)}{(5R)(2P)} - \frac{(-3V)(5R)}{(2P)(5R)}$$

$$= \frac{16V^2P}{10PR} - \frac{(-15VR)}{10PR} \qquad \text{(Use parentheses.)}$$

Step 3. Subtract numerators.

$$\frac{16V^2P}{10PR} - \frac{(-15VR)}{10PR} = \frac{16V^2P - (-15VR)}{10PR}$$

$$= \frac{16V^2P + 15VR}{10PR}$$

Hence, $\dfrac{8V^2}{5R} - \dfrac{-3V}{2P} = \dfrac{16V^2P + 15VR}{10PR}$ or $\dfrac{V(16VP + 15R)}{10PR}$.

Example 2:

Add $\dfrac{9a}{32b^3c^2z} + \dfrac{5c}{18bz^3}$.

Solution:

Step 1. Since $32b^3c^2z = 2^5 \cdot b^3 \cdot c^2 \cdot z$ and $18bz^3 = 2 \cdot 3^2 \cdot b \cdot z^3$, the LCD $= 2^5 \cdot 3^2 \cdot b^3 \cdot c^2 \cdot z^3 = 288b^3c^2z^3$.

Step 2. Expand each fraction to one which has the LCD.

$$\frac{9a}{32b^3c^2z} + \frac{5c}{18bz^3} = \frac{(9a)(9z^2)}{(32b^3c^2z)(9z^2)} + \frac{(5c)(16b^2c^2)}{(18bz^3)(16b^2c^2)} \quad \text{(See Note on next page)}$$

$$= \frac{81az^2}{288b^3c^2z^3} + \frac{80b^2c^3}{288b^3c^2z^3}$$

Step 3. Add the numerators.

$$\frac{81az^2}{288b^3c^2z^3} + \frac{80b^2c^3}{288b^3c^2z^3} = \frac{81az^2 + 80b^2c^3}{288b^3c^2z^3}$$

Hence, $\dfrac{9a}{32b^3c^2z} + \dfrac{5c}{18bz^3} = \dfrac{81az^2 + 80b^2c^3}{288b^3c^2z^3}$.

Note:

To determine the factors $9z^2$ and $16b^2c^2$ which are needed to expand the fractions, we divide the LCD by the given denominators.

Hence, $\dfrac{288b^3c^2z^3}{32b^3c^2z} = \boxed{9z^2}$ and $\dfrac{288b^3c^2z^3}{18bz^3} = \boxed{16b^2c^2}$

are the factors used to expand the two fractions in Step 2 of Example 2.

24

Example 3:

Add $\dfrac{5}{V + 1} + \dfrac{7}{R - 3}$.

Solution:

Step 1. The LCD $= (V + 1)(R - 3)$ by inspection.

Step 2. Expand each fraction to one which has the LCD.

$$\frac{5}{V + 1} + \frac{7}{R - 3} = \frac{5(R - 3)}{(V + 1)(R - 3)} + \frac{7(V + 1)}{(R - 3)(V + 1)}$$

$$= \frac{5R - 15 + 7V + 7}{(V + 1)(R - 3)}$$

$$= \frac{5R + 7V - 8}{(V + 1)(R - 3)}$$

Hence, $\dfrac{5}{V + 1} + \dfrac{7}{R - 3} = \dfrac{5R + 7V - 8}{(V + 1)(R - 3)}$.

25

CHECK POINT 6

Add or subtract as indicated.

1. $\dfrac{2m^3}{15n^3p} - \dfrac{-5m}{24np^2}$ 2. $\dfrac{9}{x - 6} + \dfrac{3x}{x + 1}$ 3. $\dfrac{1}{4a^2 - 9} - \dfrac{a}{2a + 3}$

MORE EXAMPLES

Example 1:

Subtract $\dfrac{9y^3}{3x^3 - 9x^2} - \dfrac{1}{15x^3}$.

Solution:

Step 1. Since $3x^3 - 9x^2 = 3x^2(x - 3)$ and $15x^3 = 3 \cdot 5 \cdot x^3$, the LCD $= 3 \cdot 5 \cdot x^3 \cdot (x - 3) = 15x^3(x - 3)$.

Step 2. Expand each fraction to one which has the LCD.

$$\dfrac{9y^3}{3x^3 - 9x^2} - \dfrac{1}{15x^3} = \dfrac{9y^3}{3x^2(x - 3)} - \dfrac{1}{15x^3}$$

$$= \dfrac{9y^3(5x)}{3x^2(x - 3)(5x)} - \dfrac{1(x - 3)}{15x^3(x - 3)}$$

$$= \dfrac{45xy^3}{15x^3(x - 3)} - \dfrac{x - 3}{15x^3(x - 3)}$$

Step 3. Subtract the numerators.

$$\dfrac{45xy^3}{15x^3(x - 3)} - \dfrac{x - 3}{15x^3(x - 3)} = \dfrac{45xy^3 - (x - 3)}{15x^3(x - 3)} \text{ (Use parentheses.)}$$

$$= \dfrac{45xy^3 - x + 3}{15x^3(x - 3)}$$

Hence, $\dfrac{9y^3}{3x^3 - 9x^2} - \dfrac{1}{15x^3} = \dfrac{45xy^3 - x + 3}{15x^3(x - 3)}$.

Example 2:

Add $\dfrac{7}{n^2 - 4} + \dfrac{5}{6n + 12}$.

Solution:

Step 1. Since $n^2 - 4 = n^2 - 2^2 = (n + 2)(n - 2)$ and $6n + 12 = 6(n + 2)$, the LCD $= 6(n + 2)(n - 2)$.

Step 2. Expand each fraction to one which has the LCD.

$$\dfrac{7}{n^2 - 4} + \dfrac{5}{6n + 12} = \dfrac{7}{(n + 2)(n - 2)} + \dfrac{5}{6(n + 2)}$$

$$= \dfrac{6 \cdot 7}{6(n + 2)(n - 2)} + \dfrac{5(n - 2)}{6(n + 2)(n - 2)}$$

Step 3.   Add the numerators.

$$\frac{42}{6(n + 2)(n - 2)} + \frac{5(n - 2)}{6(n + 2)(n - 2)} = \frac{42 + 5(n - 2)}{6(n + 2)(n - 2)}$$

$$= \frac{42 + 5n - 10}{6(n + 2)(n - 2)}$$

$$= \frac{5n + 32}{6(n + 2)(n - 2)}$$

Hence, $\dfrac{7}{n^2 - 4} + \dfrac{5}{6n + 12} = \dfrac{5n + 32}{6(n^2 - 4)}.$

28

Example 3:

Subtract $\dfrac{6m}{m^2 - 7m + 10} - \dfrac{4m}{m^2 - 8m + 15}.$

Solution:

Step 1.   Since $m^2 - 7m + 10 = (m - 2)(m - 5)$ and

$m^2 - 8m + 15 = (m - 3)(m - 5)$.

the LCD $= (m - 2)(m - 3)(m - 5)$.

Step 2.   Expand each fraction to one which has the LCD.

$$\frac{6m}{m^2 - 7m + 10} - \frac{4m}{m^2 - 8m + 15} = \frac{6m}{(m - 2)(m - 5)} - \frac{4m}{(m - 3)(m - 5)}$$

$$= \frac{6m(m - 3)}{(m - 2)(m - 5)(m - 3)} - \frac{4m(m - 2)}{(m - 3)(m - 5)(m - 2)}$$

Step 3.   Combine the numerators, collect like terms and reduce.

$$= \frac{6m(m - 3) - 4m(m - 2)}{(m - 2)(m - 3)(m - 5)}$$

$$= \frac{6m^2 - 18m - 4m^2 + 8m}{(m - 2)(m - 3)(m - 5)}$$

$$= \frac{2m^2 - 10m}{(m - 2)(m - 3)(m - 5)}$$

$$= \frac{2m(m - 5)^1}{(m - 2)(m - 3)(m - 5)_1}$$

$$= \frac{2m}{(m - 2)(m - 3)}$$

Hence, $\dfrac{6m}{m^2 - 7m + 10} - \dfrac{4m}{m^2 - 8m + 15} = \dfrac{2m}{(m - 2)(m - 3)}.$

Example 4:

Add or subtract as indicated $\dfrac{a}{a^2 + 8a + 15} - \dfrac{4}{a^2} + \dfrac{1}{a^3 + 3a^2}$.

Solution:

Step 1.   Since $a^2 + 8a + 15 = (a + 3)(a + 5)$ and $a^3 + 3a^2 = a^2(a + 3)$, the LCD $= a^2(a + 3)(a + 5)$.

Step 2.   Expand each fraction to one which has the LCD.

$$\dfrac{a}{a^2 + 8a + 15} - \dfrac{4}{a^2} + \dfrac{1}{a^3 + 3a^2}$$

$$= \dfrac{a}{(a + 3)(a + 5)} - \dfrac{4}{a^2} + \dfrac{1}{a^2(a + 3)}$$

$$= \dfrac{a^2 a}{a^2(a + 3)(a + 5)} - \dfrac{4(a + 3)(a + 5)}{a^2(a + 3)(a + 5)} + \dfrac{1(a + 5)}{a^2(a + 3)(a + 5)}$$

$$= \dfrac{a^3}{a^2(a + 3)(a + 5)} - \dfrac{4a^2 + 32a + 60}{a^2(a + 3)(a + 5)} + \dfrac{a + 5}{a^2(a + 3)(a + 5)}$$

Step 3.   Combine the numerators and collect like terms.

$$= \dfrac{a^3 - (4a^2 + 32a + 60) + a + 5}{a^2(a + 3)(a + 5)}$$

$$= \dfrac{a^3 - 4a^2 - 32a - 60 + a + 5}{a^2(a + 3)(a + 5)}$$

$$= \dfrac{a^3 - 4a^2 - 31a - 55}{a^2(a + 3)(a + 5)}$$

Hence, $\dfrac{a}{a^2 + 8a + 15} - \dfrac{4}{a^2} + \dfrac{1}{a^3 + 3a^2} = \dfrac{a^3 - 4a^2 - 31a - 55}{a^2(a + 3)(a + 5)}$.

CHECK POINT 7

Add and/or subtract the following fractions.

1. $\dfrac{R^2}{2R^2 + 4R} - \dfrac{5}{8R^2}$

2. $\dfrac{2a - 1}{3a - 9} - \dfrac{a + 2}{a^2 - 9}$

3. $\dfrac{5x}{3x^2 - 5x - 2} - \dfrac{3x}{x^2 + x - 6}$

4. $\dfrac{2}{n^3 - 2n^2} + \dfrac{n}{n^2 + n - 6} - \dfrac{7}{n^3}$

DRILL EXERCISES

I. Add or subtract the following fractions as indicated.

1. $\dfrac{T^2S}{2T - 1} - \dfrac{-TS^2}{2T - 1}$

2. $\dfrac{5y + 4}{7y^2} + \dfrac{2y - 3}{7y^2}$

3. $\dfrac{9x - 2}{3b} - \dfrac{x + 3}{3b}$

4. $\dfrac{6n^3 - n}{5m} - \dfrac{2n^3 - 3n}{5m}$

II. Write each of the following in prime factored exponential form.

5. $9y^3 - 36y^2$

6. $16y^2 - 36z^2$

7. $2m^2 - 18m + 36$

8. $3x^5 - x^4 - 2x^3$

III. Find the LCD of the following groups of fractions.

9. $\dfrac{4y}{x + 3},\ \dfrac{3}{x - 2}$

10. $\dfrac{z^2}{2x},\ \dfrac{4}{3y}$

11. $\dfrac{x^2}{(x + 3)^2},\ \dfrac{2x}{x + 3}$

12. $\dfrac{4}{3a},\ \dfrac{2}{9a^2}$

13. $\dfrac{7}{16ab^3c^2},\ \dfrac{11b}{28a^2c^4}$

14. $\dfrac{15z}{36x^2y^2},\ \dfrac{3z^2}{8x^4 - 24x^3}$

15. $\dfrac{a - 1}{a^2 + 4a - 21},\ \dfrac{7}{4a - 12}$

16. $\dfrac{4x - 7}{x^2 - 25},\ \dfrac{x + 2}{x^2 - 7x + 10},\ \dfrac{x - 7}{3x^2}$

IV. Add and/or subtract the following fractions.

17. $\dfrac{2}{xy} + \dfrac{3}{xz} + \dfrac{4}{yz}$

18. $\dfrac{3p^2}{10mn^2} - \dfrac{-2p}{35m^4n}$

19. $\dfrac{2}{m + 1} + \dfrac{3}{n - 2}$

20. $\dfrac{7y^2}{4x^2 - 8x} - \dfrac{3}{16x^2}$

21. $\dfrac{2x}{x^2 + 3x + 2} + \dfrac{3}{2x^2 + 4x} - \dfrac{2}{2x^2}$

22. $\dfrac{4x - 3}{x^2 - 4} - \dfrac{x - 1}{x + 2}$

23. $\dfrac{3x + 1}{5x} - \dfrac{2x + 5}{3x - 1}$

24. $\dfrac{2x + 3}{6x + 4} - \dfrac{4x - 1}{3x^2 + 5x + 2}$

ASSIGNMENT EXERCISES

I.   Add and/or subtract the following fractions as indicated.

1. $\dfrac{4z - 7}{11x^2y^2} + \dfrac{3z + 6}{11x^2y^2}$

2. $\dfrac{5a^2 + 6a}{6b} - \dfrac{4a^2 - a}{6b}$

3. $\dfrac{c}{7a} + \dfrac{2d}{3b}$

4. $\dfrac{P}{P + 1} - \dfrac{R}{R + 1}$

5. $\dfrac{5}{(x - 2)^2} + \dfrac{y}{(x - 2)}$

6. $\dfrac{4}{m - 3} - \dfrac{-n}{m - 3}$

7. $\dfrac{x^2 - x}{7y} - \dfrac{3x^2 + 2}{7y}$

8. $\dfrac{4}{a - 3} + \dfrac{7}{2a - 1}$

9. $\dfrac{2x^3}{9x^3 - 36x^2} - \dfrac{1}{72x^3}$

10. $\dfrac{-x^3y^2}{3x + 2} + \dfrac{x^2y^3}{5x + 2}$

11. $\dfrac{7z}{x^7y} + \dfrac{5}{3z}$

12. $\dfrac{6x}{5x - 20} + \dfrac{3}{x^2 - 16}$

13. $\dfrac{7}{9w^4x^2} - \dfrac{-5}{15xy^2}$

14. $\dfrac{-7b}{30a^2c^2d} + \dfrac{6}{48ad^3}$

15. $\dfrac{x}{x^2 - 4x + 3} - \dfrac{3}{2x^3} + \dfrac{4}{x^3 - 3x^2}$

16. $\dfrac{5x + 2}{2x} - \dfrac{3x - 4}{4x - 3}$

17. $\dfrac{3x + 1}{x + 1} + \dfrac{2x - 1}{x^2 - 1}$

18. $\dfrac{x - 4}{6x^2 + 13x + 6} - \dfrac{x + 3}{3x + 2}$

19. $\dfrac{3x - 4}{x^3 + 2x + x} - \dfrac{2x - 1}{x^2 + x} + \dfrac{x}{x + 1}$

SOLVING FRACTIONAL EQUATIONS AND FORMULAS, WORD PROBLEMS

Objectives: After having worked through this unit, the student should be capable of:

1. solving fractional equations;
2. recognizing extraneous solutions;
3. solving literal equations and formulas involving fractional expressions;
4. solving word problems involving the solution of fractional equations.

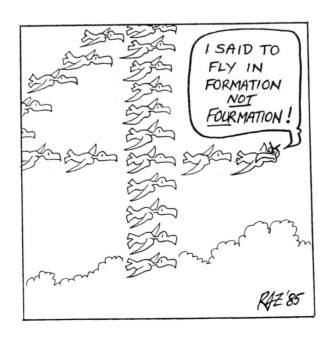

In previous units we have learned how to solve:

(a)   linear equations in one unknown with a general form
$ax + b = 0$,   where a and b are constants and $a \neq 0$,

(b)   quadratic equations in one unknown with a general form
$ax^2 + bx + c = 0$, where a, b and c are constants and $a \neq 0$.

To solve many problems in business, science and technology, we are faced with the task of solving equations which contain fractions.

Examples:

1. The number of years required when investing a principal of P to yield an amount S at a simple interest rate i is given by

$$n = \frac{S}{Pi} - \frac{1}{i}.$$

2. Analyzing a certain electric circuit, we may obtain the following equation:

$$\frac{V - 3}{2} + \frac{V - 4}{6} + \frac{V}{4} = 0.$$

REVIEW:

FRACTIONAL EQUATIONS

Equations which contain fractions are referred to as fractional equations.

To solve fractional equations, we first change the equation to one which no longer contains fractions.

This process is generally referred to as clearing an equation of fractions. The resulting equation may be a linear or quadratic equation which we then solve in the usual manner.

CLEARING AN EQUATION OF FRACTIONS

Rule:

To clear an equation of fractions, multiply each term of the equation by the lowest common denominator of all fractions in the equation.

Example:

Clear the following equation of fractions.

$$\frac{1}{2} + \frac{3x}{4} = \frac{5x}{6}$$

Solution:

1. The LCD of $\frac{1}{2}$, $\frac{3x}{4}$ and $\frac{5x}{6}$ is 12.

2. $$12\left(\frac{1}{2} + \frac{3x}{4}\right) = 12\left(\frac{5x}{6}\right)$$ (Multiply both sides of $\frac{1}{2} + \frac{3x}{4} = \frac{5x}{6}$ by the LCD.)

$$12 \cdot \frac{1}{2} + 12 \cdot \frac{3x}{4} = 12 \cdot \frac{5x}{6}$$

$$6 + 9x = 10x$$

5

CHECK POINT 1

Clear the following equations of fractions.

1. $\dfrac{7}{2} + \dfrac{2}{M} = 4$
2. $\dfrac{2}{5V} + \dfrac{3}{10} = \dfrac{1}{12}$
3. $\dfrac{3}{4} = \dfrac{f}{f - 2}$

4. $\dfrac{2}{R} = \dfrac{3}{R + 2}$
5. $\dfrac{2}{x + 1} = 5 + \dfrac{2}{x + 1}$

6

SOLVING FRACTIONAL EQUATIONS

Method:

Step 1. Find the LCD of all fractions in the equation.

Step 2. Multiply both sides of the equation by the LCD and simplify.

Step 3. Solve the resulting non-fractional equations.

Step 4. Check the solutions.

7

CLEARING OF FRACTIONS RESULTING IN LINEAR EQUATIONS

Example 1:

Solve $\dfrac{V}{6} - \dfrac{V - 3}{4} = \dfrac{1}{2}$.

Solution:

Step 1. The LCD is 12.

Step 2. Multiply both sides of the equation by the LCD.

$$12\left(\dfrac{V}{6} - \dfrac{V - 3}{4}\right) = 12\left(\dfrac{1}{2}\right)$$

$$12\left(\dfrac{V}{6}\right) - 12\left(\dfrac{V - 3}{4}\right) = 12\left(\dfrac{1}{2}\right)$$

$$2V - 3(V - 3) = 6$$

Step 3. Solve the non-fractional equation.

$$2V - 3(V - 3) = 6$$
$$2V - 3V + 9 = 6$$
$$-V = -3$$
$$V = 3$$

Step 4. Check the solution.

$$\text{L.H.S.} = \dfrac{3}{6} - \dfrac{3 - 3}{4} = \dfrac{1}{2} - \dfrac{0}{4} = \dfrac{1}{2}$$

$$\text{R.H.S.} = \dfrac{1}{2}$$

Hence, the solution is $V = 3$.

Example 2:

Solve $\dfrac{25}{3P} + \dfrac{1}{3} = \dfrac{10}{P}$.

Solution:

Step 1. LCD = 3P.

Step 2.
$$3P\left(\dfrac{25}{3P} + \dfrac{1}{3}\right) = 3P\left(\dfrac{10}{P}\right)$$

$$\overset{1}{3P} \cdot \dfrac{25}{\underset{1}{3P}} + \overset{1}{3P} \cdot \dfrac{1}{\underset{1}{3}} = \overset{1}{3P} \cdot \dfrac{10}{\underset{1}{P}}$$

$$25 + P = 30$$

Step 3. Solve the non-fractional equation.
$$25 + P = 30$$
$$P = 5$$

Step 4. Check the solution.

$$\text{L.H.S.} = \dfrac{25}{3(5)} + \dfrac{1}{3} = \dfrac{5}{3} + \dfrac{1}{3} = \dfrac{6}{3} = 2$$

$$\text{R.H.S.} = \dfrac{10}{5} = 2$$

Hence, the solution is P = 5.

EXTRANEOUS SOLUTIONS

Certain methods of solving equations may introduce solutions which will not check in the original equation. Such solutions are called extraneous solutions. In solving fractional equations, extraneous solutions are introduced when variable expressions in the denominators are eliminated by multiplying both sides of the equation by the LCD.

Example 3:

Solve $\dfrac{m + 1}{m - 3} - 6 = \dfrac{4}{m - 3}$.

Solution:

Step 1. LCD = m - 3

Step 2. Multiply both sides of the equation by the LCD.

$$(m - 3)\left(\dfrac{m + 1}{m - 3} - 6\right) = (m - 3)\left(\dfrac{4}{m - 3}\right)$$

$$\overset{1}{(m - 3)}\left(\dfrac{m + 1}{\underset{1}{m - 3}}\right) - 6(m - 3) = \overset{1}{(m - 3)}\left(\dfrac{4}{\underset{1}{m - 3}}\right)$$

$$m + 1 - 6(m - 3) = 4$$

Step 3. Solve the non-fractional equation.

$$m + 1 - 6(m - 3) = 4$$

$$m + 1 - 6m + 18 = 4$$

$$m = 3$$

Step 4. Check the solution.

L.H.S. $= \dfrac{3 + 1}{3 - 3} - 6 = \dfrac{4}{0} - 6$. Since $\dfrac{4}{0}$ is undefined, $m = 3$ cannot be

a solution but is extraneous.

Hence, the above equation has no solution.

11

Example 4:

Solve $\dfrac{3}{m - 4} - \dfrac{5}{m + 4} = \dfrac{2}{m^2 - 16}$.

Solution:

Step 1. LCD $= (m - 4)(m + 4) = m^2 - 16$.

Step 2. Multiply both sides of the equation by the LCD.

$$(m^2 - 16)\left(\dfrac{3}{m - 4} - \dfrac{5}{m + 4}\right) = (m^2 - 16)\left(\dfrac{2}{m^2 - 16}\right)$$

$$(m - 4)(m + 4)\left(\dfrac{3}{m - 4} - \dfrac{5}{m + 4}\right) = (m^2 - 16)\left(\dfrac{2}{m^2 - 16}\right)$$

$$\dfrac{3(m - 4)(m + 4)}{(m - 4)} - \dfrac{5(m - 4)(m + 4)}{(m + 4)} = (m^2 - 16)\left(\dfrac{2}{m^2 - 16}\right)$$

$$3(m + 4) - 5(m - 4) = 2$$

Step 3. Solve the non-fractional equation.

$$3(m + 4) - 5(m - 4) = 2$$

$$3m + 12 - 5m + 20 = 2$$

$$m = 15$$

Step 4. Check the solution.

L.H.S. $\dfrac{3}{15 - 4} - \dfrac{5}{15 + 4} = \dfrac{3}{11} - \dfrac{5}{19} = \dfrac{57 - 55}{209} = \dfrac{2}{209}$

R.H.S. $\dfrac{2}{15^2 - 16} = \dfrac{2}{225 - 16} = \dfrac{2}{209}$

Hence, the solution is $m = 15$.

12

CHECK POINT 2

Solve the following equations.

1. $\dfrac{2M}{3} - \dfrac{M+1}{2} = \dfrac{1}{3}$

2. $\dfrac{2}{x-3} - 5 = \dfrac{2}{x-3}$

3. $\dfrac{a+10}{a+6} = \dfrac{4}{a+6} + 3$

4. $\dfrac{6}{y^2-4} + \dfrac{3}{y-2} = \dfrac{4}{y+2}$

13

Example 1:

Solve $\dfrac{3}{x} + \dfrac{5}{x+3} = 1 - \dfrac{3}{x+3}$.

Solution:

Step 1. The LCD = $x(x+3)$.

Step 2. Multiply both sides of the equation by the LCD.

$$x(x+3)\left(\dfrac{3}{x} + \dfrac{5}{x+3}\right) = x(x+3)\left(1 - \dfrac{3}{x+3}\right)$$

$$3(x+3) + 5x = x(x+3) - 3x$$

$$3x + 9 + 5x = x^2 + 3x - 3x$$

$$-x^2 + 8x + 9 = 0$$

$$x^2 - 8x - 9 = 0 \quad \text{(Multiply both sides by } -1.)$$

Step 3. Solve the quadratic equation.

$$x^2 - 8x - 9 = 0$$

$$(x+1)(x-9) = 0$$

$$x + 1 = 0 \text{ or } x - 9 = 0$$

Hence, $x = -1$ or $x = 9$.

Step 4. Check the solution in the original equation.

1. If $x = -1$, $\dfrac{3}{-1} + \dfrac{5}{-1+3} \overset{?}{=} 1 - \dfrac{3}{-1+3}$ or $\dfrac{-1}{2} = \dfrac{-1}{2}$.

2. If $x = 9$, $\dfrac{3}{9} + \dfrac{5}{9+3} \overset{?}{=} 1 - \dfrac{3}{9+3}$ or $\dfrac{9}{12} = \dfrac{9}{12}$.

Hence, $x = -1$ and $x = 9$ are solutions of the given equation.

14

Example 2:

Solve $\dfrac{p-4}{p+2} = \dfrac{2}{p-3}$.

Solution:

Step 1. The LCD = $(p+2)(p-3)$.

Step 2. Multiply both sides of the equation by the LCD.

$$\overset{1}{(\cancel{p+2})}(p-3)\left(\frac{p-4}{\underset{1}{\cancel{p+2}}}\right) = (p+2)\overset{1}{(\cancel{p-3})}\left(\frac{2}{\underset{1}{\cancel{p-3}}}\right)$$

$$(p-3)(p-4) = 2(p+2)$$

$$p^2 - 7p + 12 = 2p + 4$$

$$p^2 - 9p + 8 = 0$$

Step 3. Solve the quadratic equation.

$$p^2 - 9p + 8 = 0$$

$$(p-1)(p-8) = 0$$

$$p - 1 = 0 \quad \text{or} \quad p - 8 = 0$$

$$\therefore \qquad\qquad p = 1 \quad \text{or} \qquad p = 8$$

Step 4. Check the solution in the original equation $\dfrac{p-4}{p+2} = \dfrac{2}{p-3}$.

If $p = 1$, $\dfrac{1-4}{1+2} = \dfrac{2}{1-3}$ or $-1 = -1$.

If $p = 8$, $\dfrac{8-4}{8+2} = \dfrac{2}{8-3}$ or $\dfrac{2}{5} = \dfrac{2}{5}$.

Hence, $p = 1$ and $p = 8$ are solutions of the given equation.

15

Example 3:

Solve $\dfrac{2t}{t-3} + \dfrac{2t-1}{t^2 - 7t + 12} = \dfrac{t+3}{t-4}$.

Solution:

Step 1. The LCD $= (t-3)(t-4)$

$$= t^2 - 7t + 12$$

Step 2. Multiply both sides of the equation by the LCD.

$$(t-3)(t-4)\left[\frac{2t}{t-3} + \frac{2t-1}{(t-3)(t-4)}\right] = (t-3)(t-4)\left(\frac{t+3}{t-4}\right)$$

$$\overset{1}{(\cancel{t-3})}(t-4)\frac{2t}{\underset{1}{\cancel{t-3}}} + \overset{1}{(\cancel{t-3})}\overset{1}{(\cancel{t-4})}\frac{2t-1}{\underset{1}{(\cancel{t-3})}\underset{1}{(\cancel{t-4})}} = (t-3)\overset{1}{(\cancel{t-4})}\left(\frac{t+3}{\underset{1}{\cancel{t-4}}}\right)$$

$$(t-4)(2t) + (2t-1) = t^2 - 9$$

$$2t^2 - 8t + 2t - 1 = t^2 - 9$$

$$t^2 - 6t + 8 = 0$$

Step 3. Solve the quadratic equation.

$(t - 2)(t - 4) = 0$

$t - 2 = 0$ or $t - 4 = 0$

$\therefore t = 2$ or $t = 4$

Step 4. Check the solution in the original equation.

$$\frac{2t}{t - 3} + \frac{2t - 1}{t^2 - 7t + 12} = \frac{t + 3}{t - 4}$$

If $t = 2,$ $\frac{2(2)}{2 - 3} + \frac{2(2) - 1}{2^2 - 7(2) + 12} = \frac{2 + 3}{2 - 4}$ or $-\frac{5}{2} = -\frac{5}{2}.$

If $t = 4,$ the R.H.S. is undefined since $\frac{4 + 3}{4 - 4} = \frac{7}{0}.$

Hence, $t = 2$ is a solution of the original equation.

However, $t = 4$ is **not** a solution and we call $t = 4$ an extraneous solution.

16

CHECK POINT 3

Solve the following equations.

1. $\frac{4}{p + 3} + \frac{6}{p + 3} = 1 - \frac{4}{p}$

2. $\frac{1}{a^2 - 16} + \frac{a}{a + 4} = \frac{2}{a - 4}$

3. $\frac{m - 1}{m^2 + 2} = \frac{2}{3m + 2}$

4. $\frac{t + 1}{t + 3} + \frac{t + 2}{t - 5} = \frac{t^2 - t + 4}{t^2 - 2t - 15}$

17

SOLVING LITERAL EQUATIONS AND FORMULAS

General Method:

Step 1. Clear the equation of fractions, if necessary.

Step 2. **Remove only those parentheses which contain the variable** for which we want to solve the equation.

Step 3. Transpose all terms containing the unknown to the left side and all other terms to the right side of the equation.

Step 4. If the terms containing the unknown cannot be combined, remove the unknown as a common factor from each term containing it.

Example: $ax + bx = 25$

$(a + b)x = 25$

Step 5. Divide both sides by the coefficient of the unknown.

Example 1:

Solve $i = \dfrac{d}{1 - nd}$ for d. (Business: Interest Rate.)

Solution:

Step 1. Clear the fraction.

$$(1 - nd)i = \dfrac{d}{\cancel{1 - nd}} \cdot \overset{1}{(\cancel{1 - nd})}$$

Step 2. Remove parentheses.

$$(1 - nd)i = d$$
$$i - ndi = d$$

Step 3. Transpose and multiply both sides by –1.

$$-d - ndi = -i$$
$$(-1)(-d - ndi) = (-1)(-i)$$
$$d + ndi = i$$

Step 4. Factor out the common factor d.

$$(1 + ni)d = i$$

Step 5. Divide both sides by (1 + ni).

$$\dfrac{\overset{1}{(\cancel{1 + ni})}d}{\underset{1}{(\cancel{1 + ni})}} = \dfrac{i}{1 + ni}$$

Hence, $d = \dfrac{i}{1 + ni}$.

Example 2:

Solve $\dfrac{E}{e} = \dfrac{R + r}{r}$ for r. (Voltage Drop.)

Solution:

Step 1. $\dfrac{E}{e} = \dfrac{R + r}{r}$

$$\overset{1}{\cancel{e}}r \cdot \dfrac{E}{\underset{1}{\cancel{e}}} = e\overset{1}{\cancel{r}} \cdot \dfrac{R + r}{\underset{1}{\cancel{r}}} \qquad (LCD = er)$$

Step 2. $rE = eR + er$ (Remove parentheses.)

Step 3. $rE - er = eR$ (Transpose.)

Step 4. $(E - e)r = eR$ (Factor out r.)

Step 5. $\dfrac{\overset{1}{(\cancel{E - e})}r}{\underset{1}{\cancel{E - e}}} = \dfrac{eR}{E - e}$ (Divide both sides by E – e.)

Hence, $r = \dfrac{eR}{E - e}$.

Example 3:

Solve $L = L_0(1 + at)$ for t. (Temperature Expansion.)

Solution:

Step 1.　Not required.

Step 2.　　　　$L = L_0(1 + at)$

　　　　　　　$= L_0 + L_0 at$　　　(Remove parentheses.)

Step 3.　$L - L_0 = L_0 at$　　　　(Transpose.)

　　　　$L_0 at = L - L_0$

Step 4.　Not required.

Step 5.　$\dfrac{L_0 at}{L_0 a} = \dfrac{L - L_0}{L_0 a}$　　　(Divide both sides by $L_0 a$.)

Hence, $t = \dfrac{L - L_0}{L_0 a}$

Example 4:

Solve $\dfrac{1}{R} = \dfrac{1}{R_1} + \dfrac{1}{R_2}$ for R_2.　　　(Electricity: Resistance.)

Solution:

Step 1.　Clear the fractions. The LCD = RR_1R_2.

$$\frac{1}{R} = \frac{1}{R_1} + \frac{1}{R_2}$$

$$RR_1R_2 \cdot \frac{1}{R} = RR_1R_2\left(\frac{1}{R_1} + \frac{1}{R_2}\right)$$

$$\frac{RR_1R_2}{R} = \frac{RR_1R_2}{R_1} + \frac{RR_1R_2}{R_2}$$

$$R_1R_2 = RR_2 + RR_1$$

Step 2.　Not required.

Step 3.　Transpose.

$$R_1R_2 - RR_2 = RR_1$$

Step 4.　Remove the common factor R_2.

$$(R_1 - R)R_2 = RR_1$$

Step 5.   Divide both sides by $R_1 - R$.

$$\frac{(R_1 - R)R_2}{R_1 - R} = \frac{RR_1}{R_1 - R}$$

Hence, $R_2 = \dfrac{RR_1}{R_1 - R}$.

22

Example 5:

To prevent uninsulated pipelines from freezing, minimum water velocity V is given by

$$V = \frac{\pi L(0.5T_w - T_a + 16)}{600D(T_w - 32)}.$$

Solve for T_w.

Solution:

Step 1.   Clear the fraction.

$$V \cdot 600D(T_w - 32) = \frac{\pi L(0.5T_w - T_a + 16)}{600D(T_w - 32)} \cdot 600D(T_w - 32)$$

Step 2.   Remove parentheses.

$$600VDT_w - 19200VD = 0.5\pi LT_w - \pi LT_a + 16\pi L$$

Step 3.   Transpose.

$$600VDT_w - 0.5\pi LT_w = 19200VD - \pi LT_a + 16\pi L$$

Step 4.   Factor out T_w.

$$(600VD - 0.5\pi L)T_w = 19200VD - \pi LT_a + 16\pi L$$

Step 5.   Divide both sides by $600VD - 0.5\pi L$.

$$\frac{(600VD - 0.5\pi L)T_w}{600VD - 0.5\pi L} = \frac{19200VD - \pi LT_a + 16\pi L}{600VD - 0.5\pi L}$$

Hence, $T_w = \dfrac{19200VD - \pi LT_a + 16\pi L}{600VD - 0.5\pi L}$

CHECK POINT 4

1. Solve $W = \dfrac{2PR}{R - r}$ for R. Mechanics: Differential Pulley.

2. Solve $S = P(1 + rt)$ for r. Business: Simple Interest.

3. Solve $Q = \dfrac{kAT(t_2 - t_1)}{d}$ for t_1. Heat Conduction.

4. Solve $A = \dfrac{m}{t}(p + t)$ for t. Thickness of Pipe.

5. Solve $f = \dfrac{r_2 - r_1}{r_1(w_1 + w_2)}$ for r_1.

WORD PROBLEMS INVOLVING FRACTIONAL EQUATIONS

When translating certain types of problems into algebra, we encounter fractional equations. These problems are somewhat more difficult to solve than those encountered previously. In the remainder of this unit, we shall deal with the types commonly referred to as work and motion problems.

WORK PROBLEMS

When calculating the time required to complete a contract, we have to be able to estimate accurately the amount of time at which certain jobs can be done. This may involve considering the use of various combinations of machines and/or people. In what follows we shall solve some simple problems of this type which are called work problems.

Example 1:

An experienced technician A can assemble a device in 10 hours. It is estimated that it will take an inexperienced technician B 15 hours to do the same job. How long will it take them together to do this job?

Discussion:

The complete job is represented by 1. This type of problem is then solved by analyzing what part or fraction of the job can be done by a person or a machine in 1 period of time.

Example:

If it takes a person 3 hours to complete a job, then in the period of 1 hour the person completes 1/3 of the job.

Solution to Example 1:

Let t = the number of hours it takes if both technicians do the job together.

Hence, $\frac{1}{t}$ = the work done in one hour by both technicians.

$\frac{1}{10}$ = the work done by the experienced technician A in one hour.

$\frac{1}{15}$ = the work done by the inexperienced technician B in one hour.

Work done in 1 hour by **A**	+	Work done in 1 hour by **B**	=	Work done in 1 hour by both **A** and **B**.
$\frac{1}{10}$	+	$\frac{1}{15}$	=	$\frac{1}{t}$

Clear fractions.
LCD = 30t.

$$3t + 2t = 30$$

$$5t = 30$$

$$t = 6$$

Check: In 6 hours technician A does $6 \times \frac{1}{10}$ or $\frac{3}{5}$ of the job.

In 6 hours technician B does $6 \times \frac{1}{15}$ or $\frac{2}{5}$ of the job.

In 6 hours both technicians do $\frac{3}{5} + \frac{2}{5}$ or $\frac{5}{5}$ of the job.

Hence, it will take the technicians 6 hours working together to do the job.

27

Example 2:

A company has two machines scheduled to produce auto parts for a given order. It is estimated that machine II would take 4 days longer than machine I to produce all parts alone. The two machines are in operation for 4 days when machine I breaks down. Machine II continues in operation and finishes the job in 2 more days. How long would it have taken machine II to do the job working alone?

Solution:

Let x = the number of days it takes machine II to do the job alone.
Then x - 4 = the number of days it takes machine I to do the job alone.

Hence, $\frac{1}{x}$ = part of the job done by machine II in one day and

$\frac{1}{x-4}$ = part of the job done by machine I in one day and

$\frac{1}{x} + \frac{1}{x-4}$ = part of the job done by both machines in one day.

| Part of job done by both machines in 4 days | + | Part of job done by machine II in 2 days | = | Complete job done |

$$4\left(\frac{1}{x} + \frac{1}{x-4}\right) \quad + \quad 2\left(\frac{1}{x}\right) \quad = \quad 1 \quad \text{Clear fractions.}$$
$$\text{LCD} = x(x-4)$$
$$4(x-4) + 4x + 2(x-4) = x(x-4)$$
$$x^2 - 14x + 24 = 0$$
$$(x-2)(x-12) = 0$$

Hence, $x = 2$ or $x = 12$.

Check:

If $x = 2$,

$\frac{1}{x-4} = \frac{1}{-2}$. The part of the job done in one day by machine I cannot be negative. Hence, $x = 2$ is not a solution.

If $x = 12$,

$\frac{1}{x} = \frac{1}{12}$ is the part of the job done in one day by machine II and

$\frac{1}{x-4} = \frac{1}{8}$ is the part of the job done in one day by machine I.

$4\left(\frac{1}{12} + \frac{1}{8}\right) = \frac{5}{6}$ is the part of the job done by both machines in 4 days.

$2\left(\frac{1}{12}\right) = \frac{1}{6}$ is the part of the job done by machine II in 2 days.

$\frac{5}{6} + \frac{1}{6} = 1$ is the complete job.

Hence, it would take machine II 12 days to do the job working alone.

28

MOTION PROBLEMS

Motion problems involve the use of the relationship $d = vt$, where

d = distance

v = velocity or rate of speed

t = time.

When solving these types of problems, we find that one of the three variables is unknown, another is known and the third variable can be written in terms of the unknown and known variable by using the above relationship.

Examples:

1. Since $d = vt$, we have $v = \frac{d}{t}$.

2. Since $d = vt$, we have $t = \frac{d}{v}$.

Example 1:

A transportation company wants to save gasoline by reducing the average speed of their trucks. If, on a standard run of 300 km, the average speed is reduced by 10 km/h, the run would take 20 minutes longer. Find the reduced speed.

Solution:

Let v = the reduced average speed in kilometres per hour.

then $v + 10$ = the faster average speed in kilometres per hour.

Since $d = vt$, we have $t = \dfrac{d}{v}$. We know that $d = 300$ km.

Let t_1 = time required to travel 300 km at the slower speed.

$$\therefore \quad t_1 = \frac{300}{v}$$

Let t_2 = time required to travel 300 km at the faster speed.

$$\therefore \quad t_2 = \frac{300}{v + 10}$$

The difference of time is given as 20 minutes and the rate is given in kilometres per hour, hence, we have to change 20 minutes to 1/3 hour.

Since it takes longer at the slower rate, $t_1 > t_2$,
we have that $t_1 - t_2 = \dfrac{1}{3}$.

$$\therefore \qquad \frac{300}{v} - \frac{300}{v + 10} = \frac{1}{3} \qquad \text{Clear the fractions. LCD} = 3v(v + 10)$$

$$900v + 9000 - 900v = v^2 + 10v$$

$$v^2 + 10v - 9000 = 0$$

$$(v + 100)(v - 90) = 0$$

Hence, $v = -100$ or $v = 90$.

Check:

If $v = -100$, $\quad t_1 = \dfrac{300}{-100} = -3$ h.

Since travel time cannot be negative, $v = -100$ is not a solution.

If $v = 90$, $\quad t_1 = \dfrac{300}{90} = 3\dfrac{1}{3}$ h $\quad$ and $\quad t_2 = \dfrac{300}{100} = 3$ h.

$3\dfrac{1}{3}$ h $- 3$ h $= \dfrac{1}{3}$ h which is the travel time saved.

Hence, the reduced speed is 90 km/h.

CHECK POINT 5

Solve the following problems. Show all steps.

1. The daily edition of a newspaper is printed by two presses. It takes one press 30 minutes longer than it takes the other press to print the daily edition. The two presses normally complete the job in 20 minutes. If the faster press has to be repaired, how long will it take the slower press to do the job alone?

2. A delivery truck has a regular run of 100 km which the driver travels regularly at a constant speed. If the driver would increase the speed by 10 km/h, he would save 1/2 hour of travelling time. Find his regular average speed.

DRILL EXERCISES

Solve the following equations.

1. $\dfrac{3}{5} + \dfrac{4}{m} = 1$

2. $\dfrac{1}{t + 4} = \dfrac{3}{5t}$

3. $\dfrac{v + 2}{v - 1} = 3$

4. $\dfrac{1}{p} - 2 = \dfrac{5}{3}$

5. $\dfrac{7}{3} = \dfrac{2}{M} - 5$

6. $\dfrac{6}{R - 2} = \dfrac{4}{R}$

7. $\dfrac{1}{2p - 3} = \dfrac{2}{11}$

8. $\dfrac{1}{x} - 1 = \dfrac{2}{3x} - \dfrac{2}{3}$

9. $\dfrac{x - 3}{x + 2} = 4 + \dfrac{x}{x + 2}$

10. $\dfrac{2}{x - 2} - \dfrac{5}{x} = \dfrac{5}{2x}$

11. $\dfrac{2}{x} + \dfrac{3}{2x} = \dfrac{1}{x - 1}$

12. $\dfrac{2}{x - 4} - \dfrac{1}{x - 3} = \dfrac{1}{x - 2}$

13. $\dfrac{1}{t + 10} + \dfrac{1}{t} = \dfrac{1}{12}$

14. $\dfrac{1}{6} - \dfrac{1}{t} = \dfrac{1}{t + 5}$

15. $3\left(\dfrac{1}{v + 3} + \dfrac{1}{v}\right) + 5\left(\dfrac{1}{v + 3}\right) = 1$

16. $\dfrac{60}{v} = \dfrac{60}{v + 6} + \dfrac{1}{3}$

17. $\dfrac{a^2}{a + 1} = \dfrac{9}{a + 1}$

18. $\dfrac{2}{x + 1} - 3 = \dfrac{4x + 6}{x + 1}$

19. $1 + \dfrac{4}{x^2 - 1} = \dfrac{2x}{x + 1}$

20. $\dfrac{2V - 3}{2V - 2} = \dfrac{1}{6} - \dfrac{1 - V}{3V - 3}$

DRILL EXERCISES (continued)

21. $\dfrac{4}{x + 2} = 1 - \dfrac{1}{2x - 5}$

22. $\dfrac{d}{d + 1} = \dfrac{d - 1}{d - 2}$

23. $\dfrac{4}{R^2 - 4} + \dfrac{1}{2 - R} = \dfrac{1}{R + 2}$

24. $\dfrac{10x}{x^2 - x - 6} = \dfrac{2}{x - 3} + 3$

25. $\dfrac{5 - m^2}{m^2 + 3m - 10} = \dfrac{m + 1}{m + 5} - \dfrac{2m + 1}{m - 2}$

Solve the following formulas as indicated.

26. Solve $C = a(R_1 + 2R_2)$ for R_1.

27. Solve $F = \dfrac{P_1 + P_2}{d(V_1 + V_2)}$ for V_1.

28. Solve $D = 2D_0 + t(2v - at)$ for a.

29. The thickness of a 2:1 ellipsoidal head for a pressure vessel is given by

$t = \dfrac{PD}{2SE - 0.2P}$. Solve for P.

30. The sludge wastage rate Q_w is related to mass of solids S in an aeration basin by the formula

$S = \dfrac{VX}{[(Q - Q_w)X_e + Q_w X_w] \cdot 1440}$. Solve for Q_w.

Solve the following word problems.

31. If a firm has two card readers and one can do a certain job in 40 minutes and the other in 50 minutes, how long will it take if both card readers are used to complete the job together?

32. An auto parts manufacturer has two machines A and B to do a certain job. Machine A, which normally can do the job alone in 150 minutes, is used but breaks down after 25 minutes. Machine B is then used and the job is completed in another 90 minutes. How long would it take if machine B would be used to do the job alone?

DRILL EXERCISES (continued)

33. A fast and a slow computer are used together to work on a certain job. After 3 minutes the fast computer is needed for another job and the slow computer finishes the job in 5 more minutes. Normally, the slow computer takes 3 minutes more than it takes the fast computer to do the job alone. How long would it take the slow computer to do the job alone?

34. A woman makes a daily trip of 20 km. If she increases the speed by 10 km per hour, she can save 6 minutes. At what speed does she normally travel?

35. A barge can travel 15 km per hour in still water. If it can travel 32 km downstream in the same period of time as it can travel 18 km upstream what is the rate of the current of the stream?

ASSIGNMENT EXERCISES

Solve the following equations.

1. $\dfrac{x + 1}{3} - \dfrac{4}{x} = \dfrac{x}{3}$

2. $\dfrac{t + 3}{4t} = 10$

3. $\dfrac{1}{5} + \dfrac{1}{2m} = \dfrac{3}{4}$

4. $\dfrac{v + 1}{v - 1} = 2 + \dfrac{v}{v - 1}$

5. $\dfrac{1}{6} + \dfrac{5}{4t} = \dfrac{4}{3t}$

6. $\dfrac{1}{v} - 1 = \dfrac{1}{3}$

7. $\dfrac{3}{d - 3} = \dfrac{2}{d}$

8. $\dfrac{5}{4} = \dfrac{1}{3a - 4}$

9. $\dfrac{2}{x + 2} - \dfrac{1}{x + 1} = \dfrac{1}{x - 1}$

10. $2\left(\dfrac{3}{3v + 2} - \dfrac{1}{2v - 3}\right) = \dfrac{1}{v - 2}$

11. $2\left(\dfrac{1}{t} - \dfrac{1}{t + 1}\right) = 3\left(\dfrac{1}{t - 1} - \dfrac{1}{t}\right)$

12. $\dfrac{255}{x} - \dfrac{5}{2} = \dfrac{240}{x + 10}$

13. $\dfrac{60}{x} = \dfrac{1}{2} + \dfrac{60}{x + 10}$

14. $290\left(\dfrac{1}{x} - \dfrac{1}{x + 18}\right) = \dfrac{9}{4}$

ASSIGNMENT EXERCISES (continued)

15. $\dfrac{R + 6}{R + 4} = \dfrac{R + 2}{R + 1}$

16. $\dfrac{3a}{a - 2} + 4 = \dfrac{a^2 + 2}{a - 2}$

17. $1 - \dfrac{2}{d - 1} = \dfrac{d}{d + 1} - \dfrac{1}{d^2 - 1}$

18. $\dfrac{2x - 10}{x} = \dfrac{15}{2} - \dfrac{2x}{5 - x}$

19. $\dfrac{1}{a + 2} + 2 = \dfrac{3}{a^2 + 2a}$

20. $\dfrac{13}{x^2 - 4} + \dfrac{x + 3}{2 - x} = \dfrac{2x - 3}{x + 2}$

21. $\dfrac{3}{p + 2} - \dfrac{1}{p - 1} = \dfrac{-3}{p^2 + p - 2}$

22. $\dfrac{1}{y + 3} - \dfrac{1}{y - 1} = \dfrac{y + 1}{y^2 + 2y - 3}$

Solve the following formulas as indicated.

23. Solve $R = \dfrac{n(E - rc)}{c}$ for c.

24. Solve $L = \dfrac{S(r - 1) + a}{r}$ for r.

25. Solve $n = \dfrac{2R}{R - r}$ for R.

26. The dependence of sound velocity measurement, c, on electronic delay time, β, is given by the formula
$$c = \dfrac{Af(1 + \alpha T)}{1 - \beta \cdot f \cdot 10^{-6}} \ .$$

Solve for f, the frequency of the pulse repetition rate.

27. Solve for E. $\quad B = \dfrac{2358 \, Bhp(100 - E)}{\Delta t \cdot E}$

(This equation is used to calculate the amount of cooling air needed for motors.)

28. The density of a slurry D_s is given by the formula $\quad D_s = \dfrac{D_p D_1}{XD_1 + (1 - X)D_p}$.

Express in terms of the weight fraction of solid in the slurry, i.e. solve for x.

ASSIGNMENT EXERCISES (continued)

Solve the following word problems.

29. An experienced mechanic can assemble a device in 6 hours. It takes a newly hired mechanic 9 hours to do the same job. How much time would be required to assemble the device if both would be assigned to work together?

30. Pump A can fill a tank in 10 hours and pump B can fill the same tank in 15 hours. After pumping together for 4 hours, pump A breaks down and pump B continues to pump on its own. How long will it take pump B to fill the rest of the tank?

31. It takes an inexperienced technician A 3 hours more to assemble a certain machine than it takes an experienced technician B. The two technicians work together for 4 hours when technician B has to leave and technician A finishes the job in 6 more hours. How long would it take technician A to do the job without any help?

32. A motor boat can travel 60 km downstream and 35 km upstream in the same period of time. If the current of the river is 5 km per hour, what is the speed of the boat in still water?

33. Courier A and courier B travel between two cities. Courier A uses a route covering 48 km and courier B uses a route covering 51 km. Courier A travels 10 km per hour faster and completes his route in 1/2 hour less time than courier B. Find the speed at which the couriers travel.

MODULE 4

**Working with Exponents and Radicals,
The Quadratic Formula**

ROOTS AND RADICALS

Objectives: After having worked through this unit, the student should be capable of:

1.   understanding the concept of a root and a principal root;
2.   using the radical symbol to indicate principal roots;
3.   indicating roots by fractional exponents;
4.   evaluating numerical expressions involving fractional exponents;
5.   using the calculator to evaluate roots.

ROOTS AND RADICALS

Finding the root of a number is the reverse of raising a number to a power. We have seen that

$$4 \times 4 = 4^2 \text{ or } 16 \qquad \text{Hence, } 16 \text{ has two equal factors of } 4.$$

$$4 \times 4 \times 4 = 4^3 \text{ or } 64 \qquad \text{Hence, } 64 \text{ has three equal factors of } 4.$$

$$4 \times 4 \times 4 \times 4 = 4^4 \text{ or } 256 \qquad \text{Hence, } 256 \text{ has four equal factors of } 4.$$

In mathematics, we often need to know the equal factors which, when multiplied, yield a given number.

THE SQUARE ROOT

The square root of a number is one of its two equal factors.

Examples:

1. +4 is a square root of 16 since $(+4)(+4) = 16$.
    −4 is another square root of 16 since $(-4)(-4) = 16$.

    Hence, the square roots of 16 are +4 and −4.

2. $\frac{2}{3}$ is a square root of $\frac{4}{9}$ since $\left(\frac{2}{3}\right)\left(\frac{2}{3}\right) = \frac{4}{9}$.

    $-\frac{2}{3}$ is another square root of $\frac{4}{9}$ since $\left(-\frac{2}{3}\right)\left(-\frac{2}{3}\right) = \frac{4}{9}$.

    Hence, the square roots of $\frac{4}{9}$ are $+\frac{2}{3}$ and $-\frac{2}{3}$.

Note:

1. A positive number has two square roots which are opposites of each other.

2. A square root which is a real number is called a real square root.

3. A negative number has **no real square** roots.
    For example, the square root of −25 is not −5 since $(-5)(-5) = 25$.
    The square roots of negative numbers are called **imaginary numbers**.

4. Zero is the only square root of zero since the product of any two non-zero numbers is **never** zero.

CHECK POINT 1

Find the square roots of the following numbers.

1. 49   2. 36   3. 64   4. $\frac{1}{4}$ 5. $\frac{9}{16}$ 6. $\frac{4}{81}$

THE PRINCIPAL SQUARE ROOT OF A NUMBER

The principal square root of a number is the **positive** square root.

Examples:

1. The principal square root of 25 is +5.

2. The principal square root of $\frac{4}{9}$ is $+\frac{2}{3}$.

6

THE RADICAL SIGN "$\sqrt{}$"

To indicate the principal square root of a number, we use the symbol $\sqrt{}$ which is called the radical sign.

Examples:

1. $\sqrt{9}$ indicates the principal square root of 9 which is +3.
   Hence, $\sqrt{9}$ = +3 or simply 3.

2. $\sqrt{\dfrac{4}{9}}$ indicates the principal square root of $\dfrac{4}{9}$ which is $+\dfrac{2}{3}$.
   Hence, $\sqrt{\dfrac{4}{9}} = +\dfrac{2}{3}$ or simply $\dfrac{2}{3}$.

3. $\sqrt{4^2}$ indicates the principal square root of 4^2 which is +4.
   Hence, $\sqrt{4^2}$ = 4.

7

If we wish to indicate the negative square root of a number, we place a negative sign in front of the radical sign.

Examples:

1. $-\sqrt{9}$ indicates the opposite or negative of the principal square root of 9.
   Hence, $-\sqrt{9}$ = -3.

2. $-\sqrt{\dfrac{16}{49}}$ indicates the opposite or negative of the principal square root of $\dfrac{16}{49}$.
   Hence, $-\sqrt{\dfrac{16}{49}} = -\dfrac{4}{7}$.

3. $-\sqrt{5^2}$ indicates the opposite or negative of the principal square root of 5^2.
   Hence, $-\sqrt{5^2}$ = -5.

8

CHECK POINT 2

Evaluate the following expressions.

1. $\sqrt{4}$

2. $\sqrt{3^2}$

3. $-\sqrt{36}$

4. $\sqrt{\dfrac{16}{25}}$

5. $-\sqrt{8^2}$

6. $\sqrt{0}$

7. $\sqrt{\dfrac{64}{81}}$

8. $-\sqrt{\dfrac{1}{4}}$

THE CUBE ROOT

The cube root of a number is one of its **three** equal factors.

Examples:

1. 4 is a cube root of 64 since $4 \times 4 \times 4 = 64$.

    Hence, a cube root of 64 or 4^3 is 4.

2. $-\frac{3}{4}$ is a cube root of $-\frac{27}{64}$ since $\left(-\frac{3}{4}\right)\left(-\frac{3}{4}\right)\left(-\frac{3}{4}\right) = -\frac{27}{64}$.

    Hence, a cube root of $-\frac{27}{64}$ or $\left(-\frac{3}{4}\right)^3$ is $-\frac{3}{4}$.

3. 2 is a cube root of 8 since $2 \times 2 \times 2 = 8$.

    Hence, a cube root of 8 or 2^3 is 2.

4. -5 is a cube root of -125 since $(-5)(-5)(-5) = -125$.

    Hence, a cube root of -125 or $(-5)^3$ is -5.

We indicate the cube root of a number by the radical sign $\sqrt[3]{}$.

Hence, $\sqrt[3]{27} = 3$. This is read as "the cube root of 27 is 3".

Examples:

1. $\sqrt[3]{-64} = -4$ 2. $\sqrt[3]{125} = 5$ 3. $\sqrt[3]{-\frac{8}{27}} = -\frac{2}{3}$

Note:

1. Every real number has exactly **one** real number as a cube root.

2. The cube root of a positive number is a positive number. $\sqrt[3]{27} = 3$.

3. The cube root of a negative number is a negative number. $\sqrt[3]{-8} = -2$.

4. The cube root of zero is zero. $\sqrt[3]{0} = 0$

CHECK POINT 3

Evaluate the following expressions.

1. $\sqrt[3]{\frac{1}{8}}$ 2. $\sqrt[3]{-64}$ 3. $\sqrt[3]{1000}$

4. $\sqrt[3]{-1}$ 5. $\sqrt[3]{\frac{8}{27}}$ 6. $\sqrt[3]{-216}$

13

THE n-th ROOT

In general, the n-th root of a number is one of its n equal factors.

1. Let n = 4. A fourth root of a number is one of its four equal factors.

   (i)   2 is a fourth root of 16 since $2 \times 2 \times 2 \times 2 = 16$.

   Hence, a fourth root of 16 or 2^4 is 2 and we write this as

   $$\sqrt[4]{2^4} = 2 \quad \text{or} \quad \sqrt[4]{16} = 2.$$

   (ii)   5 is a fourth root of 625 since $5 \times 5 \times 5 \times 5 = 625$.

   Hence, a fourth root of 625 or 5^4 is 5 and we write this as

   $$\sqrt[4]{5^4} = 5 \quad \text{or} \quad \sqrt[4]{625} = 5.$$

14

2. Let n = 5. A fifth root of a number is one of its five equal factors.

   (i)   –3 is a fifth root of –243 since $(-3)(-3)(-3)(-3)(-3) = -243$.

   Hence, a fifth root of $(-3)^5$ is –3 and we write this as

   $$\sqrt[5]{(-3)^5} = -3 \quad \text{or} \quad \sqrt[5]{-243} = -3.$$

   (ii)   10 is a fifth root of 100 000 since

   $10 \times 10 \times 10 \times 10 \times 10 = 100\ 000$.

   Hence, a fifth root of 10^5 is 10 and we write this as

   $$\sqrt[5]{10^5} = 10 \quad \text{or} \quad \sqrt[5]{100\ 000} = 10.$$

15

TERMINOLOGY

1. A **radical** is an indicated root of a number or expression.

   Examples:

   (i) $\sqrt{7}$ is a radical.

   (ii) $\sqrt[4]{10x^3}$ is a radical.

   (iii) $\sqrt[5]{19a^2}$ is a radical.

2. The **radical signs** are the symbols $\sqrt{}$, $\sqrt[3]{}$, $\sqrt[4]{}$, $\sqrt[5]{}$, and so on.

3. The **radicand** is the number or expression under the radical sign.

   Examples:

   (i)   In $\sqrt{7}$, 7 is the radicand.   (ii)   In $\sqrt[4]{10x^2}$, $10x^2$ is the radicand.

4. The **index** or order of a root is the number on the radical sign indicating which root is to be taken.

Examples:

(i)   In $\sqrt[3]{19}$, 3 is the index.

(ii)   In $\sqrt[4]{10x^2}$, 4 is the index.

(iii) In $\sqrt{10}$, the index is understood to be 2; that is, if the square root is to be taken, we do not write the index 2.

16

EVEN AND ODD ROOTS

An even root is indicated when the index of a radical sign is even.

Examples:

$\sqrt{7}$, $\sqrt[4]{129}$, $\sqrt[6]{84}$.

An odd root is indicated when the index of a radical sign is odd.

Examples:

$\sqrt[3]{10}$, $\sqrt[5]{-21}$, $\sqrt[7]{165}$.

17

Note:

1. The principal even root of a positive number is the positive root.

2. Even real roots of negative numbers cannot be found. Numbers such as $\sqrt{-1}$, $\sqrt[4]{-16}$ and $\sqrt[6]{-64}$ are called imaginary numbers.

3. Every real number has exactly **one** real odd root.

   (a) The real odd root of a positive number is a positive number.

   (b) The real odd root of a negative number is a negative number.

18

CHECK POINT 4

Evaluate the following expressions.

1. $\sqrt[5]{32}$ 2. $\sqrt[4]{81}$ 3. $\sqrt[7]{-1}$ 4. $\sqrt[8]{1}$

19

INDICATING ROOTS BY USE OF FRACTIONAL EXPONENTS

The notation $a^{1/n}$ is also used to indicate the n-th root of a number.

We define $\boxed{a^{1/n} = \sqrt[n]{a} \text{ where } a \geq 0 \text{ and } n \text{ is a positive integer.}}$

Examples:

1. $4^{1/2} = \sqrt{4}$ 2. $8^{1/3} = \sqrt[3]{8}$ 3. $\sqrt[5]{32} = 32^{1/5}$

20

In order that the above definition is consistent with the exponential law $(a^m)^n = a^{mn}$,

we define $\boxed{a^{m/n} = (a^m)^{1/n} = \sqrt[n]{a^m} \quad \text{where } a \geq 0 \text{ and } m \text{ and } n \text{ are positive integers.}}$

Examples:

1. $8^{2/3} = \sqrt[3]{8^2}$

2. $\sqrt[5]{32^3} = 32^{3/5}$

21

EVALUATING EXPRESSIONS WITH FRACTIONAL EXPONENTS

Examples:

1. $4^{1/2} = (2^2)^{1/2} = 2^{2 \cdot 1/2} = 2^1.$ Hence, $4^{1/2} = 2$ and $\sqrt{4} = 2.$

2. $(-27)^{1/3} = [(-3)^3]^{1/3} = (-3)^{3 \cdot 1/3} = -3^1.$ Hence, $(-27)^{1/3} = -3$

                      and $\sqrt[3]{-27} = -3.$

3. $64^{1/6} = (2^6)^{1/6} = 2^{6 \cdot 1/6} = 2^1.$ Hence, $64^{1/6} = 2$ and $\sqrt[6]{64} = 2.$

22

Since $a^{m/n} = (a^m)^{1/n} = (a^{1/n})^m$, we have that

$a^{m/n} = \sqrt[n]{a^m} = \left(\sqrt[n]{a}\right)^m$ where $a \geq 0$ and m and n are positive integers.

Hence, $a^{m/n}$ can be interpreted as either

1. $\sqrt[n]{a^m}$ "the n-th root of the m-th power of a" or

2. $\left(\sqrt[n]{a}\right)^m$ "the m-th power of the n-th root of a".

23

When evaluating an expression such as $32^{3/5}$, we normally **evaluate the root first.**

Example:

$\sqrt[5]{32^3} = \left(\sqrt[5]{32}\right)^3 = (2)^3 = 8.$

If we reverse the operation and first evaluate 32^3, we then have to find the 5-th root of a large number.

Example:

$\sqrt[5]{32^3} = \sqrt[5]{32\ 768} = 8.$

As can be seen, to evaluate $\sqrt[5]{32}$ and 2^3 is a lot easier than to evaluate 32^3 and $\sqrt[5]{32\ 768}.$

More Examples:

1. $8^{4/3} = \sqrt[3]{8^4} = \left(\sqrt[3]{8}\right)^4 = (2)^4 = 16$

2. $(-125)^{2/3} = \sqrt[3]{(-125)^2} = \left(\sqrt[3]{-125}\right)^2 = (-5)^2 = 25$

3. $81^{3/4} = \sqrt[4]{81^3} = \left(\sqrt[4]{81}\right)^3 = (3)^3 = 27$

PRINCIPAL ROOTS

For mental calculations, we should know the following roots.

$\sqrt{4} = 2$	$\sqrt[3]{-125} = -5$	$\sqrt[4]{625} = 5$	$\sqrt[5]{-243} = -3$
$\sqrt{9} = 3$	$\sqrt[3]{-64} = -4$	$\sqrt[4]{256} = 4$	$\sqrt[5]{-32} = -2$
$\sqrt{16} = 4$	$\sqrt[3]{-27} = -3$	$\sqrt[4]{81} = 3$	$\sqrt[5]{32} = 2$
$\sqrt{25} = 5$	$\sqrt[3]{-8} = -2$	$\sqrt[4]{16} = 2$	$\sqrt[5]{243} = 3$
$\sqrt{36} = 6$	$\sqrt[3]{8} = 2$		
$\sqrt{49} = 7$	$\sqrt[3]{27} = 3$		
$\sqrt{64} = 8$	$\sqrt[3]{64} = 4$		
$\sqrt{81} = 9$	$\sqrt[3]{125} = 5$		
$\sqrt{100} = 10$			

Note:

1. $\sqrt[n]{0} = 0$ **Examples:** $\sqrt{0} = 0,\ \sqrt[3]{0} = 0,\ \sqrt[4]{0} = 0,\ \sqrt[5]{0} = 0,$ etc.

2. $\sqrt[n]{1} = 1$ **Examples:** $\sqrt{1} = 1,\ \sqrt[3]{1} = 1,\ \sqrt[4]{1} = 1,\ \sqrt[5]{1} = 1,$ etc.

3. $\sqrt[n]{-1} = -1$ where n is an **odd** number.

          Examples: $\sqrt[3]{-1} = -1,\ \sqrt[5]{-1} = -1,$ etc.

CHECK POINT 5

I. Write the following radicals as expressions with fractional exponents.

   1. $\sqrt{3}$ 2. $\sqrt[5]{42}$ 3. $\sqrt[6]{9}$

   4. $\sqrt[3]{2^5}$ 5. $\sqrt{7^3}$ 6. $\sqrt[5]{3^4}$

II. Evaluate mentally the following.

   7. $16^{1/4}$ 8. $9^{1/2}$ 9. $(125)^{2/3}$

   10. $(-32)^{3/5}$ 11. $(-8)^{5/3}$ 12. $(81)^{3/4}$

USING THE CALCULATOR TO FIND ROOTS

So far in this unit, we have discussed the meaning of radicals and roots. We determined square roots of perfect integer squares such as 1, 4, 9. 16, 25, 36, and so on.

In addition, we determined cube roots and other roots, using whole numbers for which we can easily obtain exact roots.

However, when finding roots of most numbers, we are either not able to obtain exact values so easily or we are not able to obtain exact values of roots at all, and we have to be satisfied with approximations.

In general, to find roots of numbers other than the type discussed so far in this unit, we use calculators.

Examples:

The exact root of 2 cannot be found. However, by means of a calculator, we can approximate its value for all practical purposes to any desired degree of accuracy as shown below. The symbol $\doteq$ is used to indicate that the equality is an approximation.

$\sqrt{2} \doteq 1.4$

$\sqrt{2} \doteq 1.41$

$\sqrt{2} \doteq 1.414$

$\sqrt{2} \doteq 1.4142$

$\sqrt{2} \doteq 1.41421$

$\sqrt{2} \doteq 1.414214$ and so on.

Example:

Use the calculator to find $\sqrt{637.59}$. Round to three decimal places.

Solution:

Operation	Display
Enter: 637.59	637.59
Press: [$\sqrt{\ }$]	25.25054455

Hence, $\sqrt{637.59}$ = 25.251 (rounded to three decimal places).

SQUARE ROOTS OF A DECIMAL

If we square 0.5, the result is (0.5)(0.5) = 0.25. At this point, we should note that the squaring of 0.5 results in a **smaller** number, namely, 0.25. This is true for any number between 0 and 1.

Examples:

1. $(0.12)^2 = 0.0144$

2. $(0.03)^2 = 0.0009$

3. $(0.006)^2 = 0.000\ 036$

Hence, if we take the **square root** of a number between 0 and 1, the result will be a **larger** number.

Examples:

1. $\sqrt{0.25} = \sqrt{(0.5)^2} = 0.5$ 2. $\sqrt{0.0144} = \sqrt{(0.12)^2} = 0.12$

3. $\sqrt{0.0009} = \sqrt{(0.03)^2} = 0.03$ 4. $\sqrt{0.000\ 036} = \sqrt{(0.006)^2} = 0.006$

CHECK POINT 6

Use the calculator to approximate the following square roots to four decimal places.

1. $\sqrt{3}$ 2. $\sqrt{2.49}$ 3. $\sqrt{0.231}$ 4. $\sqrt{12\ 396}$

5. $(139)^{1/2}$ 6. $(0.847)^{1/2}$ 7. $(0.05)^{1/2}$ 8. $(54\ 168)^{1/2}$

USING THE CALCULATOR TO EVALUATE ROOTS

Scientific or Business calculators normally have a

[y^x]   and an   [$\sqrt[x]{y}$]   key.

When using a calculator, one has to know how to enter numbers into the calculator and how to instruct the calculator to perform the desired operation.

Different calculators work in different ways.

If your calculator does not perform the operation of finding a root as shown here, you should read the instructions in your calculator's manual regarding the use of the [y^x] and [$\sqrt[x]{y}$] keys.

Example 1:

Use the calculator to find $\sqrt[3]{146}$.

Solution:

Since $\sqrt[3]{146} = (146)^{1/3}$, we can use either $\sqrt[x]{y}$ or y^x on our calculator.

(a)   If we use $\sqrt[x]{y}$, then x = 3 and y = 146.

Operation		Display
Enter (y)	146	146
Press	[$\sqrt[x]{y}$]	
Enter (x)	3	3
Press	[=]	5.26563743

To check, we use the fact that if $\sqrt[3]{146}$ = 5.26563743,

then $(5.26563743)^3$ = 146

Hence, we leave 5.26563743 in our calculator and see if $(5.26563743)^3$ = 146.

Check:

	Press	[y^x]	5.26563743
	Enter (x)	3	
	Press	[=]	146

Hence, $\sqrt[3]{146}$ = 5.27 (rounded).

(b)   If we use y^x, then x = $\frac{1}{3}$ and y = 146.

	Operation		Display
	Enter (y)	146	146
	Press	[y^x]	
	Enter (x)	3	3
To get x = $\frac{1}{3}$	Press	[1/x]	0.333333333
	Press	[=]	5.26563743

We check in the same way as shown in (a) above.

Example 2:

Use the calculator to find $\sqrt[5]{10294.85}$.

Solution:

	Operation		Display
	Enter (y)	10294.85	10294.85
	Press	[$\sqrt[x]{y}$]	
	Enter (x)	5	5
	Press	[=]	6.34634978
Check:	Press	[y^x]	6.34634978
	Enter (x)	5	
	Press	[=]	10294.85

Hence, $\sqrt[5]{10294.85}$ = 6.346 (rounded).

Note:

1. We cannot find the **even** root of a negative number.
    Example:
    $\sqrt{-36} \neq -6$, since (-6)(-6) = 36.

2. **Odd** roots of negative numbers can be calculated.
    Odd roots of negative numbers are **negative**.
    Example:
    $\sqrt[3]{-27}$ = -3, since (-3)(-3)(-3) = -27.

Example 3:

Use the calculator to find $\sqrt[7]{-2569}$.

Solution:

We find the 7-th root of 2569 and then attach a minus sign to the answer since an **odd** root of a negative number is negative.

	Operation		Display
	Enter (y)	2569	2569
	Press	[$\sqrt[x]{y}$]	
	Enter (x)	7	7
	Press	[=]	3.06979347
Check:	Press	[y^x]	
	Enter (x)	7	
	Press	[=]	2569.00001

Note:

Since the calculator only displays a certain number of digits, rounding occurs. Hence, in the reverse process, digits such as 1 in the fifth decimal place may occur.

Hence, $\sqrt[7]{-2569}$ = -3.070 (rounded to four significant digits).

Example:
Use the calculator to find $(0.896)^{2/5}$.

Solution:
Since the exponent is $\frac{2}{5}$, we convert it to a decimal and use the [y^x] key.

	Operation		Display
	Enter (y)	0.896	0.896
	Press	[y^x]	
$\frac{2}{5} = 0.4 \longrightarrow$	Enter (x)	0.4	0.4
	Press	[=]	0.957024826

Check:
To check the answer in this case, we use the fact that $[(0.896)^{2/5}]^{5/2}$
$$= (0.896)^{10/10}$$
$$= (0.896)$$

	Operation		Display
$\therefore$	Press	[y^x]	
$\frac{5}{2} = 2.5 \longrightarrow$	Enter (x)	2.5	2.5
	Press	[=]	0.896

Hence, $(0.896)^{2/5} = 0.957$ (rounded to three significant digits).

CHECK POINT 7

Use the calculator to find the following radicals.
Check your answers. (Round answers to five decimal places.)

1. $\sqrt[3]{9.84}$ 2. $\sqrt[4]{8963}$ 3. $\sqrt[5]{-167}$ 4. $\sqrt[8]{0.0596}$

5. $(12.8)^{2/3}$ 6. $(159.5)^{3/4}$ 7. $(-8.6)^{3/5}$

DRILL EXERCISES

I.   Evaluate mentally the following expressions.

1. $\sqrt{9}$ 2. $\sqrt[5]{1}$ 3. $\sqrt[3]{\frac{25}{49}}$

4. $-\sqrt{\frac{1}{16}}$ 5. $\sqrt[3]{\frac{1}{8}}$ 6. $\sqrt[3]{-1}$

7. $\sqrt[5]{-32}$ 8. $\sqrt[4]{625}$ 9. $(64)^{1/6}$

10. $(32)^{4/5}$ 11. $(-27)^{1/3}$ 12. $(25)^{3/2}$

DRILL EXERCISES (continued)

II. Evaluate the following to four decimal places using a calculator. Check your answers.

13. $\sqrt{0.00146}$ 14. $\sqrt[3]{-294}$ 15. $\sqrt[6]{18.5}$

16. $\sqrt[5]{-8964}$ 17. $\sqrt[8]{0.06}$ 18. $\sqrt{1.985}$

19. $\sqrt[4]{750}$ 20. $\sqrt[7]{-0.35}$ 21. $(75.6)^{3/4}$

22. $(-2.9)^{2/3}$ 23. $(0.5)^{5/6}$ 24. $(-106)^{3/5}$

III. Rewrite the following using fractional exponents.

25. $\sqrt[3]{16}$ 26. $\sqrt[7]{3^5}$ 27. $\sqrt[5]{-9}$ 28. $\sqrt[4]{(8.5)^3}$

ASSIGNMENT EXERCISES

I. Evaluate mentally the following expressions.

1. $\sqrt{121}$ 2. $-\sqrt{\dfrac{1}{9}}$ 3. $\sqrt[4]{256}$

4. $\sqrt[5]{-1}$ 5. $\sqrt[3]{\dfrac{27}{343}}$ 6. $\sqrt[7]{1}$

7. $\sqrt{\dfrac{49}{81}}$ 8. $\sqrt[5]{-243}$ 9. $(81)^{1/4}$

10. $(243)^{2/5}$ 11. $(-32)^{3/5}$ 12. $(27)^{4/3}$

ASSIGNMENT EXERCISES (continued)

II. Evaluate the following to four decimal places using a calculator. Check your answers.

13. $\sqrt{0.0025}$

14. $\sqrt[3]{-316}$

15. $\sqrt[4]{164}$

16. $\sqrt[6]{625}$

17. $\sqrt[9]{-0.32}$

18. $\sqrt{2.425}$

19. $\sqrt[5]{749}$

20. $\sqrt[7]{749}$

21. $(-3.3)^{1/3}$

22. $(54)^{3/4}$

23. $(123)^{6/7}$

24. $(0.142)^{2/5}$

III. Rewrite the following using fractional exponents.

25. $\sqrt[4]{72}$

26. $\sqrt[6]{2^5}$

27. $\sqrt[3]{-16}$

28. $\sqrt[3]{(7.2)^2}$

NEGATIVE AND
FRACTIONAL EXPONENTS

Objectives: After having worked through this unit, the student should be capable of using the laws of exponents to:

1.   simplify algebraic expressions involving integral exponents;
2.   simplify expressions involving fractional exponents.

INTRODUCTION

Previously, we discussed laws of exponents for exponents which are integers. In this unit, we shall discuss the use of exponents which are fractions and show how to simplify algebraic expressions involving integral and fractional exponents.

REVIEW

Laws of Exponents   Examples

3.1 $a^m \cdot a^n = a^{m+n}$ $x^3 \cdot x^6 = x^9$

3.2 $\dfrac{a^m}{a^n} = a^{m-n}$ or $\dfrac{a^m}{a^n} = \dfrac{1}{a^{n-m}}$, $a \neq 0$ $\dfrac{x^5}{x^2} = x^3$ or $\dfrac{x^5}{x^2} = \dfrac{1}{x^{-3}}$

3.3 $(a^m)^k = a^{mk}$ $(x^2)^4 = x^8$

3.4 $(a^n b^m)^k = a^{nk} \cdot b^{mk}$ $(3x^3 y^4)^2 = 3^2 x^6 y^8 = 9x^6 y^8$

3.5 $\left(\dfrac{a^m}{b^n}\right)^k = \dfrac{a^{mk}}{b^{nk}}$, $b \neq 0$ $\left(\dfrac{2x^4}{y^2}\right)^3 = \dfrac{2^3 x^{12}}{y^6} = \dfrac{8x^{12}}{y^6}$

3.6 $a^0 = 1$, $a \neq 0$

1. $5^0 = 1$
2. $4x^0 = 4(1) = 4$
3. $(16x^2 yz^4)^0 = 1$
4. $(-8x)^0 = 1$

3.7   (a) $a^{-n} = \dfrac{1}{a^n}$, $a \neq 0$ $5x^{-4} = \dfrac{5}{x^4}$

(b) $\dfrac{1}{a^{-n}} = a^n$ $\dfrac{x^2 z}{y^{-4}} = x^2 z y^4$

Note:

The exponential laws are not defined whenever a base appearing in the denominator is zero. Remember, division by zero is not defined.

3

USING THE LAWS OF EXPONENTS

The following examples show how the Laws of Exponents are used to simplify and obtain expressions containing only positive exponents.

1. $a^7 \cdot a^{-3} = a^{7+(-3)} = a^4$ 　　　　　　　　　　Law　3.1

2. $\dfrac{8x^{-9}}{x^3} = \dfrac{8}{x^{3-(-9)}} = \dfrac{8}{x^{3+9}} = \dfrac{8}{x^{12}}$ 　　　　Law ___?

3. $(m^4)^{-2} = m^{-8} = \dfrac{1}{m^8}$ 　　　　　　　Laws　3.3　and　3.7(a)

4. $(3m^2n^{-3}t)^{-2} = 3^{-2}m^{-4}n^6t^{-2} = \dfrac{n^6}{9m^4t^2}$ 　　　Laws　3.4　and ___?

5. $\left(\dfrac{2a^{-2}}{5b^{-3}}\right)^{-2} = \dfrac{2^{-2}a^4}{5^{-2}b^6} = \dfrac{5^2a^4}{2^2b^6} = \dfrac{25a^4}{4b^6}$ 　　Laws ___? and ___?

6. $(9x^2y^3)^0 \div (3y)^{-2} = 1 \div (3^{-2}y^{-2}) = 9y^2$ 　　Laws ___? and ___?

Answers:

2.	Law 3.2	**4.**	Laws 3.4 and 3.7(a)
5.	Laws 3.5, 3.7(a) and 3.7(b)	**6.**	Laws 3.6 and 3.7(b)

4

CHECK POINT 1

Simplify the following and obtain expressions containing only positive exponents. State the laws of exponents you are using.

1. $15m^{-4} \cdot m^2$ 　　　　　2. $(x^{-1})^8$ 　　　　　3. $4a^8 \div a^{-6}$

4. $(2^{-3}x^{-2}y)^{-2}$ 　　　　5. $(3^{-1}m^2)^{-2} \div (16m^3n^2)^0$ 　　　6. $\left(\dfrac{3x^{-4}y^2}{2^{-1}z^3}\right)^{-2}$

More Examples

1. $(x + y)^{-1} = \dfrac{1}{x + y}$ Law   3.7(a)

2. $x^{-1} + y^{-1} = \dfrac{1}{x} + \dfrac{1}{y}$ Law   3.7(a)

Note:

$(x + y)^{-1} \neq x^{-1} + y^{-1}$ The laws of exponents 3.1 to 3.5 do not apply when we add or subtract exponential expressions.

These laws **only** apply when we multiply or divide exponential expressions

Example 3:
  Simplify $(m^{-1} - 3n)^{-2}$

Solution:

$(m^{-1} - 3n)^{-2} = \dfrac{1}{(m^{-1} - 3n)^2}$ 3.7(a)

$= \dfrac{1}{\left(\dfrac{1}{m} - 3n\right)^2}$ 3.7(a)

$= \dfrac{1}{\left(\dfrac{1 - 3mn}{m}\right)^2}$

$= \dfrac{1}{\dfrac{(1 - 3mn)^2}{m^2}}$ 3.5

$= \dfrac{m^2}{(1 - 3mn)^2}$ or $\dfrac{m^2}{1 - 6mn + 9m^2n^2}$

Note:

$(m^{-1} - 3n)^{-2} \neq m^2 - 3^{-2}n^{-2}$. Law 3.4 does not hold for $(m^{-1} - 3n)^{-2}$.

Example 4:
  Simplify $(2^{-3} + 4^{-1})^2$.

Solution:

$(2^{-3} + 4^{-1})^2 = \left(\dfrac{1}{2^3} + \dfrac{1}{4}\right)^2$ Law 3.4 **cannot** be applied. Why? See Note Frame 5. Law 3.7(a) was used.

$$= \left(\frac{1}{8} + \frac{1}{4}\right)^2$$

$$= \left(\frac{3}{8}\right)^2 = \frac{9}{64}$$

8

Example 5:

Simplify $\left(\dfrac{2m}{3t^2}\right)^{-2}\left(\dfrac{2t}{m^2}\right)^{-3}$.

Solution:

$$\left(\frac{2m}{3t^2}\right)^{-2}\left(\frac{3t}{m^2}\right)^{-3} = \frac{2^{-2}m^{-2}}{3^{-2}t^{-4}} \cdot \frac{3^{-3}t^{-3}}{m^{-6}} \qquad \text{Law 3.5 \textbf{can} be applied. Why?}$$

$$= \frac{3^2t^4}{2^2m^2} \cdot \frac{m^6}{3^3t^3} \qquad \text{Laws 3.7(a) and 3.7(b).}$$

$$= \frac{tm^4}{12}$$

9

Example 6:

Simplify $\dfrac{a^{-2} - b^{-2}}{a^2 - b^2}$.

Solution:

$$\frac{a^{-2} - b^{-2}}{a^2 - b^2} = \frac{\dfrac{1}{a^2} - \dfrac{1}{b^2}}{a^2 - b^2} \qquad \begin{array}{l}\text{Law 3.7(a).}\\ \text{Combine fractions in numerator.}\end{array}$$

$$= \frac{\dfrac{b^2 - a^2}{a^2b^2}}{a^2 - b^2} \qquad \text{Invert the denominator and multiply.}$$

$$= \frac{b^2 - a^2}{a^2b^2} \cdot \frac{1}{a^2 - b^2} \qquad \begin{array}{l}\text{Remember:}\\ (b^2 - a^2) = -a^2 + b^2 = -(a^2 - b^2)\end{array}$$

$$= \frac{-\overset{1}{\cancel{(a^2 - b^2)}}}{a^2b^2} \cdot \frac{1}{\underset{1}{\cancel{a^2 - b^2}}}$$

$$= -\frac{1}{a^2b^2} \qquad \text{or} \qquad -a^{-2}b^{-2}$$

CHECK POINT 2

Simplify the following expressions.

Remember: $(-3m)^2 = (-3m)(-3m)$ and $-3m^2 = -3mm$. Hence, $(-3m)^2 \neq -3m^2$.

1. $(m - n)^{-2}$
2. $(2a + 3b)^{-1}$
3. $(4x)^{-2} + 3y^{-2}$

4. $(a^{-1} + 2b)^{-2}$
5. $\dfrac{2m^2 - mn - n^2}{m^2(n^{-2} - m^{-2})}$ (Factor trinomial.)

6. $(4^{-2} - 2^{-3})^2$
7. $\left(\dfrac{2a^{-1}b^2}{3c^4}\right)^{-3}\left(\dfrac{4b}{9a^2c^5}\right)^2$

Problem solving in calculus frequently involves simplification of expressions such as

$$(x^2 - 1)^{-1} \cdot 2x - (x^2 + 1)(x^2 - 1)^{-2} \cdot 2x$$

These expressions can be simplified in different ways, however, for now we use the following steps:

Step 1. Change all negative exponents to positive exponents.

Step 2. Find the lowest common denominator (LCD).

Step 3. Add/Subtract the fractions after expanding all to the lowest common denominator.

Step 4. Simplify (if possible) the numerator by factoring out common factors and/or collecting like terms.

Example 1:
   Simplify $(x^2 - 1)^{-1} \cdot 2x - (x^2 + 1)(x^2 - 1)^{-2} \cdot 2x$

Solution:
Step 1. Change all negative exponents to positive exponents.

$$(x^2 - 1)^{-1} \cdot 2x - (x^2 + 1)(x^2 - 1)^{-2} \cdot 2x$$

$$= \frac{2x}{x^2 - 1} - \frac{2x(x^2 + 1)}{(x^2 - 1)^2}$$

Step 2. Find the LCD.

$$= \frac{2x(x^2 - 1)}{(x^2 - 1)(x^2 - 1)} - \frac{2x(x^2 + 1)}{(x^2 - 1)^2}$$

Step 3. Combine the fractions

$$= \frac{2x(x^2 - 1) - 2x(x^2 + 1)}{(x^2 - 1)^2}$$

Step 4. Simplify the numerator. (Factor out 2x and collect like terms.)

$$= \frac{2x[(x^2 - 1) - (x^2 + 1)]}{(x^2 - 1)^2}$$

$$= \frac{2x[-2]}{(x^2 - 1)^2}$$

$$= - \frac{4x}{(x^2 - 1)^2} \quad \text{or} \quad - \frac{4x}{x^4 - 2x^2 + 1}$$

12

Example 2:
Simplify $(x - 1)^3(-4)(x + 2)^{-3} + 3(x - 1)^2(x + 2)^{-4}$

Solution (a):

$(x - 1)^3(-4)(x + 2)^{-3} + 3(x - 1)^2(x + 2)^{-4}$

Change to positive exponents

$$= \frac{-4(x - 1)^3}{(x + 2)^3} + \frac{3(x - 1)^2}{(x + 2)^4}$$

Find the LCD

$$= \frac{-4(x - 1)^3(x + 2)}{(x + 2)^3(x + 2)} + \frac{3(x - 1)^2}{(x + 2)^4}$$

Combine fractions

$$= \frac{-4(x - 1)^3(x + 2) + 3(x - 1)^2}{(x + 2)^4}$$

Factor out the common factor $(x - 1)^2$

$$= \frac{(x - 1)^2[-4(x - 1)(x + 2) + 3]}{(x + 2)^4}$$

Multiply and collect like terms

$$= \frac{(x - 1)^2[-4x^2 - 4x + 11]}{(x + 2)^4}$$

Factor out (-1) from $-4x^2 - 4x + 11$

$$= - \frac{(x - 1)^2[4x^2 + 4x - 11]}{(x + 2)^4}$$

Note:

When simplifying expressions such as

$$(x - 1)^3(-4)(x + 2)^{-3} + 3(x - 1)^2(x + 2)^{-4}$$

it might be helpful to replace the binomial factors by single variables.

Solution (b):

Let $a = x - 1$ and $b = x + 2$

$\therefore$ $(x - 1)^3(-4)(x + 2)^{-3} + 3(x - 1)^2(x + 2)^{-4}$

$= a^3(-4)b^{-3} + 3a^2b^{-4}$

Change to positive exponents

$= \dfrac{-4a^3}{b^3} + \dfrac{3a^2}{b^4}$

Find the LCD

$= \dfrac{-4a^3b}{b^3b} + \dfrac{3a^2}{b^4}$

Combine fractions

$= \dfrac{-4a^3b + 3a^2}{b^4}$

Factor out the common factor a^2

$= \dfrac{a^2(-4ab + 3)}{b^4}$

Factor out (-1) from $-4ab + 3$

$= -\dfrac{a^2(4ab - 3)}{b^4}$

Substitute $x - 1$ and $x + 2$ back

$= -\dfrac{(x - 1)^2[4(x - 1)(x + 2) - 3]}{(x + 2)^4}$

Simplify

$= -\dfrac{(x - 1)^2[4x^2 + 4x - 11]}{(x + 2)^4}$

CHECK POINT 3

Simplify the following.

1. $3x^2(x^2 + 1)^{-1} - x^3(x^2 + 1)^{-2} \cdot 2x$

2. $2(2x + 1)^{-2} - (2x - 1)(2x + 1)^{-3}(2)$

15

OPERATIONS WITH FRACTIONAL EXPONENTS

The exponential laws listed previously are also valid, with some restrictions, when the exponents are fractions.

In the remainder of this unit, we shall show how to use exponential laws to simplify algebraic expressions involving fractional exponents.

Note:

In order to avoid difficulties with even roots of negative numbers, we shall assume that any base raised to a fractional exponent is **positive**. If we applied the laws of exponents to fractional exponents without restricting them to a positive base, contradictions as shown in the following example would occur.

Example:

$$-3 = (-3)^1 = (-3)^{2 \cdot 1/2} = [(-3)^2]^{1/2} = 9^{1/2} = 3$$

The contradiction of saying $-3 = 3$ arises by applying $a^{mn} = (a^m)^n$ in the situation where **a is negative** and n is a fraction with an even denominator.

Hence, in the following examples, every base is considered to be a positive number.

16

Using (3.1) $\boxed{a^m \cdot a^n = a^{m+n}}$ with fractional exponents.

Examples:

1. $5^{1/2} \cdot 5^{1/2} = 5^{1/2+1/2} = 5^1$ or   5

2. $4^{1/3} \cdot 4^{1/3} \cdot 4^{1/3} = 4^{1/3+1/3+1/3} = 4^1$ or   4

3. $a^{2/5} \cdot a^{-1/5} = a^{2/5+(-1/5)} = a^{1/5}$

4. $m^{1/2} \cdot m^{3/4} = m^{1/2+3/4} = m^{5/4}$

17

Using (3.2) $\boxed{\dfrac{a^m}{a^n} = a^{m-n}, \; a \neq 0}$ with fractional exponents.

Examples:

1. $\dfrac{16^{3/4}}{16^{1/2}} = 16^{3/4-1/2} = 16^{1/4} = 2$

2. $x^{1/3} \div x^{-5/9} = x^{1/3-(-5/9)} = x^{1/3+5/9} = x^{8/9}$

3. $(V - 3)^{5/6} \div (V - 3)^{1/3} = (V - 3)^{5/6-1/3} = (V - 3)^{1/2}$

18

CHECK POINT 4

Use the laws of exponents to simplify the following.

1.　$4^{1/3} \cdot 4^{2/3}$

2.　$m^{1/2} \cdot m^{-1/4}$

3.　$\dfrac{x^{3/8}}{x^{1/2}}$

4.　$(R + 3)^{3/2} \div (R + 3)^{1/2}$

19

Using　(3.3)　$\boxed{(a^m)^k = a^{mk}}$　with fractional exponents.

Examples:

1.　$(64^{2/3})^{3/4} = 64^{2/3 \cdot 3/4} = 64^{1/2} = 8$

2.　$(x^{1/2})^{4/5} = x^{1/2 \cdot 4/5} = x^{2/5}$

3.　$[(x^2 + 1)^{1/2}]^4 = (x^2 + 1)^{1/2 \cdot 4} = (x^2 + 1)^2$　or　$x^4 + 2x^2 + 1$

20

Using　(3.4)　$\boxed{(a^m \cdot b^n)^k = a^{mk} \cdot b^{nk}}$　with fractional exponents.

Examples:

1.　$(27a^{1/2})^{1/3} = 27^{1/3}a^{1/2 \cdot 1/3} = 27^{1/3}a^{1/6} = 3a^{1/6}$

2.　$(32x^{-1/2} \cdot y^{2/3})^{3/5} = 32^{3/5}x^{-1/2 \cdot 3/5} \cdot y^{2/3 \cdot 3/5}$

$$= (32^{1/5})^3 x^{-3/10} \cdot y^{2/5} = \frac{8y^{2/5}}{x^{3/10}}$$

21

Using　(3.5)　$\boxed{\left(\dfrac{a^m}{a^n}\right)^k = \dfrac{a^{mk}}{a^{nk}}}$　with fractional exponents.

Examples:

1.　$\left(\dfrac{1}{8^{4/15}}\right)^{-5/2} = \dfrac{1^{-5/2}}{8^{(4/15)(-5/2)}} = \dfrac{1}{8^{-2/3}} = 8^{2/3} = (8^{1/3})^2 = 2^2 = 4$

2.　$\left(\dfrac{2^{1/3}}{x^{1/4}}\right)^{1/2} = \dfrac{2^{1/3 \cdot 1/2}}{x^{1/4 \cdot 1/2}} = \dfrac{2^{1/6}}{x^{1/8}}$

3.　$\left(\dfrac{m^{1/2} \cdot n^{1/3}}{t^{2/3}}\right)^{3/2} = \dfrac{m^{1/2 \cdot 3/2} \cdot n^{1/3 \cdot 3/2}}{t^{2/3 \cdot 3/2}} = \dfrac{m^{3/4} \cdot n^{1/2}}{t}$

22

CHECK POINT 5

Use the laws of exponents to simplify the following.

1. $(a^{1/2})^{5/3}$

2. $(m^{2/3} \cdot n^{1/4})^{3/2}$

3. $(4x^{4/3} \cdot y^{1/2})^{1/2}$

4. $\left(\dfrac{1}{32^{6/7}}\right)^{7/10}$

5. $\left(\dfrac{x^{2/3}}{y^{1/4}}\right)^{3/2}$

6. $\left(\dfrac{16a^{4/3}b^{1/2}}{c^{3/2}}\right)^{1/4}$

23

MORE EXAMPLES:

Example 1:

Simplify $(2x^{2/3}y^{-3/5})(4x^{-1/3}y^{1/5})$.

Solution:

$$(2x^{2/3}y^{-3/5})(4x^{-1/3}y^{1/5}) = 2 \cdot 4 \cdot x^{2/3} \cdot x^{-1/3} \cdot y^{-3/5} \cdot y^{1/5}$$

$$= 8x^{2/3-1/3} \cdot y^{-3/5+1/5}$$

$$= 8x^{1/3} \cdot y^{-2/5} \quad \text{or} \quad \frac{8x^{1/3}}{y^{2/5}}$$

24

Example 2:

Simplify $\dfrac{15a^{1/3}b^{2/3}c^{-5/3}}{3a^{-2/3}b^{5/3}c^{1/3}}$.

Solution:

$$\frac{15a^{1/3}b^{2/3}c^{-5/3}}{3a^{-2/3}b^{5/3}c^{1/3}} = 5a^{1/3-(-2/3)}b^{2/3-5/3}c^{-5/3-1/3}$$

$$= 5a^{3/3}b^{-3/3}c^{-6/3}$$

$$= 5ab^{-1}c^{-2} \quad \text{or} \quad \frac{5a}{bc^2}$$

Example 3:

Simplify $\left(\dfrac{2^{-4}x^{3/2}}{8^{-2/3}y^2}\right)^{-1/2} \div \left(\dfrac{4^{1/2}x^{1/6}}{2^2y^{-2/3}}\right)^3.$

Solution:

$\left(\dfrac{2^{-4}x^{3/2}}{8^{-2/3}y^2}\right)^{-1/2} \div \left(\dfrac{4^{1/2}x^{1/6}}{2^2y^{-2/3}}\right)^3$

$= \left[\dfrac{2^{(-4)(-1/2)}x^{(3/2)(-1/2)}}{8^{(-2/3)(-1/2)}y^{2(-1/2)}}\right] \div \left[\dfrac{4^{(1/2)(3)}x^{(1/6)(3)}}{2^{(2)(3)}y^{(-2/3)(3)}}\right]$

$= \left(\dfrac{2^2x^{-3/4}}{8^{1/3}y^{-1}}\right) \div \left(\dfrac{4^{3/2}x^{1/2}}{2^6y^{-2}}\right) \qquad \text{Note:} \quad 4^{3/2} = (4^{1/2})^3 = 2^3 = 8$

$= \left(\dfrac{4y}{2x^{3/4}}\right) \div \left(\dfrac{8x^{1/2}y^2}{64}\right)$

$= \left(\dfrac{4y}{2x^{3/4}}\right) \cdot \left(\dfrac{64}{8x^{1/2}y^2}\right)$

$= \dfrac{16}{x^{5/4}y}$

CHECK POINT 6

Use the laws of exponents to simplify the following.

1. $(3^{-1}x^2y^{-2/5})(9x^{-3/4}y^3)$

2. $\dfrac{36m^{-1/2}n^{3/5}t^{-1/5}}{9m^{-3/2}n^{-7/5}t^2}$

3. $\left(\dfrac{3^2a^{-1/2}}{2b^{-3/4}}\right)^{-2} \cdot \left(\dfrac{9a^{2/3}}{16b}\right)^{1/2}$

4. $\left(\dfrac{1}{8p}\right)^{-2/3} \div \left(\dfrac{16p^2}{t^4}\right)^{3/4}$

In calculus, we frequently need to simplify algebraic expressions as shown in the following examples involving fractional exponents.

Example 1:

Simplify $(2 - x^2)^{-1/2}(-1) + (x)(-\frac{1}{2})(2 - x^2)^{-3/2}(-2x)$.

Solution:

$(2 - x^2)^{-1/2}(-1) + (x)(-\frac{1}{2})(2 - x^2)^{-3/2}(-2x)$

$$= \frac{-1}{(2 - x^2)^{1/2}} + \frac{x^2}{(2 - x^2)^{3/2}}$$ Change to positive exponents.

$$= \frac{-1(2 - x^2)^1}{(2 - x^2)^{1/2}(2 - x^2)^1} + \frac{x^2}{(2 - x^2)^{3/2}}$$ Find the common denominator.

$$= \frac{x^2 - 2}{(2 - x^2)^{3/2}} + \frac{x^2}{(2 - x^2)^{3/2}}$$ Add the fractions.

$$= \frac{2(x^2 - 1)}{(2 - x^2)^{3/2}}$$

Example 2:

Simplify $\dfrac{(3a^2 - 3x^2)^{1/2}}{(a^2 - x^2)^{3/2}}$.

Solution:

$$\frac{(3a^2 - 3x^2)^{1/2}}{(a^2 - x^2)^{3/2}} = \frac{[3(a^2 - x^2)]^{1/2}}{(a^2 - x^2)^{3/2}}$$ Factor out 3 from the numerator inside the square root.

$$= \frac{3^{1/2}(a^2 - x^2)^{1/2}}{(a^2 - x^2)^{3/2}}$$ Use exponent law 3.4.

$$= \frac{3^{1/2}}{(a^2 - x^2)^{3/2-1/2}}$$ Use exponent law 3.2.

$$= \frac{3^{1/2}}{a^2 - x^2} \quad \text{or} \quad 3^{1/2}(a^2 - x^2)^{-1}$$

As mentioned previously it might be helpful to replace the binomial factors by a single variable.

Example 3:
Simplify $\frac{1}{2}(3x + 1)^{-1/2}(3)(x^2 - 5)^{2/3} + (3x + 1)^{1/2}(\frac{2}{3})(x^2 - 5)^{-1/3}(2x)$

Solution:
Let $a = 3x + 1$ and $b = x^2 - 5$

$\therefore$ $\frac{1}{2}(3x + 1)^{-1/2}(3)(x^2 - 5)^{2/3} + (3x + 1)^{1/2}(\frac{2}{3})(x^2 - 5)^{-1/3}(2x)$

$= \frac{1}{2}a^{-1/2}(3)b^{2/3} + a^{1/2}(\frac{2}{3})b^{-1/3}(2x)$

Change to positive exponents

$= \frac{3b^{2/3}}{2a^{1/2}} + \frac{4xa^{1/2}}{3b^{1/3}}$

Find the LCD

$= \frac{3b^{2/3} \cdot 3b^{1/3}}{2a^{1/2} \cdot 3b^{1/3}} + \frac{4xa^{1/2} \cdot 2a^{1/2}}{3b^{1/3} \cdot 2a^{1/2}}$

Add and simplify

$= \frac{9b + 8xa}{6a^{1/2}b^{1/3}}$

Substitute $a = 3x + 1$ and $b = x^2 - 5$ back

$= \frac{9(x^2 - 5) + 8x(3x + 1)}{6(3x + 1)^{1/2}(x - 5)^{1/3}}$

Simplify the numerator

$= \frac{33x^2 + 8x - 45}{6(3x + 1)^{1/2}(x^2 - 5)^{1/3}}$

CHECK POINT 7

Simplify the following.

1. $3(x - 1)^{1/2} + (3x + 1)(x - 1)^{-1/2}$

2. $\dfrac{(2\pi x^2 + 2\pi y^2)^{3/2}}{(x^2 + y^2)^{1/2}}$

DRILL EXERCISES

Simplify the following and obtain expressions containing only positive exponents.

1. $3x^{-1} \cdot x^4$

2. $(a^{-4})^{-1}$

3. $(-2m^3n^{-1})^2$

4. $(5^{-1}rt^{-2})^{-3}$

DRILL EXERCISES (continued)

5. $(x^3y^0z^{-2})^{-1} \div (x^{-1}y^3z^{-2})^2$

6. $\left(\dfrac{3m^{-2}n^3}{4z}\right)^4$

7. $\dfrac{-8x^{-3}(y^2z)}{-4yz^{-2}}$

8. $\dfrac{21a^2b(a^{-1}b^{-2})}{-7a^3b^{-1}}$

9. $(a + b)^{-3}$

10. $(2m)^{-1} - 3n^{-3}$

11. $(5x - y^{-2})^{-1}$

12. $\dfrac{r^2 - rt - 2t^2}{t(t^{-1} + r^{-1})}$ (Factor trinomial)

13. $(8^{-1} - 4^{-2})^2$

14. $(2x - y^{-2})^{-3}$

15. $(3^{-2} - 2^{-3})^{-1}$

16. $\left(\dfrac{3a^2}{2b}\right)^{-3}\left(\dfrac{9a^4}{4b}\right)^2$

17. $\left(\dfrac{m^{-2}n^3}{5rt^{-4}}\right)^{-2}\left(\dfrac{3m^{-2}n^{-1}}{2rt^{-3}}\right)^3$

18. $\dfrac{a^{-1} + b^{-1}}{a^{-2} - b^{-2}}$

19. $4(1 - 2t)^3(1 + 2t) - 6(1 + 2t)^2(1 - 2t)^2$

20. $2m(m^4 - 1)^{-1} - m^2(m^4 - 1)^{-2} \cdot 4m^2$

21. $(a^{-1} - b)^{-2}$

22. $(x^2y^6)^{1/2}$

23. $3^{1/4} \cdot 3^{3/4}$

24. $a^{2/3} \cdot a^{-1/6}$

25. $(y^{-1/3})^{3/2}$

26. $[(m^2 - 1)^{1/3}]^6$

27. $(x + 1)^{3/4} \div (x + 1)^{-1/2}$

28. $t^{5/6} \div t^{2/3}$

29. $8^{5/6} \div 8^{1/2}$

30. $\dfrac{32^{3/5}}{32^{1/2}}$

31. $\left(\dfrac{1}{16^{5/8}}\right)^{-6/5}$

32. $\left(5x^{3/4} \cdot y^{1/3}\right)^2$

33. $\dfrac{m^{1/2} \cdot n^{1/3}}{m^{-3/4} \cdot n^{2/3}}$

34. $\dfrac{a^{5/6} \cdot a^{2/3}}{a^{-2}}$

DRILL EXERCISES (continued)

35. $\dfrac{x^{2/3} \cdot y^{-1/2} \cdot z^{4/5}}{x^{5/6} \cdot y^{3/8} \cdot z^{-3}}$

36. $\dfrac{t^{1/2}}{t^{1/5} \cdot t^{-1}}$

37. $\left(\dfrac{x^{1/3}z^2}{y^{-5/3}}\right)^{-3/2}$

38. $\left(\dfrac{r^{1/2}t^{2/3}}{16r^{1/5}}\right)^{3/4}$

39. $\left(\dfrac{8m^{5/6}n^{-1/4}}{m^{2/3}n^4}\right)^{-3/2}$

40. $(2^{-1}a^3b^{-2/3})(8a^{-1/6}b^{1/2})$

41. $\dfrac{12x^{1/2} \cdot y^{-2/3} \cdot z^{-1/4}}{3x^{-2} \cdot z^{1/2}} \cdot \dfrac{y^{1/6}}{x^{3/4}}$

42. $\left(\dfrac{32p}{V^{2/3}}\right)^{-1/5} \div \left(\dfrac{9V^{1/3}}{4P^{4/5}}\right)^{1/2}$

43. $\left(\dfrac{16^{-1/4}a^{2/5}b^{3/4}}{4^{-1/2}a^{-3/5}}\right)^2 \div \left(\dfrac{8ab^{-1/2}}{b^6}\right)^{-1/3}$

44. $(x^{1/2} - y^{1/2})(x^{1/2} + y^{1/2})$

45. $\dfrac{(3a^2 - 3b^2)^{5/2}}{(a^2 - b^2)^{3/2}}$

46. $7(2 - m)^{-1/2} - \dfrac{1}{2}(7m - 1)(2 - m)^{-3/2}$

47. $5(5x - 1)^{-2}(x + 3)^{1/2} + \dfrac{1}{2}(5x - 1)^{-1}(x + 3)^{-1/2}$

ASSIGNMENT EXERCISES

Simplify the following and obtain expressions containing only positive exponents.

1. $3a^5 \cdot a^{-2}$

2. $(m^{-2})^4$

3. $(5p^2) \div (2p^{-3})$

4. $(2^{-1}x^{-3}y^2z)^{-3}$

5. $\dfrac{(a^2b^0c^4)^5}{2ab^{-1}}$

6. $\left(\dfrac{m^3n^{-2}}{3t^{-2}}\right)^{-4}$

7. $(5x^2y^{-1})^{-1} \div (3x^{-3}y^4)$

8. $5(x + y)^{-2}$

9. $\dfrac{-m^3n^{-4}(4m^{-5}n^4)}{28m^{-1}n^6}$

10. $(a^{-2} - 3b)^{-1}$

11. $(a^{-1}b^2)^{-1}(5a^{-2}b^3)$

12. $(3^{-3} - 1^{-1})^{-1}$

Unit 16

ASSIGNMENT EXERCISES (continued)

13. $\left(\dfrac{5b^{-1}}{2a^2}\right)^3\left(\dfrac{-1}{8b^4}\right)^{-1}$

14. $(x^{-1} + y^{-2})^3$

15. $\dfrac{(7m^3n^{-1})^2(6m^{-3})}{-3n^{-2}}$

16. $\left(\dfrac{4x^2y^{-3}z}{x^{-1}y^{-5}}\right)^2\left(\dfrac{5y^3}{-2xz}\right)^{-3}$

17. $(2^{-3} + k)^2$

18. $5a^{-2} - (3b)^{-1}$

19. $\dfrac{x^{-1} + y^{-1}}{y^2 - x^2}$

20. $2^{1/2} \cdot 2^{-3/4}$

21. $(m^{-2}n^9)^{1/3}$

22. $(m^{-4}n^3)^{1/2}$

23. $q^{-4/3} \cdot q^{-1/6}$

24. $(t^{2/3})^{-1/2}$

25. $(r - t)^{-1/3} \div (r - t)^{1/6}$

26. $a^{1/2} \div a^{2/5}$

27. $16^{3/4} \div 16^{1/2}$

28. $\dfrac{27^{2/3}}{-3^{-1}}$

29. $(7m^{-1/2}n^{2/3})^{-2}$

30. $\dfrac{2x^2 - xy - y^2}{x(x^{-1} - y^{-1})}$

31. $\dfrac{a^{2/3}b^{-3/4}}{a^{-1/6}b^{1/3}}$

32. $\dfrac{y^{2/5} \cdot y^{-2}}{y^{1/10}}$

33. $[(x^2 - y^2)^4]^{1/8}$

34. $\left(\dfrac{a^{-2}b^{1/2}}{64}\right)^{-2/3}$

35. $5a(a^2 - 1) - 2a(a + 1)^2$

36. $rt^2(3r - 1)^{-2} + t^2(3r - 1)^{-1}$

37. $\left(\dfrac{x^{-2}y^{-2/5}}{x^{-3/2}z^3}\right)^{-1/6}$

38. $\left(\dfrac{4h^{-4}k^{3/2}}{k^{-2}}\right)^{-1/2}$

39. $\dfrac{r^{3/4}t^{-2/3}v^{-2}}{r^{1/2}t^{-1/3}v^{-4/5}}$

40. $\dfrac{64^{2/3}}{64^{1/2}}$

41. $\left(\dfrac{1}{27^{4/9}}\right)^{-5/8}$

42. $\left(\dfrac{t^{-3}r^{3/4}}{125t^{1/2}}\right)^{4/3}$

249

ASSIGNMENT EXCERCISES (continued)

43. $\left(\dfrac{9x^{5/3}y^{-1/2}}{x^2 y^{2/3}}\right)^{-3/4}$

44. $\dfrac{m^{1/4}n^{-2/3}}{6m^{-1}} \cdot \left(\dfrac{4n^{-1}}{m^{2/5}}\right)^{1/2}$

45. $(3a^{-3}b^{2/3}c^{1/4})(-4a^{1/2}b^{-1/6}c^{2/3})$

46. $\left(\dfrac{-p}{8R^{3/4}}\right)^{1/3} \div \left(\dfrac{16R^{2/3}}{25P^{6/5}}\right)^{1/4}$

47. $\dfrac{4x^{1/3}y^{-4/3}z^{1/2}}{12x^{-2}y^{-1/2}} \cdot \dfrac{x}{y^{1/6}z^{-1}}$

48. $(a^{1/2} + b^{1/2})(a^{1/2} - b^{1/2})$

49. $\dfrac{(x^2 + 2y^2)^{1/4}}{(4x^2 + 8y^2)^{3/2}}$

50. $\dfrac{4}{5}(t - 3)^{-1/3} - (t - 3)^{-4/3}(8t)$

51. $\dfrac{1}{4}(2 - 3z)^{-2}(z + 1)^{5/6} + 3(z + 1)^{-1/6}(2 - 3z)^{-1}$

52. $(x^2 + 1)^2(x^3 - 1)(3x^2) + (x^3 - 1)^2(x^2 + 1)(2x)$

53. $(x - 1)^{-1}(4)(3x^2 + 2)^3(6x) + (3x^2 + 2)^4(-1)(x - 1)^{-2}$

54. $2(x + 5)(2x^2 - 1)^{-3} + (x + 5)^2(-3)(2x^2 - 1)^{-4}(4x)$

55. $3(2x - 7)^2(2)(x - 4)^{-2} + (2x - 7)^3(-2)(x - 4)^{-3}$

56. $(x^2 + 4)^{1/2} + \dfrac{x}{2}(x^2 + 4)^{-1/2}(2x)$

57. $3(x^2 - 2)^2(2x)(x - 1)^{-1/2} + (x^2 - 2)^3(-\dfrac{1}{2})(x - 1)^{-3/2}$

58. $4x^3(x^2 - 4)^{-1/3} + x^4(-\dfrac{1}{3})(x^2 - 4)^{-4/3}(2x)$

59. $-\dfrac{1}{4}(5x - 1)^{-5/4}(5)(x^2 + 3)^{1/2} + (5x - 1)^{-1/4}(\dfrac{1}{2})(x^2 + 3)^{-1/2}(2x)$

UNIT **17**

OPERATIONS WITH RADICALS

Objectives: After having worked through this unit, the student should be capable of:

1.   using the laws of radicals to reduce a radical to simplest form;
2.   adding and subtracting radicals;
3.   multiplying and dividing radicals;
4.   simplifying expressions containing radicals, including "rationalizing the denominator".

1

REVIEW   TERMINOLOGY

1.   A **radical** is an indicated root of a number or expression.

   Examples:

   1. $\sqrt{7}$ is a radical.

   2. $\sqrt[4]{10x^3}$ is a radical.

   3. $\sqrt[5]{19a^2}$ is a radical.

2.   The **radical** signs are the symbols $\sqrt{}$, $\sqrt[3]{}$, $\sqrt[4]{}$, $\sqrt[5]{}$, and so on.

3.   The **radicand** is the number or expression under the radical sign.

   Examples:

   1.   In $\sqrt{7}$, 7 is the radicand.

   2.   In $\sqrt[4]{10x^3}$, $10x^3$ is the radicand.

4.   The **index** of a root is the number on the radical sign indicating which root is to be taken. The index is sometimes referred to as the order.

   Examples:

   1.   In $\sqrt[3]{19}$, 3 is the index.

   2.   In $\sqrt[4]{10x^3}$, 4 is the index.

   3.   In $\sqrt{10}$, the index is understood to be 2; that is, if the square root is to be taken, we do not write the index 2.

2

When we introduced the concept of radicals, we learned that they can be written using fractional exponents.

We defined $\sqrt[n]{a} = a^{1/n}$

$\sqrt[n]{a^m} = a^{m/n}$

Examples:

1. $\sqrt[5]{32} = 32^{1/5}$ 2. $\sqrt[4]{81^3} = 81^{3/4}$

Hence, the laws of exponents can be used in operations involving radicals in the same way as they were used with fractional exponents. In what follows, **we assume that all radicands are positive** when the index is an even number.

Using the above definitions and the laws of exponents, we shall present, in what follows, the laws of radicals.

$$(17.1) \qquad \boxed{\sqrt[n]{a^n} = \left(\sqrt[n]{a}\right)^n = a}$$

Examples:

1. $\sqrt[3]{2^3} = (2^3)^{1/3} = 2 = (2^{1/3})^3 = \left(\sqrt[3]{2}\right)^3$

2. $\sqrt{5^2} = (5^2)^{1/2} = 5 = (5^{1/2})^2 = (\sqrt{5})^2$

In general, we have $\sqrt[n]{a^n} = (a^n)^{1/n} = a = (a^{1/n})^n = \left(\sqrt[n]{a}\right)^n$

$$(17.2) \qquad \boxed{\sqrt[n]{ab} = \sqrt[n]{a}\ \sqrt[n]{b}}$$

Examples:

1. $\sqrt{36} = \sqrt{4 \cdot 9} = (4 \cdot 9)^{1/2} = 4^{1/2} \cdot 9^{1/2} = \sqrt{4}\ \sqrt{9}$

2. $\sqrt[3]{64} = \sqrt[3]{8 \cdot 8} = (8 \cdot 8)^{1/3} = 8^{1/3} \cdot 8^{1/3} = \sqrt[3]{8}\ \sqrt[3]{8}$

In general, we have $\sqrt[n]{ab} = (ab)^{1/n} = a^{1/n} \cdot b^{1/n} = \sqrt[n]{a}\ \sqrt[n]{b}.$

$$(17.3) \qquad \boxed{\sqrt[n]{\frac{a}{b}} = \frac{\sqrt[n]{a}}{\sqrt[n]{b}}} \qquad b \neq 0$$

Examples:

1. $\sqrt{\dfrac{25}{49}} = \left(\dfrac{25}{49}\right)^{1/2} = \dfrac{25^{1/2}}{49^{1/2}} = \dfrac{\sqrt{25}}{\sqrt{49}}$

2. $\sqrt[4]{\dfrac{16}{81}} = \left(\dfrac{16}{81}\right)^{1/4} = \dfrac{16^{1/4}}{81^{1/4}} = \dfrac{\sqrt[4]{16}}{\sqrt[4]{81}}$

In general, we have $\sqrt{\dfrac{a}{b}} = \left(\dfrac{a}{b}\right)^{1/n} = \dfrac{a^{1/n}}{b^{1/n}} = \dfrac{\sqrt[n]{a}}{\sqrt[n]{b}} \qquad b \neq 0.$

(17.4)
$$\sqrt[m]{\sqrt[n]{a}} = \sqrt[mn]{a}$$

Example:

$$\sqrt{\sqrt[3]{64}} = (64^{1/3})^{1/2} = 64^{1/6} = \sqrt[6]{64}$$

In general, we have $\sqrt[m]{\sqrt[n]{a}} = (a^{1/n})^{1/m} = a^{1/mn} = \sqrt[mn]{a}.$

CHECK POINT 1

As shown in the previous examples, use the laws of exponents to show that the following equations are true.

1. $\sqrt{4^2} = (\sqrt{4})^2 = 4$

2. $\sqrt[5]{2^5} = \left(\sqrt[5]{2}\right)^5 = 2$

3. $\sqrt{81} = \sqrt{9}\,\sqrt{9}$

4. $\sqrt[3]{216} = \sqrt[3]{8}\,\sqrt[3]{27}$

5. $\sqrt{\dfrac{4}{25}} = \dfrac{\sqrt{4}}{\sqrt{25}}$

6. $\sqrt[3]{\dfrac{27}{64}} = \dfrac{\sqrt[3]{27}}{\sqrt[3]{64}}$

7. $\sqrt[3]{\sqrt{729}} = \sqrt[6]{729}$

SUMMARY   LAWS OF RADICALS

17.1 $\sqrt[n]{a^n} = \left(\sqrt[n]{a}\right)^n = a$

17.2 $\sqrt[n]{ab} = \sqrt[n]{a}\,\sqrt[n]{b}$

17.3 $\sqrt[n]{\dfrac{a}{b}} = \dfrac{\sqrt[n]{a}}{\sqrt[n]{b}}$ $b \neq 0$

17.4 $\sqrt[m]{\sqrt[n]{a}} = \sqrt[mn]{a}$

Note: 1. The above laws only apply to operations involving multiplication or division.

$$\sqrt{a + b} \neq \sqrt{a} + \sqrt{b} \qquad \sqrt{a - b} \neq \sqrt{a} - \sqrt{b}$$

2. All letters a, b, m and n are assumed to be positive.

9

SIMPLIFYING RADICALS
Radicals contained in answers should be reduced, if possible, to simplest form. To simplify radicals, we use the laws listed previously.

Example 1:
Simplify $\sqrt{48}$.

Solution:

Step 1.   Factor 48 into two factors so that one is the largest possible perfect square.

$$\sqrt{48} = \sqrt{16 \times 3}$$

Step 2.   Use the laws of radicals to reduce the radicals to simplest form.

$$\sqrt{48} = \sqrt{16 \times 3}$$

$$= \sqrt{16}\ \sqrt{3} \qquad (17.2)$$

$$= 4\ \sqrt{3} \qquad (17.1)$$

Hence, $\sqrt{48} = 4\sqrt{3}$.

10

Example 2:
Simplify $\sqrt[3]{-250}$.

Solution:

Step 1.   Factor -250 into two factors so that one is the largest possible negative perfect cube.

$$\sqrt[3]{-250} = \sqrt[3]{-125 \cdot 2}$$

Step 2.   Use the laws of radicals to reduce the radical to simplest form.

$$\sqrt[3]{-250} = \sqrt[3]{-125 \cdot 2}$$

$$= \sqrt[3]{-125}\ \sqrt[3]{2} \qquad (17.2)$$

$$= -5\ \sqrt[3]{2} \qquad (17.1)$$

Hence, $\sqrt[3]{-250} = -5\ \sqrt[3]{2}$

11

More Examples:

1. $\sqrt{12} = \sqrt{4 \cdot 3} = \sqrt{4}\ \sqrt{3} = 2\ \sqrt{3}$ 3. $\sqrt{128} = \sqrt{64 \cdot 2} = \sqrt{64}\ \sqrt{2} = 8\ \sqrt{2}$

2. $\sqrt[3]{32} = \sqrt[3]{8 \cdot 4} = \sqrt[3]{8}\ \sqrt[3]{4} = 2\ \sqrt[3]{4}$ 4. $\sqrt[3]{81} = \sqrt[3]{27 \cdot 3} = \sqrt[3]{27}\ \sqrt[3]{3} = 3\ \sqrt[3]{3}$

12

CHECK POINT 2

Reduce the following radicals to simplest form.

1. $\sqrt{50}$
2. $\sqrt[3]{24}$
3. $\sqrt[3]{54}$
4. $\sqrt{108}$

13

In general, when simplifying radicals, we remove from the radicand of a radical of order n all perfect n-th power factors.

Method:

Step 1.   Factor the radicand of index n into perfect n-th powers.

Step 2.   Use the laws of radicals to reduce the radical to simplest form.

14

Example 1:

Simplify $\sqrt{8x^3}$

Solution:

Step 1. $\sqrt{8x^3} = \sqrt{4 \cdot 2 \cdot x^2 \cdot x}$ Factor radicand into perfect squares.

$= \sqrt{4 \cdot x^2 \cdot 2 \cdot x}$ Place perfect square factors first.

Step 2. $= \sqrt{4x^2}\sqrt{2x}$ Use (17.2)

$= \sqrt{4}\sqrt{x^2}\sqrt{2x}$

$= 2x\sqrt{2x}$ Use (17.1)

Hence, $\sqrt{8x^3} = 2x\sqrt{2x}$.

Example 2:

Simplify $\sqrt[5]{128a^8b^7}$.

Solution:

Step 1.

$$\sqrt[5]{128a^8b^7} = \sqrt[5]{32 \cdot 4 \cdot a^5 \cdot a^3 \cdot b^5 \cdot b^2} \qquad \text{Factor into perfect 5-th powers.}$$

$$= \sqrt[5]{32 \cdot a^5 \cdot b^5 \cdot 4 \cdot a^3 \cdot b^2} \qquad \text{Place perfect 5-th powers first.}$$

Step 2.

$$= \sqrt[5]{32 \cdot a^5 \cdot b^5} \sqrt[5]{4 \cdot a^3 \cdot b^2} \qquad \text{Use (17.2)}$$

$$= \sqrt[5]{32} \sqrt[5]{a^5} \sqrt[5]{b^5} \sqrt[5]{4a^3b^2} \qquad \text{Use (17.1)}$$

$$= 2ab \sqrt[5]{4a^3b^2}$$

Hence, $\sqrt[5]{128a^8b^7} = 2ab \sqrt[5]{4a^3b^2}$.

Example 3:

Simplify $\sqrt[3]{x^7y^{10}z^4}$.

Solution:

Step 1.

$$\sqrt[3]{x^7y^{10}z^4} = \sqrt[3]{x^6 \cdot x \cdot y^9 \cdot y \cdot z^3 \cdot z} \qquad \begin{array}{l}\text{Factor into exponents which can}\\ \text{be changed to perfect cubes.}\end{array}$$

$$= \sqrt[3]{(x^2)^3(y^3)^3z^3xyz}$$

Step 2.

$$= \sqrt[3]{(x^2)^3(y^3)^3z^3} \sqrt[3]{xyz} \qquad \text{Use (17.2)}$$

$$= \sqrt[3]{(x^2)^3} \sqrt[3]{(y^3)^3} \sqrt[3]{z^3} \sqrt[3]{xyz}$$

$$= x^2y^3z \sqrt[3]{xyz} \qquad \text{Use (17.1)}$$

Hence, $\sqrt[3]{x^7y^{10}z^4} = x^2y^3z \sqrt[3]{xyz}$.

17

To simplify radicals means also to reduce the index of the radical if possible. This can be accomplished by changing radicals to fractional exponents.

Examples:

1. $\sqrt[4]{49} = \sqrt[4]{7^2} = (7^2)^{1/4} = 7^{1/2} = \sqrt{7}$

2. $5\sqrt[6]{125} = 5\sqrt[6]{5^3} = 5(5^3)^{1/6} = 5\sqrt{5}$

3. $\dfrac{\sqrt[6]{128}}{\sqrt{2}} = \dfrac{\sqrt[6]{(2)^7}}{\sqrt{2}} = \dfrac{2^{7/6}}{2^{1/2}} = 2^{2/3} = \sqrt[3]{4}$

18

CHECK POINT 3

I.   Reduce the following.

1. $\sqrt{12a^3}$ 2. $\sqrt[3]{16x^5y^4}$ 3. $\sqrt[6]{m^7n^{13}t^{19}}$ 4. $\sqrt{8x^5y^9z}$

II. Reduce the index of the following radicals.

5. $\sqrt[4]{36}$ 6. $\sqrt[6]{16}$ 7. $\sqrt[9]{64}$ 8. $\dfrac{\sqrt[4]{25}}{\sqrt{5}}$

19

ADDITION AND SUBTRACTION OF RADICALS

We have learned that expressions such as $x^3 - y$, $2x + 5y - 3$ and $2x^3 + 4x^2 - 3x$ cannot be simplified by addition and subtraction. We can only add or subtract like terms. A similar situation exists when we add or subtract expressions containing radicals.

Radicals which have the **same index** and the **same radicand** are called **like radicals**.

We can add expressions containing radicals only if they contain like radicals.

Examples:

1. $3\sqrt{2} + 5\sqrt{2} = 8\sqrt{2}$ The terms $3\sqrt{2}$ and $5\sqrt{2}$ contain the identical or **like** radical $\sqrt{2}$.

2. $8\sqrt[4]{xy} + \sqrt[4]{xy} - 3\sqrt[4]{xy} = 6\sqrt[4]{xy}$ The terms $8\sqrt[4]{xy}$, $\sqrt[4]{xy}$, $3\sqrt[4]{xy}$ contain the identical or like radical $\sqrt[4]{xy}$.

3. $8\sqrt{6} - 5\sqrt{7}$ These terms cannot be combined by subtraction since $\sqrt{6}$ and $\sqrt{7}$ are not like radicals; that is, the radicands are different.

4. $6\sqrt[3]{x} - \sqrt[3]{y}$ $\sqrt[3]{x}$ and $\sqrt[3]{y}$ are not like radicals. Hence, we cannot combine the terms.

5. $7\sqrt[3]{5} + 3\sqrt{5}$ $7\sqrt[3]{5}$ and $3\sqrt{5}$ are not like radicals, that is, the indices are different. One term contains the cube root $\sqrt[3]{5}$, the other term contains the square root.

MORE EXAMPLES

The following are examples showing how we can simplify by adding and/or subtracting expressions containing radicals.

Example 1:

Simplify $3\sqrt{6} - \sqrt{10} + 5\sqrt{6} + 4\sqrt{10} - 6\sqrt{6}$.

Solution:

Add and/or subtract those terms containing like radicals.

$$3\sqrt{6} - \sqrt{10} + 5\sqrt{6} + 4\sqrt{10} - 6\sqrt{6} = (3\sqrt{6} + 5\sqrt{6} - 6\sqrt{6}) + (-\sqrt{10} + 4\sqrt{10})$$

$$= 2\sqrt{6} + 3\sqrt{10}$$

Example 2:

Simplify $6\sqrt[3]{x^2} + 11\sqrt{y} - 4\sqrt[3]{x^2} - 8\sqrt{y}$.

Solution:

Combine the terms containing like radicals.

$$6\sqrt[3]{x^2} + 11\sqrt{y} - 4\sqrt[3]{x^2} - 8\sqrt{y} = (6\sqrt[3]{x^2} - 4\sqrt[3]{x^2}) + (11\sqrt{y} - 8\sqrt{y})$$

$$= 2\sqrt[3]{x^2} + 3\sqrt{y}$$

Normally, when simplifying expressions containing radicals, we **reduce** first all radicals to simplest form as shown previously; then we add and/or subtract the like radicals.

For example, $\sqrt[3]{48}$ is written as $\sqrt[3]{8 \times 6} = \sqrt[3]{8}\sqrt[3]{6} = 2\sqrt[3]{6}$ and

$\sqrt[4]{49}$ is written as $\sqrt[4]{49} = \sqrt[4]{7^2} = (7^2)^{1/4} = 7^{1/2} = \sqrt{7}$.

Unit 17

Example 3:

Simplify $5\sqrt[3]{48} + 2\sqrt{7} - 4\sqrt[3]{6} - \sqrt[4]{49}$.

Solution:

Step 1.   Reduce radicals to simplest form.

$$5\sqrt[3]{48} + 2\sqrt{7} - 4\sqrt[3]{6} - \sqrt[4]{49} = 5\sqrt[3]{8 \times 6} + 2\sqrt{7} - 4\sqrt[3]{6} - \sqrt[4]{7^2}$$

$$= 5\sqrt[3]{8}\,\sqrt[3]{6} + 2\sqrt{7} - 4\sqrt[3]{6} - (7^2)^{1/4}$$

$$= 5(2\sqrt[3]{6}) + 2\sqrt{7} - 4\sqrt[3]{6} - 7^{1/2}$$

$$= 10\sqrt[3]{6} + 2\sqrt{7} - 4\sqrt[3]{6} - \sqrt{7}$$

Step 2.   Add and/or subtract like radicals.

$$5\sqrt[3]{48} + 2\sqrt{7} - 4\sqrt[3]{6} - \sqrt[4]{49} = 10\sqrt[3]{6} + 2\sqrt{7} - 4\sqrt[3]{6} - \sqrt{7}$$

$$= 6\sqrt[3]{6} + \sqrt{7}$$

Example 4:

Simplify $2\sqrt{32} + 4\sqrt{24} - 3\sqrt{8} - 5\sqrt{54}$.

Solution:

Step 1.   Reduce radicals to simplest form.

$$2\sqrt{32} + 4\sqrt{24} - 3\sqrt{8} - 5\sqrt{54} = 2\sqrt{16 \times 2} + 4\sqrt{4 \times 6} - 3\sqrt{4 \times 2} - 5\sqrt{9 \times 6}$$

$$= 2(4\sqrt{2}) + 4(2\sqrt{6}) - 3(2\sqrt{2}) - 5(3\sqrt{6})$$

$$= 8\sqrt{2} + 8\sqrt{6} - 6\sqrt{2} - 15\sqrt{6}$$

Step 2.   Combine the radicals.

$$2\sqrt{32} + 4\sqrt{24} - 3\sqrt{8} - 5\sqrt{54} = 8\sqrt{2} + 8\sqrt{6} - 6\sqrt{2} - 15\sqrt{6}$$

$$= 2\sqrt{2} - 7\sqrt{6}$$

26

Example 5:

Simplify $\sqrt[3]{2x^7y^9} + 3\sqrt[3]{16x^4y^6}$.

Solution:

$$\sqrt[3]{2x^7y^9} + 3\sqrt[3]{16x^4y^6} = \sqrt[3]{2(x^2)^3 \cdot x \cdot (y^3)^3} + 3\sqrt[3]{2^3 \cdot 2 \cdot x^3 \cdot x \cdot (y^2)^3}$$

(Reduce radicals.)

$$= \sqrt[3]{(x^2)^3 \cdot (y^3)^3 \cdot 2 \cdot x} + 3\sqrt[3]{2^3 \cdot x^3 \cdot (y^2)^3 \cdot 2 \cdot x}$$

$$= \sqrt[3]{(x^2)^3(y^3)^3} \sqrt[3]{2x} + 3\sqrt[3]{2^3x^3(y^2)^3} \sqrt[3]{2x}$$

$$= x^2y^3 \sqrt[3]{2x} + 3(2xy^2) \sqrt[3]{2x}$$

(Factor out the like radical.)

$$= \sqrt[3]{2x} (x^2y^3 + 6xy^2)$$

27

Example 6:

Simplify $\sqrt{24a^5b} - 4\sqrt{5ab^3} + \sqrt{54a^3b}$.

Solution:

$$\sqrt{24a^5b} - 4\sqrt{5ab^3} + \sqrt{54a^3b} = \sqrt{4 \cdot 6(a^2)^2ab} - 4\sqrt{5ab^2b} + \sqrt{9 \cdot 6a^2ab}$$

Reduce radicals.

$$= \sqrt{4(a^2)^2} \sqrt{6ab} - 4\sqrt{b^2} \sqrt{5ab} + \sqrt{9a^2} \sqrt{6ab}$$

$$= 2a^2\sqrt{6ab} - 4b\sqrt{5ab} + 3a\sqrt{6ab} \qquad \text{or}$$

$$= (2a^2 + 3a)\sqrt{6ab} - 4b\sqrt{5ab}$$

Factor out from the first and last term the like radical.

28

CHECK POINT 4

Perform the indicated operations and write the answers in simplest form.

1. $3\sqrt{5} - 6\sqrt{10}$

2. $\sqrt{x^3} + 2\sqrt{x}$

3. $\sqrt[4]{9} - 2\sqrt{27} + \sqrt{50}$

4. $\sqrt[3]{54} + 2\sqrt[3]{2} - \sqrt[3]{81}$

5. $\sqrt{3x^3y^5} - \sqrt{12xy^3}$

6. $\sqrt[3]{2a^4b} - \sqrt[3]{16ab^7} + \sqrt[3]{a^7b^4}$

MULTIPLICATION OF RADICALS

To multiply radicals, we use

(17.2) $\quad \sqrt[n]{a} \; \sqrt[n]{b} = \sqrt[n]{ab}.$

Note:

1. This law applies only to radicals which have the **same index** — n.

2. If possible, we reduce radicals to simplest form **after** multiplication.

Examples

1. $\sqrt[3]{7} \; \sqrt[3]{5} = \sqrt[3]{35}$ Since both $\sqrt[3]{7}$ and $\sqrt[3]{5}$ have the **same index**, that is, both are cube roots, we can multiply them.

2. $\sqrt[4]{6} \; \sqrt[3]{7} \neq \sqrt[?]{42}$ We cannot multiply $\sqrt[4]{6}$ and $\sqrt[3]{7}$ using (17.2) since $\sqrt[4]{6}$ is a fourth root and $\sqrt[3]{7}$ is a cube root.

Note:

If we change $\sqrt[4]{6}$ and $\sqrt[3]{7}$ to fractional exponents, we can multiply them as follows:

$\sqrt[4]{6} \; \sqrt[3]{7} = (6)^{1/4}(7)^{1/3}$

$\qquad = (6)^{3/12}(7)^{4/12}$ Change the fractional exponents to equivalent fractions with common denominator.

$\qquad = (6^3)^{1/12}(7^4)^{1/12}$ Use $a^{nm} = (a^n)^m$

$\qquad = (6^3 \cdot 7^4)^{1/12}$ Use $a^n b^n = (ab)^n$

$\qquad = \sqrt[12]{6^3 \cdot 7^4}$

More Examples:

1. $\sqrt{2} \; \sqrt{8} = \sqrt{16} = 4$

2. $\sqrt[3]{9} \; \sqrt[3]{3} = \sqrt[3]{27} = 3$

3. $\sqrt{32} \; \sqrt{6} = \sqrt{192} = \sqrt{64 \times 3} = 8\sqrt{3}$

4. $\sqrt[5]{x^3} \; \sqrt[5]{x^4} = \sqrt[5]{x^7} = \sqrt[5]{x^5 \cdot x^2} = x\sqrt[5]{x^2}$

5. $\sqrt{\dfrac{3a}{2b}} \; \sqrt{\dfrac{5a}{8b}} = \sqrt{\dfrac{3a \cdot 5a}{2b \cdot 8b}} = \dfrac{\sqrt{15a^2}}{\sqrt{16b^2}} = \dfrac{a\sqrt{15}}{4b}$

MULTIPLICATION OF EXPRESSIONS CONTAINING RADICALS

Expressions containing radicals are multiplied in the same way as other algebraic expressions, but we include the use of

$$\sqrt[n]{a} \ \sqrt[n]{b} = \sqrt[n]{ab}.$$

Example:

Multiply $(5\sqrt{2x})(3\sqrt{xy})$.

Solution:

$$(5\sqrt{2x})(3\sqrt{xy}) = 5 \cdot 3 \cdot \sqrt{2x} \cdot \sqrt{xy}$$

$$= 5 \cdot 3 \cdot \sqrt{2x^2y}$$

$$= 15x\sqrt{2y}$$

Example 2:

Multiply $\sqrt[5]{a} \ (3 - \sqrt[5]{2a^4})$.

Solution:

$$\sqrt[5]{a} \ (3 - \sqrt[5]{2a^4}) = 3\sqrt[5]{a} - \sqrt[5]{a} \ \sqrt[5]{2a^4}$$

$$= 3\sqrt[5]{a} - \sqrt[5]{2a^5}$$

$$= 3\sqrt[5]{a} - a\sqrt[5]{2}$$

When multiplying expressions containing radicals, we can use the shortcut methods for special products.

(a) $(a + b)(a - b) = a^2 - b^2$

(b) $(a + b)(c + d) = ac + ad + bc + bd$

               F   O   I   L

(c) $(a + b)^2 = a^2 + 2ab + b^2$

Example 3:

Multiply $(4 + \sqrt{5})(4 - \sqrt{5})$.

Solution:

$$(4 + \sqrt{5})(4 - \sqrt{5}) = 4^2 - (\sqrt{5})^2$$

$$= 16 - 5 = 11$$

Example 4:

Multiply $(3x + 2\sqrt{y})(3x - 2\sqrt{y})$.

Solution:

$$(3x + 2\sqrt{y})(3x - 2\sqrt{y}) = (3x)^2 - (2\sqrt{y})^2$$

$$= 9x^2 - 4y$$

36

Example 5:
Multiply $(2 + \sqrt{2x})(3 + \sqrt{18y})$.

Solution:

$$(2 + \sqrt{2x})(3 + \sqrt{18y}) = \overset{F}{6} + \overset{O}{2\sqrt{18y}} + \overset{I}{3\sqrt{2x}} + \overset{L}{\sqrt{2x}\sqrt{18y}}$$

$$= 6 + 2\sqrt{9 \cdot 2y} + 3\sqrt{2x} + \sqrt{36xy}$$

$$= 6 + 6\sqrt{2y} + 3\sqrt{2x} + 6\sqrt{xy}$$

37

CHECK POINT 5

Perform the indicated operations and write the answer in simplest form.

1. $\sqrt{32}\,\sqrt{2}$

2. $\sqrt[3]{5}\,\sqrt[3]{25}$

3. $\sqrt[4]{x^3}\,\sqrt[4]{ax}$

4. $\sqrt{\dfrac{5m}{2t^3}}\,\sqrt{\dfrac{3m^5}{18t}}$

5. $(3\sqrt[5]{4x^2})(6y\sqrt[5]{8x^4})$

6. $(4y - 3\sqrt{x})(4y + 3\sqrt{x})$

7. $(5a + \sqrt{2b})(a + 2\sqrt{8b})$

38

DIVISION OF RADICALS
To divide radicals, we use

(17.3)
$$\frac{\sqrt[n]{a}}{\sqrt[n]{b}} = \sqrt[n]{\frac{a}{b}}$$

Note:
1. This law applies only to radicals which have the **same index** — n.
2. If possible, we reduce radicals, after division, to simplest form.

39

Examples:

1. $\dfrac{\sqrt[5]{128}}{\sqrt[5]{4}} = \sqrt[5]{\dfrac{128}{4}} = \sqrt[5]{32} = 2$

2. $\dfrac{\sqrt[3]{48x^4y}}{\sqrt[3]{6x}} = \sqrt[3]{\dfrac{48x^4y}{6x}} = \sqrt[3]{8x^3y} = 2x\sqrt[3]{y}$

3. $\dfrac{\sqrt{x^2 - 1}}{\sqrt{x + 1}} = \sqrt{\dfrac{x^2 - 1}{x + 1}} = \sqrt{\dfrac{(x - 1)(x + 1)}{x + 1}} = \sqrt{x - 1}$

4. $\dfrac{\sqrt{50a^5}}{\sqrt{4a^3b}} = \sqrt{\dfrac{50a^5}{4a^3b}} = \sqrt{\dfrac{25a^2}{2b}} = \dfrac{\sqrt{25a^2}}{\sqrt{2b}} = \dfrac{5a}{\sqrt{2b}}$

5. $\dfrac{\sqrt{x^2 - 5x - 6}}{\sqrt{4x - 24}} = \sqrt{\dfrac{x^2 - 5x - 6}{4x - 24}} = \sqrt{\dfrac{(x - 6)(x + 1)}{4(x - 6)}} = \dfrac{1}{2}\sqrt{x + 1}$

40

RATIONALIZING THE DENOMINATOR

Expressions such as

$$\frac{10}{\sqrt{2}}, \quad \frac{3a}{\sqrt{b}}, \quad \frac{7}{1 + \sqrt{3}}, \quad \frac{19}{\sqrt{x} - \sqrt{y}}$$

are normally written in an equivalent form having no radicals in the denominator. We refer to this process as **rationalizing the denominator.**

Examples:

1. $\dfrac{10}{\sqrt{2}} = \dfrac{10}{\sqrt{2}} \cdot \dfrac{\sqrt{2}}{\sqrt{2}} = \dfrac{10\sqrt{2}}{2} = 5\sqrt{2}$

2. $\dfrac{3a}{\sqrt{b}} = \dfrac{3a}{\sqrt{b}} \cdot \dfrac{\sqrt{b}}{\sqrt{b}} = \dfrac{3a\sqrt{b}}{b}$

Note:

$$\sqrt{2}\sqrt{2} = 2^{1/2} \cdot 2^{1/2} = 2^{1/2 + 1/2} = 2$$

$$\sqrt{b}\sqrt{b} = b^{1/2} \cdot b^{1/2} = b^{1/2 + 1/2} = b$$

41

In general, when the rationalization involves a radical of index n, we proceed as follows.

Step 1.   Determine the factor by which the radicand has to be multiplied so that we obtain a new radicand which is a perfect n-th power.

Step 2.   Multiply the denominator and numerator by a radical which has the radicand determined in Step 1.

Example:

Rationalize $\dfrac{6x}{\sqrt[3]{x}}$

Solution:

Step 1.   Determine the factor by which x has to be multiplied so that we obtain a perfect cube.

The factor is x^2 since $x \cdot x^2 = x^3$.

Step 2.   Multiply the numerator and denominator by $\sqrt[3]{x^2}$.

$$\frac{6x}{\sqrt[3]{x}} \cdot \frac{\sqrt[3]{x^2}}{\sqrt[3]{x^2}} = \frac{6x\sqrt[3]{x^2}}{\sqrt[3]{x^3}} = \frac{6x\sqrt[3]{x^2}}{x} = 6\sqrt[3]{x^2}$$

More Examples

1. $\dfrac{10y}{\sqrt[4]{2y}} = \dfrac{10y}{\sqrt[4]{2y}} \cdot \dfrac{\sqrt[4]{8y^3}}{\sqrt[4]{8y^3}} = \dfrac{10y\sqrt[4]{8y^3}}{\sqrt[4]{16y^4}} = \dfrac{10y\sqrt[4]{8y^3}}{\sqrt[4]{(2y)^4}} = \dfrac{\overset{5}{\cancel{10}}y\sqrt[4]{8y^3}}{\underset{1}{\cancel{2}}y} = 5\sqrt[4]{8y^3}$

2. $\dfrac{1}{\sqrt[5]{x^2}} = \dfrac{1}{\sqrt[5]{x^2}} \cdot \dfrac{\sqrt[5]{x^3}}{\sqrt[5]{x^3}} = \dfrac{\sqrt[5]{x^3}}{\sqrt[5]{x^5}} = \dfrac{\sqrt[5]{x^3}}{x}$

When rationalizing expressions such as

$$\frac{7}{1 + \sqrt{3}} \quad \text{and} \quad \frac{19}{\sqrt{x} - \sqrt{y}}$$

we use the fact that $(a + b)(a - b) = a^2 - b^2$.

$\therefore \quad (1 + \sqrt{3})(1 - \sqrt{3}) = 1^2 - (\sqrt{3})^2 = 1 - 3 = -2$

and $(\sqrt{x} - \sqrt{y})(\sqrt{x} + \sqrt{y}) = (\sqrt{x})^2 - (\sqrt{y})^2 = x - y.$

Hence, to rationalize a sum we multiply it by the difference.
Similarly, to rationalize a difference we multiply it by the sum.

NOTE:

Expressions of the form x + y and x - y are called a pair of **conjugates**.

Example: $\sqrt{z} + \sqrt{3}$ and $\sqrt{z} - \sqrt{3}$ are a pair of conjugates.

As we discussed in unit 1, a complex number is an expression **a + bj**, where **a** is called the **real part** and **bj** the **imaginary part** of the complex number. The conjugate of the complex number a + bj is defined as the number a - bj.

Example: $5 - j\sqrt{17}$ and $5 + j\sqrt{17}$ are a pair of **complex conjugates**.

Example 1:

Rationalize $\dfrac{3}{1 + \sqrt{5}}$.

Solution:

Since the denominator is a sum, we multiply both numerator and denominator by the difference $(1 - \sqrt{5})$.

$$\therefore \quad \frac{3}{1 + \sqrt{5}} = \frac{3}{1 + \sqrt{5}} \cdot \frac{1 - \sqrt{5}}{1 - \sqrt{5}}$$

$$= \frac{3(1 - \sqrt{5})}{1^2 - (\sqrt{5})^2}$$

$$= \frac{3(1 - \sqrt{5})}{1 - 5}$$

$$= \frac{3(1 - \sqrt{5})}{-4}$$

$$= -\frac{3}{4}(1 - \sqrt{5}) \quad \text{or} \quad \frac{3\sqrt{5}}{4} - \frac{3}{4}$$

Example 2:

Rationalize $\dfrac{19}{\sqrt{x} - \sqrt{y}}$.

Solution:

Since the denominator is a difference, we multiply both denominator and numerator by the sum $(\sqrt{x} + \sqrt{y})$.

$$\frac{19}{\sqrt{x} - \sqrt{y}} = \frac{19}{\sqrt{x} - \sqrt{y}} \cdot \frac{\sqrt{x} + \sqrt{y}}{\sqrt{x} + \sqrt{y}}$$

$$= \frac{19(\sqrt{x} + \sqrt{y})}{(\sqrt{x})^2 - (\sqrt{y})^2}$$

$$= \frac{19(\sqrt{x} + \sqrt{y})}{x - y}$$

In general, to simplify expressions containing radicals, the following should be observed.

1. Remove all perfect n-th powers from a radical of index n.

2. Reduce, if possible, the index of the radical.

3. Rationalize all denominators.

CHECK POINT 6

Perform the indicated operations and write the answer in simplest form.

1. $\dfrac{\sqrt{32}}{\sqrt{2}}$

2. $\dfrac{\sqrt[3]{16x^4 y}}{\sqrt[3]{2x}}$

3. $\dfrac{\sqrt[4]{a^2 - b^2}}{\sqrt[4]{a - b}}$

4. $\dfrac{\sqrt{m^2 + m - 6}}{\sqrt{9m - 18}}$

5. $\dfrac{7}{\sqrt{a}}$

6. $\dfrac{3}{1 + 2\sqrt{y}}$

7. $\dfrac{x}{2\sqrt{y} - 5}$

8. $\dfrac{2m^2}{\sqrt[3]{4m}}$

DRILL EXERCISES

Reduce the following radicals to simplest form.

1. $\sqrt{32}$

2. $\sqrt[5]{64}$

3. $\sqrt[3]{48}$

4. $\sqrt{75}$

5. $\sqrt[4]{96}$

6. $\sqrt[3]{16}$

7. $\sqrt{24}$

8. $\sqrt[4]{162}$

9. $\sqrt{8x^3}$

10. $\sqrt[3]{a^5 b^7}$

11. $\sqrt[5]{128m^{10}n^8 t^{12}}$

12. $\sqrt[3]{54x^4 y^9}$

13. $\sqrt{p^3 r^6 t^5}$

14. $\sqrt[6]{64a^{11}b^{10}c^7}$

Reduce the index of the following.

15. $\sqrt[4]{49}$

16. $\sqrt[6]{25}$

17. $\sqrt[9]{64}$

18. $\dfrac{\sqrt[4]{81}}{\sqrt{3}}$

Perform the indicated operations and write the answer in simplest form.

19. $2\sqrt{3} + 5\sqrt{6}$

20. $5\sqrt{m} - 2\sqrt{m}$

21. $\sqrt{8t} + 4\sqrt{2t}$

22. $3\sqrt{a^5} + 2\sqrt{a}$

23. $7\sqrt[3]{x} + 3\sqrt[3]{xy} - 2\sqrt[3]{x}$

24. $\sqrt{75} + \sqrt[4]{9} - \sqrt{54}$

25. $\sqrt{3a^3 b} + a\sqrt{12ab} - \sqrt{27ab^3}$

26. $\sqrt[5]{64m^7 n^6} - \sqrt[5]{2m^2 n} + 3m\sqrt[5]{32m^2 n^6}$

27. $\dfrac{\sqrt{2t}}{5} + \sqrt{\dfrac{8t^3}{25}}$

28. $2x\sqrt{98x^5 y} + x^3\sqrt{18xy} - x^2\sqrt{50x^3 y}$

29. $m\sqrt{\dfrac{m}{4n^2}} - n\sqrt{\dfrac{m^3}{9n^4}}$

30. $7\sqrt[3]{a^6 b^4} - a^2 b\sqrt[3]{ab^2} - a^2 b\sqrt[3]{8b} + a\sqrt[3]{125a^4 b^5}$

268

Unit 17

DRILL EXERCISES (continued)

31. $\sqrt{2}\sqrt{18}$

32. $\sqrt{a}\;\sqrt[3]{a^2}$

33. $\sqrt{5x}\sqrt{5x}$

34. $(6\sqrt{t})^2$

35. $(-2\sqrt{27})(4\sqrt{3})$

36. $\left(\sqrt[3]{2}\right)\left(2\sqrt[3]{4}\right)$

37. $\left(\sqrt[5]{t^3}\right)\left(\sqrt[3]{t^2}\right)$

38. $\sqrt[5]{16}\;\sqrt[5]{4}$

39. $(3\sqrt{4})^2$

40. $\left(2\sqrt[3]{4}\right)^3$

41. $(a\sqrt{3})^2$

42. $\left(x\sqrt[5]{y}\right)^5$

43. $\sqrt[4]{2m^2}\cdot\sqrt[4]{8m^6}$

44. $3\sqrt{xy}\cdot\sqrt{x}$

45. $\sqrt[3]{4a^2}\cdot\sqrt[3]{16a^2}$

46. $\sqrt{\dfrac{3}{7}}\cdot\sqrt{\dfrac{7}{12}}$

47. $\sqrt{2}\cdot\sqrt{\dfrac{1}{162}}$

48. $\sqrt{\dfrac{3x}{5y}}\cdot\sqrt{\dfrac{12x^3}{5y^3}}$

49. $2\sqrt[3]{ab}\cdot\sqrt[3]{\dfrac{a^2}{b}}$

50. $\dfrac{1}{\sqrt{3(x+1)}}\cdot\sqrt{\dfrac{x^2-1}{12}}$

51. $\sqrt[4]{\dfrac{2m-1}{2m^3}}\cdot\sqrt[4]{\dfrac{2m+1}{8m}}$

52. $\left(t^2\sqrt[5]{2t^3r^2}\right)\left(r\sqrt[5]{3t^4r^4}\right)$

53. $\left(2v\sqrt[3]{3p^2}\right)\left(3p\sqrt[3]{18pv^2}\right)$

54. $\sqrt{3}(\sqrt{8}-2\sqrt{3})$

55. $\sqrt{x}(\sqrt{2}+3\sqrt{xy})$

56. $2\sqrt{t}(3\sqrt{2t}-5t\sqrt{2})$

57. $(2-3\sqrt{5})(2+3\sqrt{5})$

58. $(4-\sqrt{x})^2$

59. $(\sqrt{m}-3\sqrt{n})(\sqrt{m}+3\sqrt{n})$

60. $(2\sqrt{y}-1)(3\sqrt{2}+\sqrt{y})$

61. $(5\sqrt{2a}+\sqrt{b})(\sqrt{8a}+2\sqrt{b})$

62. $(2-\sqrt{6a})(3+\sqrt{7a})$

63. $(\sqrt{v}-\sqrt{t})^2$

64. $\dfrac{12}{\sqrt{3}}$

65. $\dfrac{2a}{\sqrt[3]{a}}$

66. $\dfrac{6xy}{\sqrt{2y}}$

67. $\sqrt{\dfrac{9x^3y}{4z}}$

68. $\sqrt[3]{\dfrac{9m}{3n^2}}$

69. $\sqrt{\dfrac{5a^3b^5}{8c^3}}$

70. $\dfrac{\sqrt[4]{3x^2-27}}{\sqrt[4]{3}\;\sqrt[4]{x+3}}$

71. $\dfrac{10}{2+3\sqrt{5}}$

269

DRILL EXERCISES (continued)

72. $\dfrac{5}{1 - \sqrt{3a}}$

73. $\dfrac{m - 2\sqrt{3}}{m + 2\sqrt{3}}$

74. $\dfrac{2a + 5\sqrt{b}}{3\sqrt{a}}$

75. $\dfrac{\sqrt{v}}{1 - \sqrt{v}}$

76. $\dfrac{\sqrt{t} - \sqrt{p}}{\sqrt{t} + \sqrt{p}}$

77. $\dfrac{3\sqrt{t} - 1}{3\sqrt{t} + 1}$

78. $\dfrac{\sqrt{a - b}}{b - \sqrt{a - b}}$

79. $\dfrac{\sqrt{x}}{2\sqrt{x} - 3\sqrt{y}}$

80. $\dfrac{\sqrt{m + 1}}{\sqrt{m + 1} - \sqrt{m}}$

81. $\dfrac{\sqrt{x^2 + 2x - 24}}{\sqrt{3x + 18}}$

ASSIGNMENT EXERCISES

Reduce the following to simplest form.

1. $\sqrt{18}$

2. $\sqrt[3]{128}$

3. $\sqrt[4]{48}$

4. $\sqrt{54}$

5. $\sqrt[5]{96}$

6. $\sqrt[3]{108}$

7. $\sqrt{200}$

8. $\sqrt[4]{243}$

9. $\sqrt{12y^3}$

10. $\sqrt[4]{p^6 t^9}$

11. $\sqrt[3]{48a^4 b^8 c^3}$

12. $\sqrt[5]{64x^7 y^{10} z^2}$

13. $\sqrt{a^4 b^3 c^6}$

14. $\sqrt[7]{128m^9 n^5 p^{14}}$

Reduce the index of the following.

15. $\sqrt[6]{27}$

16. $\sqrt[4]{64}$

17. $\dfrac{\sqrt[3]{125}}{\sqrt{5}}$

18. $\sqrt[12]{81}$

19. $2\sqrt[9]{27}$

20. $\dfrac{\sqrt[6]{343}}{\sqrt[4]{49}}$

Perform the indicated operations and write the answer in simplest form.

21. $3\sqrt{3b} - \sqrt{3b}$

22. $9\sqrt{2x} + \sqrt{50x}$

23. $2\sqrt{5} - 7\sqrt{10}$

24. $\sqrt{27a^5} - a\sqrt{12a^3}$

25. $3\sqrt{t} + 2\sqrt{t^7}$

26. $\sqrt[3]{m^2 n} - \sqrt[3]{mn} + 2\sqrt[3]{mn^2}$

ASSIGNMENT EXERCISES (continued)

27. $\sqrt[3]{54x^7y} - \sqrt{8xy} + x\sqrt[3]{16x^4y}$

28. $\sqrt{12} - \sqrt[6]{27} + \sqrt[6]{25}$

29. $r\sqrt{\dfrac{18r}{49}} + \dfrac{\sqrt{2r^3}}{7}$

30. $3b\sqrt[4]{32a^3b^6} - \sqrt[4]{2a^7b^2c^4}$

31. $4y\sqrt[3]{x^4y^2} - \sqrt[6]{x^{15}y^3} - x\sqrt{9x^3y} + x\sqrt[3]{xy^5}$

32. $\sqrt[3]{\dfrac{y^2}{27x^3}} - \sqrt[3]{\dfrac{8y^5}{x^6}}$

33. $\sqrt[3]{4} \cdot \sqrt[3]{12}$

34. $\sqrt[4]{a^3}\ \sqrt[4]{a}$

35. $\sqrt{3y} \cdot \sqrt{3y}$

36. $\left(2\ \sqrt[3]{m}\right)^3$

37. $(4\sqrt{t})^2$

38. $(6\sqrt{20})(-3\sqrt{5})$

39. $\left(-\sqrt[3]{16}\right)\left(2\ \sqrt[3]{4}\right)$

40. $\sqrt{a}\ \sqrt[3]{a^2}$

41. $\sqrt[5]{8}\ \sqrt[5]{12}$

42. $(5\sqrt{3})^2$

43. $\left(-3\ \sqrt[3]{4}\right)^3$

44. $\left(y\ \sqrt[4]{xy}\right)^4$

45. $(a^2\ \sqrt{7})^2$

46. $\sqrt[5]{m^6n^4} \cdot \sqrt[5]{64mn}$

47. $\sqrt[4]{27n^7} \cdot \sqrt[4]{3n^2}$

48. $\sqrt{9a} \cdot \sqrt{ab}$

49. $\sqrt{\dfrac{5}{3}} \cdot \sqrt{60}$

50. $\sqrt{\dfrac{2}{5}} \cdot \sqrt{\dfrac{5}{18}}$

51. $\sqrt{\dfrac{b^3}{3a}} \cdot \sqrt{\dfrac{8}{27ab}}$

52. $\sqrt{\dfrac{5}{x}} \cdot \sqrt[6]{\dfrac{64}{x^3}}$

ASSIGNMENT EXERCISES (continued)

53. $\sqrt{t^2 - r^2} \cdot \sqrt{\dfrac{20}{r + t}}$

54. $\sqrt[3]{\dfrac{6n^2}{m + 5}} \cdot \sqrt[3]{\dfrac{4n^2}{m^2 + 10m + 25}}$

55. $7\sqrt[5]{8x^4y^2} \cdot y\sqrt[5]{8x^3y^4}$

56. $\sqrt[4]{9x^6y^2} \cdot 5x\sqrt{12xy}$

57. $\sqrt{5}(3\sqrt{5} - \sqrt{8})$

58. $\sqrt{v}(\sqrt{7} + 2\sqrt{v^3y})$

59. $3\sqrt{t}(\sqrt{2t} - t^2\sqrt{8})$

60. $(x\sqrt{6} - 2)(x\sqrt{6} + 2)$

61. $(8 + \sqrt{d})^2$

62. $(3\sqrt{v} + \sqrt{t})(\sqrt{v} - 2\sqrt{t})$

63. $(2\sqrt{5a} - \sqrt{3b})(\sqrt{a} - 2\sqrt{15b})$

64. $(4\sqrt{y} - 2)(1 - \sqrt{y})$

65. $(\sqrt{2a} - \sqrt{3})(\sqrt{6a} + 8)$

66. $(\sqrt{c} - \sqrt{d})^2$

67. $\dfrac{20}{\sqrt{5}}$

68. $\dfrac{3x}{\sqrt[3]{x^2}}$

69. $\dfrac{6xy}{\sqrt{3y}}$

70. $\sqrt{\dfrac{36c}{25ab^3}}$

71. $\sqrt[3]{\dfrac{7w}{4v^2}}$

72. $\sqrt{\dfrac{11q^7t^2}{12r^3}}$

73. $\dfrac{\sqrt[5]{2t^2 - 32}}{\sqrt[5]{t - 4}\,\sqrt[5]{2}}$

74. $\dfrac{8}{3 - 2\sqrt{6}}$

75. $\dfrac{9}{\sqrt{2a} + 1}$

76. $\dfrac{\sqrt{x} + \sqrt{y}}{\sqrt{x} - \sqrt{y}}$

77. $\dfrac{5\sqrt{2} - b}{5\sqrt{2} + b}$

78. $\dfrac{4\sqrt{b} - 5a}{2\sqrt{b}}$

79. $\dfrac{\sqrt{m}}{\sqrt{m} + 3}$

80. $\dfrac{1 + 4\sqrt{x}}{1 - 4\sqrt{x}}$

81. $\dfrac{\sqrt{x + y} - x}{\sqrt{x + y}}$

82. $\dfrac{\sqrt{y}}{4\sqrt{x} + \sqrt{y}}$

83. $\dfrac{\sqrt{t - 1}}{\sqrt{t} + \sqrt{t - 1}}$

84. $\sqrt{\dfrac{4x^2 - 4x + 1}{25x + 100}}$

SOLVING QUADRATIC EQUATIONS BY THE METHOD OF COMPLETING THE SQUARE

Objectives: After having worked through this unit, the student should be capable of:

1.   solving quadratic equations by use of the method of "completing the square".

1

REVIEW

Previously, we have learned that the standard form for a quadratic equation in x is

$$ax^2 + bx + c = 0, \text{ where a, b and c are constants and } a \neq 0.$$

We call these equations quadratic because 2 is the highest exponent present when the equation is written in standard form.

Note:

The restriction $a \neq 0$ is necessary since, if $a = 0$ and $b \neq 0$, we have

$$0x^2 + bx + c = 0 \quad \text{or}$$
$$bx + c = 0$$

which is the general linear equation in x written in standard form.

2

Examples:

1. $2x^2 - 6x + 5 = 0$ is a quadratic equation in x, written in standard form where $a = 2$, $b = -6$ and $c = 5$.

2. $y^2 - 9 = 0$ is a quadratic equation in y, written in standard form where $a = 1$, $b = 0$ and $c = -9$.

3. $7t^2 + 2t = 0$ is a quadratic equation in t, written in standard form where $a = 7$, $b = 2$ and $c = 0$.

3

In a previous unit, we learned how to solve some quadratic equations by factoring. However, the quadratic equations we find when solving practical problems, in most cases, cannot be solved by factoring.

Hence, we shall discuss in this unit the method called "completing the square" which will enable us to solve any quadratic equation. Moreover, we shall use this method in the next unit to derive the quadratic formula. This formula is generally used to find the solution of any quadratic equation which cannot be solved by factoring.

4

THE SQUARE ROOT PROPERTY

In order to solve quadratic equations by the method of completing the square, we will use the following square root property:

If b is any number such that $b \geq 0$ and $x^2 = b$, then $x = \sqrt{b}$ or $x = -\sqrt{b}$.

To show that this is true, we solve by factoring the quadratic equation $x^2 = b$.

$$x^2 - b = 0$$
$$x^2 - (\sqrt{b})^2 = 0 \qquad b = \sqrt{b} \cdot \sqrt{b} = (\sqrt{b})^2$$
$$(x - \sqrt{b})(x + \sqrt{b}) = 0 \qquad \text{Factors of the difference of squares.}$$
$$\therefore \quad x - \sqrt{b} = 0 \quad \text{or} \quad x + \sqrt{b} = 0$$
$$x = \sqrt{b} \quad \text{or} \qquad x = -\sqrt{b}$$

This property tells us that we can take the square root of both sides of an equation as long as the numbers involved are positive or zero.

We normally write expressions such as "$x = \sqrt{b}$ or $x = -\sqrt{b}$" as $x = \pm\sqrt{b}$ which is read as "x is equal to plus or minus the square root of b".

Examples:

1. If $x^2 = 7$, then $x = \pm\sqrt{7}$ which means $x = \sqrt{7}$ or $x = -\sqrt{7}$.

2. If $t^2 = 64$, then $t = \pm 8$ which means $t = 8$ or $t = -8$.

3. If $(y + 4)^2 = 10$, then $y + 4 = \pm\sqrt{10}$ which means $y = -4 + \sqrt{10}$ or $y = -4 - \sqrt{10}$.

4. If $(5t - 2)^2 = 13$, then $5t - 2 = \pm\sqrt{13}$ which means $t = \dfrac{2 + \sqrt{13}}{5}$ or $t = \dfrac{2 - \sqrt{13}}{5}$.

CHECK POINT 1

Solve the following equations by taking the square root of both sides.

1. $V^2 = 5$

2. $R^2 = 121$

3. $(x - 6)^2 = 1$

4. $(3y + 4)^2 = 7$

PERFECT SQUARE TRINOMIAL

We call a trinomial which has two equal binomial factors a perfect square trinomial.

Examples:

1. $x^2 + 2x + 1 = (x + 1)^2$ Hence, $x^2 + 2x + 1$ is a perfect square trinomial.

2. $t^2 - 16t + 64 = (t - 8)^2$ Hence, $t^2 - 16t + 64$ is a perfect square trinomial.

3. $25m^2 - 10m + 1 = (5m - 1)^2$ Hence, $25m^2 - 10m + 1$ is a perfect square trinomial.

4. $y^2 + y + \dfrac{1}{4} = (y + \dfrac{1}{2})^2$ Hence, $y^2 + y + \dfrac{1}{4}$ is a perfect square trinomial.

Let us examine how the terms of a perfect square trinomial relate to each other.

$$(x + d)^2 = (x + d)(x + d)$$
$$= x^2 + dx + dx + d^2$$
$$= x^2 + 2dx + d^2$$

Note:

1. The coefficient of x^2 is 1.

2. The coefficient of x is 2d and the last term is d^2.

3. Half of the coefficient of x is $\frac{2d}{2}$ = d. If we square d we have d^2 which is the last term of the perfect square trinomial $x^2 + 2dx + d^2$. Hence, if we square half of 2d, the coefficient of the middle term, we obtain d^2, the last term of the perfect square trinomial.

Examples:

1. $x^2 + 8x + 16 = (x + 4)^2$

   $(\frac{1}{2} \cdot 8)^2 \longrightarrow 4^2$

2. $t^2 - 10t + 25 = (x - 5)^2$

   $\left(\frac{1}{2} \cdot (-10)\right)^2 \longrightarrow (-5)^2$

3. $y^2 + y + \frac{1}{4} = \left(y + \frac{1}{2}\right)^2$

   $\left(\frac{1}{2} \cdot 1\right)^2 \longrightarrow \left(\frac{1}{2}\right)^2$

Note:

This relationship is only valid when the coefficient of the squared term such as x^2, t^2 and y^2 is 1.

Example:

Construct a perfect square trinomial from $x^2 - \frac{3x}{2}$ by adding a constant term and write it as the square of a binomial.

Solution:

Note:

The coefficient of the squared term is 1.

Step 1. Identify the coefficient of x.

The coefficient of x is $-\frac{3}{2}$.

Step 2. Add to $x^2 - \dfrac{3x}{2}$ the square of half of $-\dfrac{3}{2}$ which is

$$\left[\dfrac{1}{2} \cdot \left(-\dfrac{3}{2}\right)\right]^2 = \left(-\dfrac{3}{4}\right)^2.$$

$$\therefore \quad x^2 - \dfrac{3x}{2} + \left(-\dfrac{3}{4}\right)^2 = x^2 - \dfrac{3x}{2} + \dfrac{9}{16}$$

Step 3. Write $x^2 - \dfrac{3x}{2} + \left(-\dfrac{3}{4}\right)^2$ as the square of a binomial.

$$x^2 - \dfrac{3x}{2} + \left(-\dfrac{3}{4}\right)^2 = \left(x - \dfrac{3}{4}\right)^2$$

Note:
Half of the coefficient of x is the second term in the squared binomial.

10

CHECK POINT 2

Construct perfect square trinomials from the given expressions by adding constant terms and write each as the square of a binomial.

1. $x^2 + 6x$ 2. $t^2 - 8t$ 3. $m^2 + m$ 4. $p^2 - \dfrac{p}{2}$

11

THE METHOD OF COMPLETING THE SQUARE
When the coefficient of the squared variable is 1, we can solve quadratic equations by following the given steps.

Step 1. Isolate the constant term on the right side of the equation.

Step 2. Construct a perfect square trinomial on the left side by adding to both sides of the equation $(\tfrac{1}{2}d)^2$. See Frame 8.

Step 3. Write the left side as a square binomial and solve the equation by taking the square root of both sides.

Example 1:

Solve $x^2 - 6x + 2 = 0$

Solution:

Step 1.   Isolate the constant term on the right side of the equation.

$$x^2 - 6x + 2 = 0$$
$$x^2 - 6x = -2$$

Step 2.   Construct a perfect square trinomial on the left side by adding to both sides of the equation

$$\left(\frac{1}{2} \cdot (-6)\right)^2$$

$$x^2 - 6x + (-3)^2 = -2 + (-3)^2$$

$$\left(\frac{1}{2} \cdot (-6)\right)^2 \uparrow\underline{\hspace{3cm}}\uparrow$$

Step 3.   Write the left side as a square of a binomial and solve the equation by taking the square root of both sides.

$$x^2 - 6x + (-3)^2 = -2 + 9$$
$$(x - 3)^2 = 7$$
$$x - 3 = \pm\sqrt{7}$$
$$x = 3 \pm \sqrt{7}$$

Hence, the solutions of $x^2 - 6x + 2 = 0$ are $x = 3 + \sqrt{7}$ or $x = 3 - \sqrt{7}$.

These solutions may also be written in decimal form: $x = 5.646$ or $x = 0.354$ (rounded to three decimal places).

Example 2:

Solve $t^2 + 12t - 9 = 0$.

Solution:

Step 1.   Isolate the constant term on the right side of the equation.

$$t^2 + 12t - 9 = 0$$
$$t^2 + 12t = 9$$

Step 2.   Construct a perfect square trinomial on the left side by adding to both sides of the equation

$$\left(\frac{1}{2} \cdot 12\right)^2 .$$

$$t^2 + 12t + 6^2 = 9 + 6^2$$

$$\left(\frac{1}{2} \cdot 12\right)^2 \uparrow\underline{\hspace{3cm}}\uparrow$$

Step 3. Write the left side as a square of a binomial and solve the equation by taking the square root of both sides.

$$t^2 + 12t + 6^2 = 9 + 36$$

$$(t + 6)^2 = 45$$

$$t + 6 = \pm\sqrt{45}$$

$$t = -6 \pm \sqrt{45}$$

$$= -6 \pm 3\sqrt{5}$$

Hence, the solutions of $t^2 + 12t - 9 = 0$ are $t = -6 + 3\sqrt{5}$ or $t = -6 - 3\sqrt{5}$.

This solution may also be written in decimal form: $t = 0.708$ or $t = -12.708$ (rounded to three decimal places).

14

CHECK POINT 3

Solve the following equations by the method of completing the square.

1. $x^2 + 8x - 9 = 0$ 2. $t^2 - 4t - 3 = 0$ 3. $p^2 + 18p - 7 = 0$

15

MORE EXAMPLES:

The preceding examples gave explanations and showed the detailed steps involved when solving quadratic equations by completing the square.

However, normally we display the solutions as shown in the following examples.

Example 1:
Solve $p^2 - 7p + 2 = 0$

Solution:
Step 1. $p^2 - 7p + 2 = 0$

$$p^2 - 7p = -2$$

Step 2. $p^2 - 7p + \left(-\dfrac{7}{2}\right)^2 = -2 + \left(-\dfrac{7}{2}\right)^2$

Step 3. $\left(p - \dfrac{7}{2}\right)^2 = -\dfrac{8}{4} + \dfrac{49}{4}$

$$\left(p - \dfrac{7}{2}\right)^2 = \dfrac{41}{4}$$

$$p - \dfrac{7}{2} = \pm\sqrt{\dfrac{41}{4}}$$

$$p = \dfrac{7}{2} \pm \sqrt{\dfrac{41}{4}}$$

Hence, $p = \dfrac{7 + \sqrt{41}}{2}$ or $p = \dfrac{7 - \sqrt{41}}{2}$.

In decimal form: $p = 6.702$ or $p = 0.298$ (rounded, three decimal places).

279

Note:

Solutions of some quadratic equations are complex numbers which were discussed in Unit 1.

As mentioned, to evaluate square roots of negative numbers, we define a number j such that $j = \sqrt{-1}$ and $j^2 = -1$. The number j is called the imaginary unit. Using this imaginary unit, we define the square root of a negative number as follows: $\sqrt{-b} = j\sqrt{b}$ where b is a positive real number. Numbers such as $\sqrt{-4} = 2j$ and $\sqrt{-17} = j\sqrt{17}$ are called imaginary numbers.

Example 2:
Solve $m^2 + 5m + 12 = 0$.

Solution:

Step 1. $m^2 + 5m + 12 = 0$

$$m^2 + 5m = -12$$

Step 2. $m^2 + 5m + (\frac{5}{2})^2 = -12 + (\frac{5}{2})^2$

Step 3. $(m + \frac{5}{2})^2 = -\frac{48}{4} + \frac{25}{4}$

$$(m + \frac{5}{2})^2 = -\frac{23}{4}$$

$$m + \frac{5}{2} = \pm\sqrt{\frac{-23}{4}}$$

$$m = -\frac{5}{2} \pm \frac{\sqrt{23}\,j}{2} \qquad\qquad \text{Note: } \sqrt{\frac{-23}{4}} = \sqrt{\frac{23\,j}{4}} = \frac{\sqrt{23}\,j}{2}$$

Hence, $m = \dfrac{-5 + \sqrt{23}\,j}{2}$ or $m = \dfrac{-5 - \sqrt{23}\,j}{2}$.

This solution may be written in decimal form: $m = -2.5 + 2.398j$ or $m = -2.5 - 2.398j$ (rounded to three decimal places).

Example 3:
Solve $K^2 - K + 1 = 0$.

Solution:

Step 1. $K^2 - K + 1 = 0$

$$K^2 - K = -1$$

Step 2. $K^2 - K + \left(-\frac{1}{2}\right)^2 = -1 + \left(-\frac{1}{2}\right)^2$ The coefficient of K is -1.

$$\left(K - \frac{1}{2}\right)^2 = \frac{-4}{4} + \frac{1}{4}$$

Step 3. $\left(K - \dfrac{1}{2}\right)^2 = -\dfrac{3}{4}$

$$K - \frac{1}{2} = \pm \sqrt{\frac{-3}{4}}$$

$$K = \frac{1}{2} \pm \sqrt{\frac{-3}{4}}$$

$$K = \frac{1}{2} \pm \frac{\sqrt{3}\,j}{2}$$

Hence, $K = \dfrac{1 + \sqrt{3}\,j}{2}$ or $K = \dfrac{1 - \sqrt{3}\,j}{2}$.

In decimal form: $K = 0.5 + 0.866j$ or $K = 0.5 - 0.866j$ (rounded to three decimal places).

18

CHECK POINT 4

Solve the following equations by the method of completing the square.

1. $x^2 - 5x + 1 = 0$ 2. $t^2 - t + 2 = 0$ 3. $E^2 + 3E + 10 = 0$

19

Examples where the coefficient of the squared variable is not 1.

Example 1:
  Solve $5x^2 - 6x - 11 = 0$.

Solution:
  The method of completing the square as shown previously can only be used if the coefficient of x^2 is 1. Hence, we first divide both sides of the equation by 5.

$$5x^2 - 6x - 11 = 0$$

$$\frac{5x^2}{5} - \frac{6x}{5} - \frac{11}{5} = \frac{0}{5}$$

$$x^2 - \frac{6x}{5} - \frac{11}{5} = 0$$

We now proceed using the method of completing the square.

  Step 1. $x^2 - \dfrac{6}{5}x - \dfrac{11}{5} = 0$ Isolate the constant.

$$x^2 - \frac{6}{5}x = \frac{11}{5}$$

  Step 2. $x^2 - \dfrac{6x}{5} + \left(-\dfrac{3}{5}\right)^2 = \dfrac{11}{5} + \left(-\dfrac{3}{5}\right)^2$ Complete the square.

Note: $\dfrac{1}{2} \cdot \left(-\dfrac{6}{5}\right) = -\dfrac{3}{5}$

Step 3. $\left(x - \dfrac{3}{5}\right)^2 = \dfrac{11}{5} + \dfrac{9}{25}$

$\left(x - \dfrac{3}{5}\right)^2 = \dfrac{64}{25}$

$x - \dfrac{3}{5} = \pm\sqrt{\dfrac{64}{25}}$ Take the square root of both sides.

$x - \dfrac{3}{5} = \pm\dfrac{8}{5}$

$x = \dfrac{3}{5} \pm \dfrac{8}{5}$

Hence, $x = \dfrac{11}{5}$ or $x = -1$.

20

Example 2:

Solve $-2n^2 + n - 1 = 0$.

Solution:
Divide both sides of $-2n^2 + n - 1 = 0$ by -2.

$$\dfrac{-2n^2}{-2} + \dfrac{n}{-2} - \dfrac{1}{-2} = \dfrac{0}{-2}$$

$$n^2 - \dfrac{n}{2} + \dfrac{1}{2} = 0$$

Use the method of completing the square to solve $n^2 - \dfrac{n}{2} + \dfrac{1}{2} = 0$.

Step 1. $n^2 - \dfrac{n}{2} = -\dfrac{1}{2}$

Step 2. $n^2 - \dfrac{n}{2} + \left(-\dfrac{1}{4}\right)^2 = -\dfrac{1}{2} + \left(-\dfrac{1}{4}\right)^2$ Note: $\dfrac{1}{2} \cdot \left(-\dfrac{1}{2}\right) = -\dfrac{1}{4}$.

Step 3. $\left(n - \dfrac{1}{4}\right)^2 = \dfrac{-8}{16} + \dfrac{1}{16}$

$\left(n - \dfrac{1}{4}\right)^2 = \dfrac{-7}{16}$

$n - \dfrac{1}{4} = \pm\sqrt{\dfrac{-7}{16}}$

$n = \dfrac{1}{4} \pm \sqrt{\dfrac{-7}{16}}$ or $n = \dfrac{1}{4} \pm \dfrac{\sqrt{7}j}{4}$

Hence, $n = \dfrac{1 + \sqrt{7}j}{4}$ or $n = \dfrac{1 - \sqrt{7}j}{4}$. (This may be written in decimal form as $0.25 + 0.661j$ or $0.25 - 0.661j$, rounded to three decimal places.)

Quadratic equations appear in many forms. Before they can be solved, we have to write them in standard form.

Example 3:
Solve $2t^2 - (5t + 2) = 4(2 - 3t) - t^2$.

Solution:

Step 1. Write the equation in standard form.

$$2t^2 - (5t + 2) = 4(2 - 3t) - t^2$$

$$2t^2 - 5t - 2 = 8 - 12t - t^2$$

$$3t^2 + 7t - 10 = 0$$

Step 2. Divide both sides by the coefficient of t^2.

$$3t^2 + 7t - 10 = 0$$

$$\frac{3t^2}{3} + \frac{7t}{3} - \frac{10}{3} = \frac{0}{3}$$

$$t^2 + \frac{7}{3}t - \frac{10}{3} = 0$$

Step 3. Use the method of completing the square to solve

$$t^2 + \frac{7}{3}t - \frac{10}{3} = 0.$$

$$t^2 + \frac{7}{3}t = \frac{10}{3} \qquad \textbf{Note: } \frac{1}{2} \cdot \frac{7}{3} = \frac{7}{6}.$$

$$t^2 + \frac{7}{3}t + \left(\frac{7}{6}\right)^2 = \frac{10}{3} + \left(\frac{7}{6}\right)^2$$

$$\left(t + \frac{7}{6}\right)^2 = \frac{120}{36} + \frac{49}{36}$$

$$\left(t + \frac{7}{6}\right)^2 = \frac{169}{36} \qquad \text{Take the square root of both sides.}$$

$$t + \frac{7}{6} = \pm \sqrt{\frac{169}{36}}$$

$$t = -\frac{7}{6} \pm \frac{13}{6}$$

Hence, $t = 1$ or $t = \frac{-10}{3} = -3.333$ (rounded to three decimal places.).

CHECK POINT 5

Solve the following equations by the method of completing the square.

1. $3x^2 - 6x - 5 = 0$

2. $-6t^2 + t + 1 = 0$

3. $-m^2 + 2(5m + 1) = 4m - 3(m^2 - 1)$

DRILL EXERCISES

I. Solve the following equations by taking the square root of both sides.

1. $t^2 = 1$

2. $x^2 = 4\pi p$

3. $(m - 3)^2 = \dfrac{1}{9}$

4. $\left(y + \dfrac{1}{2}\right)^2 = \dfrac{49}{4}$

5. $\left(2n - \dfrac{1}{6}\right)^2 = \dfrac{7}{36}$

6. Solve for R: $(R + \pi)^2 = \dfrac{h}{2}$

II. Construct perfect square trinomials from the given expressions.

7. $x^2 - 4x$

8. $y^2 + 10y$

9. $m^2 - m$

10. $t^2 + \dfrac{t}{2}$

11. $p^2 - \dfrac{2p}{3}$

12. $x^2 + ax$

13. $y^2 - ky$

14. $n^2 + n$

II. Solve the following quadratic equations by the method of completing the square.

15. $x^2 + 5x - 24 = 0$

16. $y^2 - 20y + 51 = 0$

17. $m^2 + 7m + 3 = 0$

18. $t^2 + t - 1 = 0$

19. $E^2 + 2E + \dfrac{1}{2} = 0$

20. $x^2 - 3x + 7 - 0$

21. $y^2 + 6y + 2 = 0$

22. $n^2 - 5n + 1 = 0$

23. $R^2 - \dfrac{R}{2} - 5 = 0$

24. $x^2 + \dfrac{7x}{3} + 6 = 0$

25. $2y^2 - 6y + 3 = 0$

26. $4t^2 + 20t - 11 = 0$

27. $9m^2 - 12m - 1 = 0$

28. $5E^2 + 8E + 3 = 0$

DRILL EXERCISES (continued)

29. $3n^2 - 5n + 9 = 0$ 30. $4x^2 = 5x + 7$

31. $9y^2 + 2 = 6y$ 32. $11t - 5 = 3t^2$

33. $3 = 3m^2 + 5m$ 34. $3x = 2(x^2 - 5)$

35. $5(y^2 + 1) = 4(2 - y)$

36. $7x^2 - 2(x + 3) = 3(x^2 - 1) + 2x$

37. $11p - 4(1 - p^2) = 2(1 - p) + 2p^2$

38. $t(3t - 7) = 2(3t^2 + 2) - t$

39. $(V - 2)(V + 3) + 9 = 4V(V + 1)$

ASSIGNMENT EXERCISES

I. Solve the following equations by taking the square root of both sides.

1. $k^2 = 8$ 2. $y^2 = 36r\theta$

3. $\left(2 + x\right)^2 = \dfrac{25}{4}$ 4. $\left(n - \dfrac{1}{3}\right)^2 = \dfrac{16}{9}$

5. $\left(\dfrac{t}{2} + \dfrac{4}{5}\right)^2 = \dfrac{11}{25}$ 6. $(P - 1)^2 = \dfrac{\pi}{3}$

II. Construct perfect square trinomials from the given expressions.

7. $a^2 + 6a$ 8. $n^2 - 5n$

9. $y^2 + y$ 10. $r^2 - \dfrac{r}{3}$

11. $x^2 - \dfrac{5x}{2}$ 12. $b^2 - b$

13. $p^2 - tp$ 14. $x^2 + kx$

ASSIGNMENT EXERCISES (continued)

III. Solve the following quadratic equations by the method of completing the square.

15. $x^2 + 3x - 54 = 0$

16. $b^2 - 10b - 3 = 0$

17. $a^2 - 4a = -8$

18. $n^2 - 22n + 117 = 0$

19. $m^2 - 3m + 1 = 0$

20. $x^2 - 7x - 2 = 0$

21. $b^2 + 1 = -b$

22. $t^2 - \dfrac{2t}{5} - 1 = 0$

23. $R^2 + \dfrac{3R}{2} + \dfrac{9}{64} = 0$

24. $d^2 + \dfrac{15d}{2} = 4$

25. $3P^2 + 10P - 8 = 0$

26. $6a^2 - 13a + 6 = 0$

27. $2y^2 - 11y + 12 = 0$

28. $2E^2 - 4E - 3 = 0$

29. $9m^2 - 3m + 12 = 0$

30. $6k^2 + 5k - 4 = 0$

31. $3t^2 - 4t - \dfrac{4}{3} = 0$

32. $5b^2 = 2b + 3$

33. $3x - 4(x^2 + 1) = 2(1 - x) - 5x^2$

34. $y(y + 3) - 4y = 4(y + 3)$

35. $0 = 2d(d - 2) + 1$

36. $6t^2 + 7(t + 1) = t(3 + 5t) + 13$

37. $(R - 1)(R + 4) = 3R(R + 2)$

38. $-4(2 - x^2) = 6(x^2 - x) + 1$

THE QUADRATIC FORMULA SOLVING EQUATIONS WITH RADICALS

Objectives: After having worked through this unit, the student should be capable of:

1.   using the quadratic formula to solve quadratic equations;
2.   solving certain equations containing radicals.

1

THE QUADRATIC FORMULA

In the previous unit we discussed how to solve any quadratic equation by the method of completing the square.

This method is used to obtain a general formula which provides in a simple way the solutions to any quadratic equation. To derive the formula we solve the general quadratic equation $ax^2 + bx + c = 0$ for x by the method of completing the square.

2

DERIVATION OF THE QUADRATIC FORMULA

If a, b and c are constants, $a \neq 0$ and $a > 0$, solve $ax^2 + bx + c = 0$ for x.

Step 1.   Divide both sides of the equation by a, the coefficient of x^2.

$$ax^2 + bx + c = 0$$

$$\frac{ax^2}{a} + \frac{bx}{a} + \frac{c}{a} = \frac{0}{a}$$

$$x^2 + \frac{bx}{a} + \frac{c}{a} = 0$$

Step 2.   Subtract $\frac{c}{a}$ from both sides.

$$x^2 + \frac{b}{a} x = -\frac{c}{a}$$

Step 3.   Add to both sides $\left(\frac{1}{2} \cdot \frac{b}{a}\right)^2$ to obtain a perfect square trinomial on the left side.

$$x^2 + \frac{b}{a} x + \left(\frac{b}{2a}\right)^2 = \left(\frac{b}{2a}\right)^2 - \frac{c}{a}$$

Step 4.   Rewrite the left side as the square of a binomial and add the fractions on the right side.

$$x^2 + \frac{b}{a} x + \left(\frac{b}{2a}\right)^2 = \left(\frac{b}{2a}\right)^2 - \frac{c}{a}$$

$$\downarrow$$

$$\left(x + \frac{b}{2a}\right)^2 = \frac{b^2 - 4ac}{4a^2}$$

Step 5.   Take the square root of both sides.

$$x + \frac{b}{2a} = \pm \sqrt{\frac{b^2 - 4ac}{4a^2}}$$

Step 6.   Add $-\frac{b}{2a}$ to both sides and simplify.

$$x = -\frac{b}{2a} \pm \frac{\sqrt{b^2 - 4ac}}{2a}$$

$$x = \frac{-b \pm \sqrt{b^2 - 4ac}}{2a}$$

Hence, the solutions to any quadratic equation written in the standard form $ax^2 + bx + c = 0$ are

$$x = \frac{-b + \sqrt{b^2 - 4ac}}{2a} \quad \text{or} \quad x = \frac{-b - \sqrt{b^2 - 4ac}}{2a}$$

where a is the coefficient of x^2,
         b is the coefficient of x,
         c is the constant term.

3

USING THE QUADRATIC FORMULA

When using the quadratic formula to solve quadratic equations, we follow the given steps;

Step 1. Write the given equation in standard form.

Step 2. Identify the values of a, b and c.

Step 3. Substitute the values of a, b and c into the formula and evaluate.

4

Example 1:

Solve $x^2 = 2x + 15$.

Solution:

Step 1. Write the equation in standard form: $x^2 - 2x - 15 = 0$.

Step 2. Identify: a, the coefficient of x^2, a = 1
               b, the coefficient of x, b = -2
               c, the constant term, c = -15.

Step 3. Substitute and evaluate.

$$x = \frac{-b \pm \sqrt{b^2 - 4ac}}{2a}$$

$$= \frac{-(-2) \pm \sqrt{(-2)^2 - 4(1)(-15)}}{2(1)}$$

$$= \frac{2 \pm \sqrt{4 + 60}}{2}$$

$$= \frac{2 \pm 8}{2}$$

Hence, the solutions of $x^2 = 2x + 15$ are $x = \frac{2 + 8}{2}$ or $x = \frac{2 - 8}{2}$

$$= 5 \qquad\qquad = -3$$

289

Note:

This equation could have been solved more quickly by using the factoring method.

$$x^2 - 2x - 15 = 0$$

$$(x + 3)(x - 5) = 0$$

$$\therefore \qquad x = -3 \quad \text{or} \quad x = 5$$

However, the equations in the following example cannot be solved by factoring

---
5

Example 2:

Solve $x^2 + 9x + 2 = 0$.

Solution:

Step 1.   The equation is in standard form.

Step 2.   Identify: a, the coefficient of x^2, a = 1,

b, the coefficient of x,   b = 9,

c, the constant term,   c = 2.

Step 3.   Substitute and evaluate.

$$x = \frac{-b \pm \sqrt{b^2 - 4ac}}{2a}$$

$$= \frac{-9 \pm \sqrt{9^2 - 4(1)(2)}}{2(1)}$$

$$= \frac{-9 \pm \sqrt{81 - 8}}{2}$$

$$= \frac{-9 \pm \sqrt{73}}{2}$$

Hence, the solutions of $x^2 + 9x + 2 = 0$ are $x = \dfrac{-9 + \sqrt{73}}{2}$ or $x = \dfrac{-9 - \sqrt{73}}{2}$
The solutions in decimal form are $x = -0.23$ or $x = -8.77$ (rounded to two decimal places.)

6

Example 3:

Solve $p(3p - 1) = 10$.

Solution:

Step 1.   Write the equation in standard form.

$$p(3p - 1) = 10$$

$$3p^2 - p - 10 = 0$$

Step 2.   Identify: a, the coefficient of p^2, a = 3,

b, the coefficient of p, b = -1,

c, the constant term,   c = -10.

Step 3.   Substitute and evaluate.

$$p = \frac{-b \pm \sqrt{b^2 - 4ac}}{2a}$$

$$= \frac{-(-1) \pm \sqrt{(-1)^2 - 4(3)(-10)}}{2(3)}$$

$$= \frac{1 \pm \sqrt{121}}{6} \quad or \quad \frac{1 \pm 11}{6}$$

Hence, the solutions of $p(3p - 1) = 10$ are $p = 2$ or $p = -\frac{5}{3}$.

7

CHECK POINT 1

Use the quadratic formula to solve the following:

1. $2m^2 + 5m + 3 = 0$ 2. $E^2 - E = 1$ 3. $3x = 1 - 10x^2$

8

MORE EXAMPLES

As discussed in the previous unit, solutions of some quadratic equations are complex numbers. We use the imaginary unit $j = \sqrt{-1}$ to change numbers such as $\sqrt{-16}$ to $4j$ or $\sqrt{-31}$ to $\sqrt{31}j$.

The preceding examples give explanations and show the detailed steps involved in using the quadratic formula. However, normally we display the solutions as shown in the following examples.

Example 1:
Solve $I^2 + 10I + 3000 = 0$.

Solution:
a = 1, b = 10, c = 3000

$$I = \frac{-b \pm \sqrt{b^2 - 4ac}}{2a}$$

$$= \frac{-10 \pm \sqrt{(10)^2 - 4(1)(3000)}}{2(1)}$$

$$= \frac{-10 \pm \sqrt{-11\,900}}{2}$$

$$= \frac{-10 \pm \sqrt{11\,900}\,j}{2}$$

Hence, $I = \frac{-10 + \sqrt{11\,900}\,j}{2}$ or $= \frac{-10 - \sqrt{11\,900}\,j}{2}$.

The approximate values are $-5 + 54.5j$ or $-5 - 54.5j$.

Example 2:

Solve $80 = -16t^2 + 90t$.

Solution:

Since $80 = -16t^2 + 90t$, we have $16t^2 - 90t + 80 = 0$.

Hence, $a = 16$, $b = -90$, $c = 80$.

$$t = \frac{-b \pm \sqrt{b^2 - 4ac}}{2a}$$

$$= \frac{-(-90) \pm \sqrt{(-90)^2 - 4(16)(80)}}{2(16)} = \frac{90 \pm \sqrt{2980}}{32}$$

Hence, $t = \dfrac{90 + \sqrt{2980}}{32}$ or $t = \dfrac{90 - \sqrt{2980}}{32}$

The approximate values in decimal form are 4.518 or 1.107 (rounded to three decimal places).

Example 3:

Solve $\dfrac{1}{3} = \dfrac{3u^2}{5} - \dfrac{u}{2}$.

Solution:

Since $\dfrac{1}{3} = \dfrac{3u^2}{5} - \dfrac{u}{2}$, we have $\dfrac{-3u^2}{5} + \dfrac{u}{2} + \dfrac{1}{3} = 0$.

$$\frac{3u^2}{5} - \frac{u}{2} - \frac{1}{3} = 0 \qquad \text{Multiply both sides by } -1.$$

Note:

Since it is generally easier to work with whole numbers than fractions, we first clear the above equation of fractions.

The LCD of 5, 2 and 3 is 30.

$$\therefore \quad 30\left(\frac{3u^2}{5} - \frac{u}{2} - \frac{1}{3}\right) = 30(0) \quad \text{Multiply both sides by the LCD of 5, 2 and 3.}$$

$$18u^2 - 15u - 10 = 0$$

$a = 18$, $b = -15$, $c = -10$

$$u = \frac{-b \pm \sqrt{b^2 - 4ac}}{2a}$$

$$= \frac{-(-15) \pm \sqrt{(-15)^2 - 4(18)(-10)}}{2(18)} \qquad = \frac{15 \pm \sqrt{945}}{36}$$

Hence, $u = \dfrac{15 + 3\sqrt{105}}{36} = \dfrac{\cancel{3}^{1}(5 + \sqrt{105})}{\cancel{3}_{1}(12)} = \dfrac{5 + \sqrt{105}}{12}$ or

$$u = \dfrac{15 - 3\sqrt{105}}{36} = \dfrac{\cancel{3}^{1}(5 - \sqrt{105})}{\cancel{3}_{1}(12)} = \dfrac{5 - \sqrt{105}}{12}$$

The approximate values in decimal form are 1.27 or −0.44 (rounded to two decimal places).

11

CHECK POINT 2

Solve the following equations. Give the approximate answers in decimal form rounded to three decimal places.

1. $3R^2 - 50R + 1200 = 0$ 2. $145 = 100t - 16t^2$ 3. $\dfrac{E}{6} = \dfrac{4}{3} + \dfrac{2E^2}{9}$

12

When solving practical problems involving quadratic equations, we encounter mostly coefficients that are not whole numbers. Using the quadratic formula as shown previously is the normal way of obtaining solutions for these problems.

Example 1:
Solve $0.12C^2 + 0.9C - 3.4 = 0$

Solution:
$a = 0.12$, $b = 0.9$, $c = -3.4$

$$c = \frac{-b \pm \sqrt{b^2 - 4ac}}{2a}$$

$$= \frac{-0.9 \pm \sqrt{(0.9)^2 - 4(0.12)(-3.4)}}{2(0.12)} = \frac{-0.9 \pm \sqrt{2.442}}{0.24}$$

Hence, $c = \dfrac{-0.9 + \sqrt{2.442}}{0.24}$ or $c = \dfrac{-0.9 - \sqrt{2.442}}{0.24}$.

The approximate values in decimal form are 2.8 or -10.3 (rounded to one decimal place).

13

Example 2:
Solve $7 = 2\pi r^2 + 5\pi r$.

Solution:
Since $7 = 2\pi r^2 + 5\pi r$, we have $-2\pi r^2 - 5\pi r + 7 = 0$.

$2\pi r^2 + 5\pi r - 7 = 0$ Multiply both sides by -1.

$a = 2\pi$, $b = 5\pi$, $c = -7$

$$r = \frac{-b \pm \sqrt{b^2 - 4ac}}{2a}$$

$$= \frac{-5\pi \pm \sqrt{(5\pi)^2 - 4(2\pi)(-7)}}{2(2\pi)}$$

$$= \frac{-5\pi \pm \sqrt{25\pi^2 + 56\pi}}{4\pi}$$

Hence, $r = \dfrac{-5\pi + \sqrt{25\pi^2 + 56\pi}}{4\pi}$ or $r = \dfrac{-5\pi - \sqrt{25\pi^2 + 56\pi}}{4\pi}$

Using $\pi = 3.14$, the approximate values in decimal form are 0.39 or -2.89 (rounded to two decimal places).

Example 3:

   Solve $1 = \sqrt{6}\, x - \sqrt{2}\, x^2$.

Solution:

   Since $1 = \sqrt{6}\, x - \sqrt{2}\, x^2$, we have $\sqrt{2}\, x^2 - \sqrt{6}\, x + 1 = 0$.

   $a = \sqrt{2}, \; b = -\sqrt{6}, \; c = 1$

$$x = \frac{-b \pm \sqrt{b^2 - 4ac}}{2a}$$

$$= \frac{-(-\sqrt{6}) \pm \sqrt{(-\sqrt{6})^2 - 4(\sqrt{2})(1)}}{2(\sqrt{2})}$$

$$= \frac{\sqrt{6} \pm \sqrt{6 - 4\sqrt{2}}}{2\sqrt{2}}$$

Hence, $x = \dfrac{\sqrt{6} + \sqrt{6 - 4\sqrt{2}}}{2\sqrt{2}}$ or $x = \dfrac{\sqrt{6} - \sqrt{6 - 4\sqrt{2}}}{2\sqrt{2}}$

The approximate values in decimal form are 1.073 or 0.659 (rounded to three decimal places).

Example:

   Solve $LI^2 + RI + \dfrac{1}{C} = 0$ for I.

Solution:

   Clear the equation of the fraction $\dfrac{1}{C}$.

$$LI^2 + RI + \frac{1}{C} = 0$$

$$C\left(LI^2 + RI + \frac{1}{C}\right) = C \cdot 0$$

$$CLI^2 + CRI + 1 = 0$$

   $a = CL, \; b = CR, \; C = 1$

$$I = \frac{-b \pm \sqrt{b^2 - 4ac}}{2a}$$

$$= \frac{-CR \pm \sqrt{(CR)^2 - 4(CL)(1)}}{2CL}$$

$$= \frac{-CR \pm \sqrt{C^2R^2 - 4CL}}{2CL}$$

Hence, $I = \dfrac{-CR + \sqrt{C^2R^2 - 4CL}}{2CL}$ or $I = \dfrac{-CR - \sqrt{C^2R^2 - 4CL}}{2CL}$.

CHECK POINT 3

Solve the following equations. Give the approximate values in decimal form rounded to three decimal places.

1. $0.015 = 2.5V - 12.6V^2$ 2. $128.6 = 7.5\pi r + 2\pi r^2$

3. $\sqrt{5}\ R^2 - 7 = \sqrt{7}\ R$ 4.   Solve for t: $d = vt - 16t^2$

SOLVING EQUATIONS CONTAINING RADICALS

In the remainder of this unit we shall learn how to solve some equations with radicals.

I   **Examples of equations containing one radical.**

1. $\sqrt{x + 1} = 7$ 2. $5 - \sqrt[3]{2y - 1} = 0$ 3. $\sqrt{2p - 1} + 2 = p$

II   **Examples of equations containing two square root terms.**

1. $\sqrt{3x + 5} - \sqrt{2x + 7} = 0$ 2. $\sqrt{t} = \sqrt{8t + 9} - 3$

To solve equations shown in the preceding examples, we make use of the following property:

(19.1) | If $x = a$,   then $x^n = a^n$ where $a > 0$

REVIEW: THE INDEX OF A RADICAL

The index or order of a radical is the number on the radical sign indicating which root is to be taken.

SOLVING EQUATIONS CONTAINING ONE RADICAL

To solve equations containing one radical, we use the following steps.

Step 1.   Isolate the radical term on one side of the equation.

Step 2.   Raise both sides of the equation to an exponent equal to the index of the radical.

Step 3.   Solve the resulting equation and check for extraneous roots.

EXTRANEOUS ROOTS

As mentioned previously, certain methods of solving equations may introduce solutions which will not check in the original equation.

Such solutions are called extraneous solutions. When solving equations with radicals, extraneous solutions are introduced when we raise both sides of the equation to an exponent in order to eliminate the radical sign.

Hence, it is important to check all solutions in the original equation when we solve equations containing radicals.

Example 1:

Solve $2\sqrt{x + 1} - 3 = 0$.

Solution:

Step 1. Isolate the radical term on one side of the equation.

$$2\sqrt{x + 1} - 3 = 0$$
$$2\sqrt{x + 1} = 3$$

Step 2. Since the index of a square root is 2, we raise both sides to the exponent 2.

$$(2\sqrt{x + 1})^2 = 3^2$$
$$4(x + 1) = 9$$

Step 3. Solve the equation and check the solution.

$$4x + 4 = 9$$
$$4x = 5$$
$$x = \frac{5}{4}$$

Check: If $x = \frac{5}{4}$, $2\sqrt{x + 1} - 3 = 2\sqrt{\frac{5}{4} + 1} - 3 = 2(\frac{3}{2}) - 3 = 0$.

Hence, $x = \frac{5}{4}$ is the solution of $2\sqrt{x + 1} - 3 = 0$.

Example 2:

Solve $\sqrt[4]{5m - 9} = 2$.

Solution:

Step 1. Isolate the radical term on one side of the equation.

$$\sqrt[4]{5m - 9} = 2$$

Step 2. Since the index of the radical is 4, we raise both sides to the exponent of 4.

$$\left(\sqrt[4]{5m - 9}\right)^4 = (2)^4$$
$$5m - 9 = 16$$

Step 3. Solve the equation and check the solution. $5m - 9 = 16$

$$m = 5$$

Check: If $m = 5$, $\sqrt[4]{5m - 9} = \sqrt[4]{25 - 9} = \sqrt[4]{16} = 2$.

Hence, $m = 5$ is the solution of $\sqrt[4]{5m - 9} = 2$.

23

Example 3:

Solve $\sqrt{2p - 1} - p + 2 = 0$.

Solution:

Step 1.   Isolate the radical term on one side of the equation.

$$\sqrt{2p - 1} - p + 2 = 0$$
$$\sqrt{2p - 1} = p - 2$$

Step 2.   Raise both sides to the exponent 2.

$$(\sqrt{2p - 1})^2 = (p - 2)^2$$
$$2p - 1 = p^2 - 4p + 4$$

Step 3.   Solve the equation and check the solution.

$$2p - 1 = p^2 - 4p + 4$$

$-p^2 + 6p - 5 = 0$ Multiply both sides by –1.

$p^2 - 6p + 5 = 0$ Factor.

$(p - 1)(p - 5) = 0$

$\therefore \quad p = 1 \quad$ or $\quad p = 5$

Check: If $p = 1$, $\sqrt{2p - 1} - p + 2 = \sqrt{2 - 1} - 1 + 2 = 1 - 1 + 2 = 2 \neq 0$.

If $p = 5$, $\sqrt{2p - 1} - p + 2 = \sqrt{10 - 1} - 5 + 2 = 3 - 5 + 2 = 0$.

Hence, $p = 1$ is an extraneous solution and $p = 5$ is the only solution of $\sqrt{2p - 1} - p + 2 = 0$.

24

CHECK POINT 4

Solve the following.

1. $\sqrt{2V - 7} - 7 = 0$ 2. $\sqrt[3]{4x + 25} = 5$ 3. $1 = 2\sqrt{x + 4} - x$

25

SOLVING EQUATIONS WITH TWO SQUARE ROOT TERMS

To solve equations with two square root terms, we use the following steps.

1.   Write one square root term on each side of the equation.

2.   Square both sides of the equation.

3.   If the result contains a square root term, isolate this term on one side of the equation and square both sides again.

4.   Solve the resulting equation and check for extraneous solutions.

Example 1:

Solve $\sqrt{3x + 5} - \sqrt{2x + 7} = 0$.

Solution:

Step 1.   Write one square root term on each side of the equation.

$$\sqrt{3x + 5} - \sqrt{2x + 7} = 0$$
$$\sqrt{3x + 5} = \sqrt{2x + 7}$$

Step 2.   Square both sides.

$$(\sqrt{3x + 5})^2 = (\sqrt{2x + 7})^2$$
$$3x + 5 = 2x + 7$$

Step 3.   Solve.

$$3x + 5 = 2x + 7$$
$$x = 2$$

Check: If $x = 2$, $\sqrt{3x + 5} - \sqrt{2x + 7} = \sqrt{6 + 5} - \sqrt{4 + 7} = \sqrt{11} - \sqrt{11} = 0$

Hence, $x = 2$ is the solution of $\sqrt{3x + 5} - \sqrt{2x + 7} = 0$.

Example 2:

Solve $2\sqrt{t} - \sqrt{8t + 9} + 3 = 0$.

Solution:

Step 1.   Write square root terms on opposite sides of the equation.

$$2\sqrt{t} - \sqrt{8t + 9} + 3 = 0$$
$$2\sqrt{t} + 3 = \sqrt{8t + 9}$$

Step 2.   Square both sides of the equation.

$$(2\sqrt{t} + 3)^2 = (\sqrt{8t + 9})^2$$
$$4t + 12\sqrt{t} + 9 = 8t + 9$$

Step 3.   Isolate the square root term $12\sqrt{t}$ on one side of the equation and again square both sides.

$$4t + 12\sqrt{t} + 9 = 8t + 9$$
$$12\sqrt{t} = 4t$$
$$(12\sqrt{t})^2 = (4t)^2$$
$$144t = 16t^2$$

Step 4. Write the quadratic equation in standard form, solve it and check the solutions.

$$16t^2 - 144t = 0$$

$$16t(t - 9) = 0$$

$$t = 0 \quad \text{or} \quad t = 9.$$

Check: If $t = 0$, $2\sqrt{t} - \sqrt{8t + 9} + 3 = 2\sqrt{0} - \sqrt{8(0) + 9} + 3 = 0 - 3 + 3 = 0$.

If $t = 9$, $2\sqrt{t} - \sqrt{8t + 9} + 3 = 2\sqrt{9} - \sqrt{8(9) + 9} + 3 = 6 - 9 + 3 = 0$.

Hence, both $t = 0$ and $t = 9$ are solutions of $2\sqrt{t} - \sqrt{8t + 9} + 3 = 0$.

28

Solutions are normally presented as shown in the following example.

Example 3:
Solve $\sqrt{3p - 5} - \sqrt{p + 7} = 2$.

Solution:

$$\sqrt{3p - 5} - \sqrt{p + 7} = 2$$

$$\sqrt{3p - 5} = 2 + \sqrt{p + 7} \qquad \text{write square root terms on opposite sides.}$$

$$(\sqrt{3p - 5})^2 = (2 + \sqrt{p + 7})^2 \qquad \text{square both sides.}$$

$$3p - 5 = 4 + 4\sqrt{p + 7} + p + 7$$

$$2p - 16 = 4\sqrt{p + 7} \qquad \text{isolate square root terms on one side.}$$

$$2(p - 8) = 4\sqrt{p + 7} \qquad \text{simplify by dividing both sides by 2.}$$

$$p - 8 = 2\sqrt{p + 7}$$

$$(p - 8)^2 = (2\sqrt{p + 7})^2 \qquad \text{square both sides again.}$$

$$p^2 - 16p + 64 = 4(p + 7)$$

$$p^2 - 20p + 36 = 0$$

$$(p - 2)(p - 18) = 0$$

$$\therefore \quad p = 2 \quad \text{or} \quad p = 18$$

Check: If $p = 2$, $\sqrt{3p - 5} - \sqrt{p + 7} = \sqrt{6 - 5} - \sqrt{2 + 7} = 1 - 3 = -2 \neq 2$.

If $p = 18$, $\sqrt{3p - 5} - \sqrt{p + 7} = \sqrt{54 - 5} - \sqrt{18 + 7} = 7 - 5 = 2$.

Hence, $p = 2$ is an extraneous root and $p = 18$ is the only solution of $\sqrt{3p - 5} - \sqrt{p + 7} = 2$.

CHECK POINT 5

Solve the following.

1. $\sqrt{2x + 3} - \sqrt{x + 8} = 0$

2. $\sqrt{m + 27} + 2\sqrt{m} = 0$

3. $\sqrt{6t + 13} - 3 = t$

4. $\sqrt{7 - p} - \sqrt{p + 13} + 2 = 0$

DRILL EXERCISES

I   Use the quadratic formula to solve the following equations. Give approximate answers in decimal form rounded to two decimal places.

1. $x^2 - 5x + 2 = 0$

2. $3t^2 - t = 3$

3. $v^2 + 4v + 1 = 0$

4. $5E^2 = 7 - E$

5. $1 = p(p - 1)$

6. $2R^2 - 3R = 6$

7. $35 = 6d^2 - 23d$

8. $2(3y^2 + 1) = y$

9. $8x(2x - 1) = -1$

10. $V^2 + 100V + 3000 = 0$

11. $250 = 500t - 16t^2$

12. $\dfrac{d}{5} = \dfrac{3d^2}{4} + \dfrac{1}{2}$

13. $\dfrac{p^2}{2} - \dfrac{2p}{7} + 1 = 0$

14. $\dfrac{3x^2}{5} + \dfrac{x}{2} - \dfrac{3}{4} = 0$

15. $\dfrac{e - 1}{3} = \dfrac{e^2}{2}$

16. $0.12Z^2 - 0.5Z + 0.2 = 0$

17. $0.95 = 0.5\pi r + 2\pi r^2$

18. $0.5R^2 = 2(1 - 0.5R)$

19. Solve for t: $K(1 - t)^2 = d$

20. $\sqrt{2}(x^2 - 2x) - \sqrt{7} = 0$

21. $t^2 - \sqrt{3}t + \sqrt{6} = 0$

22. $\dfrac{V^2}{2} + \sqrt{5}V = \dfrac{3}{4}$

23. $(p - 3)^2 + (p - 3) + 1 = 0$

24. $\dfrac{y^2}{2} = 2.4y + 1.2$

25. $\pi r(2r + 8.9) = 16.5$

26. $\dfrac{E^2}{5(1 + 0.2)} = 8(E - 0.2)$

DRILL EXERCISES (continued)

II Solve the following equations.

27. $\sqrt{x + 5} = 3$

28. $\sqrt{3y - 1} = 5$

29. $\sqrt{2p - 1} - 10 = 0$

30. $\sqrt[3]{t - 1} - 5 = 0$

31. $2\sqrt[4]{3p - 1} = 3$

32. $\sqrt[5]{y - 7} - 2 = 0$

33. $2\sqrt{x} - 1 = 9$

34. $\sqrt{10t - 1} - \sqrt{4t + 7} = 0$

35. $\sqrt{x} - \sqrt{3x - 4} = 0$

36. $\sqrt{3p + 4} + \sqrt{9p - 7} = 0$

37. $x = \sqrt{2x + 7} - 2$

38. $\sqrt{5t - 26} - t + 4 = 0$

39. $2 + \sqrt{R + 4} - \sqrt{4R + 5} = 0$

40. $\sqrt{E + 1} = 1 - 2\sqrt{E}$

41. $\sqrt{p + 7} + 1 - \sqrt{3p - 2} = 0$

42. $1 + \sqrt{4m + 3} = \sqrt{2m + 5}$

43. $2 + \sqrt{2n - 4} - \sqrt{3n + 4} = 0$

44. $\sqrt{2y - 1} = 2 - \sqrt{y}$

45. $\sqrt{x + 4} - \sqrt{x + 7} = 1$

III Solve the following word problems.

46. A rectangular name plate for a machine has an area of 44 cm^2 and a perimeter of 27 cm. Calculate the dimensions of the plate.

47. The hypotenuse of a right-angled triangle is 34.0 m. Find the lengths of the remaining two sides if one of these remaining sides is 14.0 m longer than the other.

DRILL EXERCISES (continued)

48. In an electrical circuit, we must connect two resistors in parallel so that their total resistance is $2\tilde{0}$ ohms. One of the two resistors must be $1\tilde{0}$ ohms greater than the other. Find the size of the two resistors.

The formula relating total resistance to two resistances in parallel is: $\dfrac{1}{R_t} = \dfrac{1}{R_1} + \dfrac{1}{R_2}$

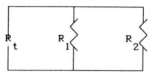

49. The pressure and volume of a gas are related by $p = \dfrac{100}{v}$.

If the volume is increased by 2, the pressure is diminished by 2. Find p and v.

50. The figure at the right shows an electric circuit with resistance (R ohms), an inductance coil (L henrys), and a capacitor (C farads) connected in series to a voltage source. In determining the current in the circuit, it is necessary to solve first the equation for m:

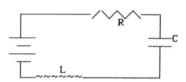

$$Lm^2 + Rm + \frac{1}{C} = 0 \ .$$

(a) Find m if R = 10 ohms, L = 1 henry, and C = 1/10025 farad.
(b) Solve for m in terms of the litaeral coefficients L, R and C.

51. The figure below is a cross-section of an L-shaped steel bar. The outside dimensions are 4.00 cm and 2.50 cm. Because of the load which the bar must support, the area of the cross-section must be 1.75 cm^2. Find the width "w".

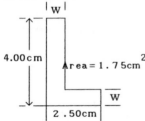

52. Ice on a power transmission line may be melted off by increasing the voltage. The following equation gives the relation between the voltage E in the line, the temperature t ($^\circ$F) of the wire and the air temperature t_0 ($^\circ$F):

$$\frac{E^2}{a(1 + bt)} = K(t - t_0)$$

If E is 580 volts, t_0 is 24°F, K(the coefficient of heat transfer) is 350 units, a (the resistance of the line at 0°F) is 52 ohms and b (the temperature coefficient of resistance of the wire) is 0.003 units, find t.

ASSIGNMENT EXERCISES

I   Use the quadratic formula to solve the following equations. Give approximate answers in decimal form rounded to two decimal places.

1. $y^2 + 3y - 10 = 0$ 2. $2x^2 - 7 = x$

3. $-R^2 + 2R + 1 = 0$ 4. $b^2 - 6b + 10 = 0$

5. $a(2a - 3) = -2$ 6. $E = 8 - E^2$

7. $20p = 4p^2 + 3$ 8. $-2(1 - x^2) = x$

9. $3t(4t - 2) = -1$ 10. $30y - 4y^2 = 12$

11. $\dfrac{x^2}{2} - 3x = \dfrac{5}{4}$ 12. $\dfrac{3d^2}{5} = \dfrac{1}{2}d - 1$

13. $\dfrac{2k + 1}{7} = \dfrac{k^2}{3}$ 14. $\dfrac{m}{4} - \dfrac{2}{3} = \dfrac{5m^2}{6}$

15. $0.78a^2 + 1.5a - 6.4 = 0$ 16. $2.7\pi - 0.49x = 8\pi x^2$

17. $3(5 - 1.6z) = 0.4z^2$ 18. $3(y - 1)^2 = 2$

19.   Solve for E: $\dfrac{1}{2}K = (2 - E)^2$ 20. $\sqrt{3}(1 - 2x^2) = \sqrt{6}x$

21. $\sqrt{5}R^2 + R - \sqrt{2} = 0$ 22. $\dfrac{t^2}{3} + \sqrt{7}t = \dfrac{2}{5}$

23. $y - 6 = 1 - (y + 2)^2$ 24. $\dfrac{3d^2}{4} = 2.3 - 1.7d$

25. $\pi x(2.8 - 3x) = 14.5$ 26. $0.2(P + 1) = \dfrac{P^2}{P - 0.5}$

ASSIGNMENT EXERCISES (continued)

II   Solve the following equations.

27. $6 = \sqrt{5t - 4}$ 28. $\sqrt{2R - 7} = 3$

29. $\sqrt{9 - 8p} + 1 = 0$ 30. $\sqrt[3]{E + 2} + 3 = 0$

31. $3\sqrt[4]{4b + 1} = 5$ 32. $\sqrt[6]{x - 9} - 2 = 0$

33. $5 - 4\sqrt{m} = 3$ 34. $\sqrt{2y - 5} = \sqrt{1 - 7y}$

35. $\sqrt{6k - 4} - \sqrt{k} = 0$ 36. $\sqrt{7t + 1} + \sqrt{3t - 2} = 0$

37. $\sqrt{5E + 10} + 1 = 2E$ 38. $2b - 1 - \sqrt{11b + 5} = 0$

39. $\sqrt{p - 4} + 2 - \sqrt{2p - 1} = 0$ 40. $\sqrt{3R + 4} = 2 + \sqrt{R}$

41. $\sqrt{n - 2} = \sqrt{2n + 3} + 2$ 42. $\sqrt{t} - \sqrt{3t - 2} = -2$

43. $\sqrt{3m + 1} + \sqrt{3m - 2} = 3$ 44. $\sqrt{7y - 2} = \sqrt{y + 1} + \sqrt{3}$

45. $\sqrt{k + 4} + 1 - \sqrt{3k + 1} = 0$

III Solve the following word problems.

46.   A circular swimming pool is surrounded by a walk 2 m wide. The area
        of the walk is 11/25 of the area of the pool. Find the radius of the
        pool to the nearest centimetre.

47.   The reactance of the circuit
        shown is given by the equation

        $X = wL - \dfrac{1}{wC}$ where w is the angular frequency in rad/s,

        L is in henries, C is in farads and X is in ohms.

        Calculate w if $X = 50 \ \Omega$ when $L = 0.15$ henries   and

        $C = 42 \times 10^{-6}$ farads.

48.   The hypotenuse of a right angled triangle is 80 cm. The length of
        one leg exceeds the length of the other leg by 10 cm. Find the
        length of the two legs to the nearest centimetre.

ASSIGNMENT EXERCISES (continued)

49. A picture frame of uniform width has outer dimensions 12 cm by 15cm. Find the width of the frame if 100 cm^2 of picture shows. Round to one decimal place.

50. A square lawn is surrounded by a walk 4.00 m wide. The area of the lawn is equal to the area of the walk. Find the length of the side of the lawn correct to the nearest tenth of a metre.

51. In calculating the depth of a column footing, the following equation was encountered:

    $(0.400 \times 0.6 \times \sqrt{25})\ (4 \times d \times [0.400 + d]) \times 10^3 = 353[3.1^2 - (0.400 + d)^2]$

    Solve for d.

52. For a certain electrical circuit, the power delivered to a device is given by the equation

    $P = EI - I^2R$

    where P is the power in watts (w), E is the applied voltage in volts (v) and R is a series resistance in ohms (Ω).

    (a) Calculate I if P = 60 w, E = 120 v and R = 46 Ω.
    (b) Calculate I if P = 1200 w, E = 120 v and R = 2 Ω.

53. The temperature t of explosion (in degrees centigrade) of nitroglycerin is given by a solution of the equation:

    $(\alpha + bt)t = 1000\ Q_{mv} + 15(\alpha + 15b)$

    where, for nitroglycerin, $\alpha = 43.705$; $b = 0.01775$; $Q_{mv} = 346.5$.

    Determine the temperature t. Also give the result in degrees absolute T, where $T = t + 273^\circ$.

54. For ammonium picrate the data for the formula in Exercise 53 are as follows: $\alpha = 54.04$; $b = 0.01212$; $Q_{mv} = 175.3$.

    Determine the temperature of explosion.

55. The "L-shape" figure below has uniform width and an area of 6.25 cm^2. Determine the width of the figure.

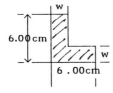

ASSIGNMENT EXERCISES (continued)

56. A contractor has to put a swimming pool into a backyard 20.0 m by 30.0 m. He is to use the entire yard for the swimming pool, leaving a uniform strip of grass around the pool. If the pool is to have an area equal to that of the grassed area around it, what will be the dimensions of the pool?

57. If two resistors R_1 and R_2 are connected in parallel,

the resultant parallel resistance $R_p = \dfrac{R_1 R_2}{R_1 + R_2}$.

If they are connected in series,

the resultant series resistance $R_s = R_1 + R_2$.

(a) Calculate values of R_1 and R_2 if the resultant parallel resistance is 20 Ω and the resultant series resistance is 80 Ω.

(b) Calculate values of R_1 and R_2 if $R_p = 10$ Ω and $R_s = 60$ Ω.

58. Determine the dimensions of a rectangle whose perimeter is 30.0 m and whose area is 55.25 m .

MODULE 5

Logarithms and Graphs of Functions

Table of contents   **Page**

UNIT 20

LOGARITHMS
BASIC CONCEPTS AND LAWS

Objectives: After having worked through this unit, the student should be capable of:

1.   understanding the concept of logarithms;
2.   changing expressions from exponential to logarithmic form and vice versa;
3.   using the laws of logarithms to rewrite expressions;
4.   understanding the terms common logarithms and natural logarithms;
5.   using the laws of logarithms to solve literal equations;
6.   changing the form of logarithms and exponential equations.

1

INTRODUCTION

Logarithm is another word for exponent.

When working with exponential expressions, we encounter the following situations.

1.	Evaluate the exponential expression.	$10^3 = x$ (Powers)
2.	Evaluate the base.	$x^3 = 27$ (Roots)
3.	Evaluate the exponent.	$10^x = 100$ (Logarithms)

2

POWERS

We refer to 10, $10^2 = 100$, $10^3 = 1000$, $10^4 = 10\ 000$, etc, as powers of ten. Hence, when evaluating exponential expressions, we find powers of a given base.

Examples:

1. $10^3 = 10 \times 10 \times 10 = 1000$ is a power of ten.

2. $2^5 = 2 \times 2 \times 2 \times 2 \times 2 = 32$ is a power of two.

3. $5^2 = 5 \times 5 = 25$ is a power of five.

3

ROOTS

Evaluating the base of an exponential expression is the same as finding a root.

Examples:

1. $x^3 = 1000$

   $(x^3)^{1/3} = (1000)^{1/3}$ Raise both sides to the exponent 1/3.

   $x^1 = \sqrt[3]{1000}$

   $x = 10$

2. $a^5 = 32$

   $(a^5)^{1/5} = (32)^{1/5}$ Raise both sides to the exponent 1/5.

   $a^1 = \sqrt[5]{32}$

   $a = 2$

3. $m^2 = 25$

   $(m^2)^{1/2} = (25)^{1/2}$ Raise both sides to the exponent 1/2.

   $m^1 = \sqrt{25}$

   $m = 5$

LOGARITHMS

Evaluating the exponent of an exponential expression is the same as finding a logarithm.

Example:

Find the exponent (logarithm) in $10^x = 1000$.

Discussion:

The question in this logarithm problem is:

"To what exponent (x) do we have to raise the base 10 to obtain the number 1000?"

This question can also be written in mathematics as

$$\log_{10} 1000 = x$$

We read this as

"The log to the base 10 of 1000 is x."

Hence, $\log_{10} 1000 = x$ means the same as $10^x = 1000$.

Since $10^3 = 1000$, we have $\log_{10} 1000 = 3$.

When we write $10^x = 1000$, we say the question is written in **exponential form**.

When we write $\log_{10} 1000 = x$. we say the question is written in **logarithmic form**.

CHANGING FROM EXPONENTIAL TO LOGARITHMIC FORM

Examples:

1. $3^4 = 81$ means $\log_3 81 = 4$ (The log to the base 3 of 81 is 4.)

2. $16^{1/4} = 2$ means $\log_{16} 2 = \frac{1}{4}$ (The log to the base 16 of 2 is 1/4.)

3. $5^3 = 125$ means $\log_5 125 = 3$ (The log to the base —— of 125 is ——.)

4. $2^6 = 64$ means   ————————   (The log to the base —— of 64 is ——.)

Answers: 3.   The log to the base 5 of 125 is 3.

4. $\log_2 64 = 6$, the log to the base 2 of 64 is 6.

6

More Examples

1. $7^{-2} = \dfrac{1}{49}$ means $\log_7(1/49) = -2$ (The log to the base 7 of 1/49 is -2.)

2. $10^{-4} = 0.0001$ means $\log_{10}0.0001 = -4$ (The log to the base 10 of 0.0001 is -4.)

3. $2^{-3} = \dfrac{1}{8}$ means $\log_2(1/8) = -3$ (The log to the base 2 of 1/8 is -3.)

7

CHECK POINT 1

Change the following to logarithmic form.

1. $2^4 = 16$ 2. $100^{1/2} = 10$ 3. $5^4 = 625$

4. $3^{-2} = \dfrac{1}{9}$ 5. $10^{-3} = 0.001$ 6. $4^7 = 16\ 384$

8

CHANGING FROM LOGARITHMIC TO EXPONENTIAL FORM

Examples:

1. $\log_2 32 = 5$ means $2^5 = 32$ (2 to the exponent 5 is 32)

2. $\log_{10}10\ 000 = 4$ means $10^4 = 10\ 000$ (10 to the exponent 4 is 10 000)

3. $\log_8 2 = \dfrac{1}{3}$ means $8^{1/3} = 2$ (____ to the exponent ____ is 2)

4. $\log_{10}0.01 = -2$ means $10^{-2} = 0.01$ (____ to the exponent ____ is ____)

5. $\log_5(1/125) = -3$ means   ____ = ____   (____ to the exponent ____ is $\dfrac{1}{125}$)

Answers:

3. 8 to the exponent 1/3 is 2.

4. 10 to the exponent -2 is 0.01.

5. $5^{-3} = \dfrac{1}{125}$, 5 to the exponent -3 is $\dfrac{1}{125}$.

9

CHECK POINT 2

Change the following to exponential form.

1. $\log_3 81 = 4$ 2. $\log_{10} 100 = 2$ 3. $\log_2 128 = 7$

4. $\log_{10} 0.1 = -1$ 5. $\log_{81} 9 = \dfrac{1}{2}$ 6. $\log_8 (1/64) = -2$

10

Definition

If $a^x = N$, where $a > 0$ and $a \neq 1$, then x is the logarithm to the base a of N.

Hence, the logarithm of the number N is the exponent to which we have to raise the base to obtain the number N.

$\therefore$ $\log_a N = x$ means $a^x = N$ where $a > 0$, $a \neq 1$ and $N > 0$.

Note:

1. Logarithms are **not** defined for negative numbers (N).

2. Logarithms are only defined when the base a is positive and not equal to 1.

3. Both $\log_a N = x$ and $a^x = N$ have the same meaning.

    We refer to
    $\log_a N = x$ as the **logarithmic form** and

    $a^x = N$ as the **exponential form**.

11

Examples:

1. $\log_2 (-128) = x$ is **not defined** since N = -128.

2. $\log_{-6} 100 = x$ is **not defined** since a = -6.

3. $\log_1 25 = x$ is **not defined** since a = 1.

Note:
Since $a^0 = 1$ for $a > 0$, we have $\log_a 1 = 0$.

4. $\log_6 1 = 0$ since $6^0 = 1$.

5. $\log_{10} 1 = 0$ since $10^0 = 1$.

6. $\log_{1/2} 1 = 0$ since $(\frac{1}{2})^0 = 1$.

12

CHECK POINT 3

Indicate which of the following expressions **are** defined. For those which are **not** defined, state the reason.

1. $\log_{10}(-100) = x$ 2. $\log_2 1 = x$ 3. $\log_{-5} 125 = x$

4. $\log_1 10 = x$ 5. $\log_0 1 = x$ 6. $\log_5 0 = x$

Change the following expressions to exponential form.

7. $\log_2 V = 0.5$ 8. $\log_e(x + 1) = \pi$ 9. $\log_{10} P = 2y$

Change the following expressions to logarithmic form.

10. $e^{2\pi} = R$ 11. $10^{2K} = V + 1$ 12. $2^{x+1} = 64$

13

Since logarithms are exponents, they have certain properties which are based on the laws of exponents.

In what follows, we shall refer to the following laws of exponents.

(3.1) $a^p \cdot a^q = a^{p+q}$

(3.2) $\dfrac{a^p}{a^q} = a^{p-q}$

(3.3) $(a^p)^k = a^{pk}$

14

LAWS OF LOGARITHMS

Using the fact that $a^p \cdot a^q = a^{p+q}$, we can show that

$\log_a(m \cdot n) = \log_a m + \log_a n$,   that is, the logarithm of a product equals the sum of the logarithms of its factors.

To prove this, we let $x = \log_a m$ and $y = \log_a n$.

Since $x = \log_a m$ means $a^x = m$ and $y = \log_a n$ means $a^y = n$,

$$m \cdot n = a^x \cdot a^y$$
$$= a^{x+y}$$

Changing $m \cdot n = a^{x+y}$ to logarithmic form, we obtain

$$\log_a(m \cdot n) = x + y$$

Since we let $x = \log_a m$ and $y = \log_a n$, we have

(20.1) $\boxed{\log_a(m \cdot n) = \log_a m + \log_a n}$

Example:

Since $2^5 = 32$ means $\log_2 32 = 5$

$2^2 = 4$ means $\log_2 4 = 2$

and $2^3 = 8$ means $\log_2 8 = 3$

we have $\log_2 (4 \times 8) = \log_2 32 = 5$

and $\log_2 4 + \log_2 8 = 2 + 3 = 5$

Hence, $\log_2 (4 \times 8) = \log_2 4 + \log_2 8$.

Using $\dfrac{a^m}{a^n} = a^{m-n}$, it can be shown in a similar way as before that

(20.2) $\boxed{\log_a \left(\dfrac{m}{n}\right) = \log_a m - \log_a n}$

that is, the logarithm of a quotient equals the logarithm of the numerator minus the logarithm of the denominator.

Example:

Since $2^5 = 32$ means $\log_2 32 = 5$

$2^2 = 4$ means $\log_2 4 = 2$

and $2^3 = 8$ means $\log_2 8 = 3$

we have $\log_2 \left(\dfrac{32}{4}\right) = \log_2 8 = 3$

and $\log_2 32 - \log_2 4 = 5 - 2 = 3$

Hence, $\log_2 \left(\dfrac{32}{4}\right) = \log_2 32 - \log_2 4$.

Using $(a^m)^k = a^{mk}$, it can be shown that

(20.3) $\boxed{\log_a m^k = k \log_a m}$

that is, the logarithm of a number to the exponent k equals the product of the exponent k times the logarithm of the number.

Example:

Since $2^6 = 64$ means $\log_2 64 = 6$

$2^3 = 8$ means $\log_2 8 = 3$

we have $\log_2 8^2 = \log_2 64 = 6$

and $2 \log_2 8 = 2(3) = 6$

Hence, $\log_2 8^2 = 2 \log_2 8$.

18

Since $a = a^1$, we have that $\log_a a = 1$

and $\log_a a^K = K \cdot \log_a a = K(1) = K$

Hence, (20.4) $\boxed{\log_a a^K = K}$

19

SUMMARY LAWS OF LOGARITHMS

(20.1) $\log_a (m \cdot n) = \log_a m + \log_a n$

(20.2) $\log_a \left(\dfrac{m}{n}\right) = \log_a m - \log_a n$

(20.3) $\log_a m^k = k \log_a m$

(20.4) $\log_a a^K = K$

Note: (a) In each equation, all logarithms must have the **same** base.
(b) Law (20.1) can be extended to more than two factors.

20

USING LAWS OF LOGARITHMS TO REWRITE EXPRESSIONS
It is often desirable to rewrite sums and differences of logarithms as the logarithm of one expression or vice versa. However, to do so, the logarithms involved **must** have the **same** base.

Examples using (20.1) $\log_a (m \cdot n) = \log_a m + \log_a n$

1. $\log_{10}(7x) = \log_{10}7 + \log_{10}x$

2. $\log_2 5 + \log_2 6 = \log_2(5 \times 6) = \log_2 30$

3. $\log_5(2\pi r) = \log_5 2 + \log_5 \pi + \log_5 r$

4. $\log_8 \sqrt{3} + \log_8 y = \log_8(\sqrt{3}y)$

5. $\log_e(x^2 - 1) = \log_e[(x - 1)(x + 1)] = \log_e(x - 1) + \log_e(x + 1)$

6. $\log_3 7 + \log_5 3$ cannot be rewritten since the logarithms **do not have the same base.**

Examples using (20.2) $\log_a\left(\dfrac{m}{n}\right) = \log_a m - \log_a n$

1. $\log_2 45 - \log_2 15 = \log_2\left(\dfrac{45}{15}\right) = \log_2 3$

2. $\log_e\left(\dfrac{6}{z}\right) = \log_e 6 - \log_e z$

3. $\log_{10}(xy) - \log_{10}y = \log_{10}\left(\dfrac{xy}{y}\right) = \log_{10}x$

4. $\log_4\left(\dfrac{y + 1}{y - 1}\right) = \log_4(y + 1) - \log_4(y - 1)$

Examples using (20.3) $\log_a m^k = k \log_a m$

and (20.4) $\log_a a^K = K$

1. $\log_{10}7^5 = 5 \log_{10}7$

2. $\log_e z^{0.5} = 0.5 \log_e z$

3. $1.75 \log_2 V = \log_2 V^{1.75}$

4. $\log_{10}10^x = x$

5. $(x - 2) \log_e 10 = \log_e 10^{x-2}$

6. $\log_e e = 1$

7. $\log_5(0.12)^y = y \log_5(0.12)$

8. $3.5 \log_6 6 = 3.5$

CHECK POINT 4

Use the laws of logarithms to write each of the following as a difference, a sum or a product. (Simplify, if possible.)

1. $\log_{10}(2 \times 5)$

2. $\log_e\left(\dfrac{4}{y}\right)$

3. $\log_2 5^3$

4. $\log_3 3^z$

5. $\log_5(3xy)$

6. $\log_{10}\left(\dfrac{x}{10}\right)$

7. $\log_e(4e)$

8. $\log_8\left(\dfrac{1}{x}\right)$

Write the sums and the differences of logarithms as the logarithms of one expression. (Simplify, if possible.)

9. $\log_2 6 + \log_2 3$

10. $\log_5(8x) - \log_5 8$

11. $\log_e 5 + \log_e y$

Evaluate the following:

12. $7 \log_5 5$

13. $\log_e e^{0.5}$

14. $\log_{10}10^{x+1} = 3$

(Solve for x.)

COMMON AND NATURAL LOGARITHMS

In our previous examples, we discussed logarithms and the laws of logarithms referring to many different logarithm bases.

In what follows, we shall restrict our discusion to logarithms to the base 10 and the base e (e = 2.718...).

Logarithms to base 10 are called **common logarithms**.
Logarithms to base e are called **natural logarithms**.

When we write log x, we mean $\log_{10} x$, the common logarithm.

When we write ln x, we mean $\log_e x$, the natural logarithm.

Hence, $\log_{10} x = \log x$ and $\log_e x = \ln x$.

REVIEW

LITERAL EQUATIONS AND FORMULAS

An equation which contains two or more **different** letters representing variables or constants is called a literal equation.

A formula is a literal equation which expresses a relationship between two or more mathematical or physical quantities.

USING LAWS OF LOGARITHMS TO SOLVE LITERAL EQUATIONS

When solving literal equations for a particular variable or expression, we often make use of the laws of logarithms.

Example 1:
Solve $a = b \log(5x)$ for log x.

Solution:

$a = b \log(5x)$	Divide both sides by b.
$\dfrac{a}{b} = \log(5x)$	Use Law 20.1 to obtain separate log terms for the constant 5 and the variable x.
$\dfrac{a}{b} = \log 5 + \log x$	Solve for log x.

$$-\log x = \log 5 - \frac{a}{b}$$

$$\log x = -\log 5 + \frac{a}{b} \quad \text{or} \quad \frac{a - b \log 5}{b}$$

Example 2:

Solve $\ln \sqrt{x} = \ln (5y^3)$ for $\ln y$.

Solution:

$\ln \sqrt{x} = \ln (5y^3)$ Use Law 20.1 to separate the constant 5 and the variable y^3.

$\ln \sqrt{x} = \ln 5 + \ln y^3$

$\ln \sqrt{x} = \ln 5 + 3 \ln y$ Use Law 20.3 and solve for $\ln y$.

$-3 \ln y = \ln 5 - \ln \sqrt{x}$

$\ln y = \dfrac{\ln 5 - \ln \sqrt{x}}{-3}$ or $\dfrac{\ln \sqrt{x} - \ln 5}{3}$

Example 3:

Solve $10 = \log \sqrt{\dfrac{R}{V}}$ for $\log V$.

Solution:

$10 = \log \sqrt{\dfrac{R}{V}}$ <u>Note</u>: $\sqrt{\dfrac{R}{V}} = \left(\dfrac{R}{V}\right)^{1/2}$

$10 = \dfrac{1}{2} \log \left(\dfrac{R}{V}\right)$ Use Law 20.3

$10 = \dfrac{1}{2}[\log R - \log V]$ Use Law 20.2 to obtain a separate log term for V.

$10 = \dfrac{1}{2} \log R - \dfrac{1}{2} \log V$ Solve for $\log V$.

$\dfrac{1}{2} \log V = \dfrac{1}{2} \log R - 10$ Multiply by LCD = 2.

$\log V = 2(\dfrac{1}{2} \log R - 10) = \log R - 20$

Example 4:

Solve $\log(x + 3) - \log(x^2 + 3x) = 3 \log a$ for $\log x$.

Solution:

$\log(x + 3) - \log(x^2 + 3x) = 3 \log a$

$\log\left(\dfrac{x + 3}{x^2 + 3x}\right) = 3 \log a$ Use Law 20.2 to combine $x + 3$ and $x^2 + 3x$ for simplification.

$\log\left(\dfrac{\overset{1}{\cancel{x + 3}}}{x\underset{1}{(\cancel{x + 3})}}\right) = 3 \log a$ Factor and simplify.

$$\log \left(\frac{1}{x}\right) = 3 \log a \qquad \text{Use Law 20.2 to obtain separate log terms for the constant and the variable.}$$

$$\log 1 - \log x = 3 \log a \qquad \underline{\text{Remember}}: \log 1 = 0$$

$$0 - \log x = 3 \log a \qquad \text{Solve for } \log x.$$

$$\log x = -3 \log a$$

37

CHECK POINT 6

Use the appropriate laws of logarithms to solve the given equations.

1. Solve for log x: $\log \sqrt[3]{x^2} = 10$

2. Solve for log a: $2 \log(am) = 5 \log n$

3. Solve for ln R: $3 = \ln\left(\pi \sqrt{\frac{P}{R}}\right)$

4. Solve for ln x: $\ln(\pi x^2) - \ln(3x) = 2 \ln \pi$

38

REVIEW

If $a^x = N$, where $a > 0$ and $a \neq 1$, then x is the logarithm to the base a of N.

Hence, the logarithm of the number N is the exponent to which we have to raise the base to obtain the number N.

$\therefore \quad \log_a N = x \quad$ means $\quad a^x = N \quad$ and

$\qquad a^x = N \qquad$ means $\quad \log_a N = x.$

39

Examples:

1. $\log R = 3.5$ means $10^{3.5} = R$

2. $e^{2\pi} = K$ means $\ln K = 2\pi.$

3. $0.5x = \log V$ means $10^{0.5x} = V$

4. $e^{-1.75+y} = x^2 + 1$ means $-1.75 + y = \ln(x^2 + 1)$

40

CHECK POINT 7

Change exponential forms to logarithmic forms and vice versa.

1. $10^{2x} = 529$ 2. $e^{y+1} = \pi + K$ 3. $2^{x+5} = 128$

4. $\log_{10} V = z - 1$ 5. $-1.2 + x = \ln R$ 6. $2\pi = \log_2(p + 2)$

41

When changing more complex expressions from logarithmic to exponential form or vice versa, we first change the expressions to the form

$$\log_a N = x \quad \text{or} \quad a^x = N.$$

This may involve combining logarithmic or exponential terms using the appropriate laws.

42

Example 1:
   Change $T = \ln \pi + a \ln P$ to exponential form.

Solution:

$T = \ln \pi + a \ln P$ Change to the form $x = \log_a N.$

$T = \ln \pi + \ln P^a$ Law 20.3

$T = \ln(\pi P^a)$ Law 20.1   Note: $\ln x = \log_e x$

        Change to exponential form.

$e^T = \pi P^a$

43

Example 2:

Change $V = Ke^{-P/t}$ to logarithmic form and solve for P.

Solution:

$V = Ke^{-P/t}$

$\dfrac{V}{K} = e^{-P/t}$ Divide both sides by K to obtain the form $N = a^x$.
        Note: $\log_e x = \ln x.$

$\ln\left(\dfrac{V}{K}\right) = -\dfrac{P}{t}$ Change to logarithmic form.

$-\dfrac{P}{t} = \ln\left(\dfrac{V}{K}\right)$ Solve for P.

$\therefore \quad P = -t \ln\left(\dfrac{V}{K}\right) \quad \text{or} \quad P = -t(\ln V - \ln K)$

$P = t(\ln K - \ln V)$

44

Example 3:
   Change $P = 2 \log\left(\dfrac{R}{M}\right)$ to exponential form and solve for R.

Solution:
   $P = 2 \log\left(\dfrac{R}{M}\right)$ Divide both sides by 2 to obtain the form $x = \log_a N.$

$\dfrac{P}{2} = \log\left(\dfrac{R}{M}\right)$ Change to exponential form.   **Note:** $\log x = \log_{10} x.$

$10^{P/2} = \dfrac{R}{M}$ Solve for R.

$R = M \cdot 10^{P/2}$

45

Example 4:
Change $10^{(P-\pi x)} = y - K$ to logarithmic form and solve for P.

Solution:
The expression $10^{(P-\pi x)} = y - K$ is in the form $a^x = N$.

Hence, $P - \pi x = \log(y - K)$

$$P = \pi x + \log(y - K)$$

46

Example 5:
Given $\log V = 3 - \log P$, solve for V by changing the exponential form.

Solution:

$\log V = 3 - \log P$	Change to the form $\log_a N = x$.
$\log V + \log P = 3$	Add $\log P$ to both sides.
$\log(VP) = 3$	Law 20.1
$VP = 10^3$	Change to exponential form.
$V = \dfrac{10^3}{P}$	Solve for V.

47

Example 6:
Given $e^{IR} = 2\pi e^{0.5k}$, solve for R by changing to logarithmic form.

Solution:

$e^{IR} = 2\pi e^{0.5k}$	Change to the form $a^x = N$.
$\dfrac{e^{IR}}{e^{0.5k}} = 2\pi$	Divide both sides by $e^{0.5k}$.
$e^{IR-0.5k} = 2\pi$	Use 3.2 and change to logarithmic form.
$IR - 0.5k = \ln(2\pi)$	Solve for R.
$R = \dfrac{\ln(2\pi) + 0.5k}{I}$	

48

CHECK POINT 8

I   Change the following to exponential form.

1. $y = \dfrac{1}{3} \log\left(\dfrac{x}{a}\right)$ 2. $\pi \ln x - \ln a = -1.25K$

3. $\log (u - a) = v + 3$ 4. $\ln R + 5 = 2 \ln P$

II   Change the following to logarithmic form.

5. $2\pi e^{-Kt} = e^{0.1x}$ 6. $M - N = 10^{(V+3)}$

    Solve for t.   Solve for V.

7. $5E = ae^{\pi R}$ 8. $10^{x+1} = 1000\ K$

    Solve for R.   Solve for x.

DRILL EXERCISES

Note: All variable expressions are restricted to positive values since the logarithms are not defined for negative numbers.

I   Change the following to logarithmic form.

1. $2^3 = 8$ 2. $4^{-2} = \dfrac{1}{16}$ 3. $6^4 = 1296$

4. $36^{1/2} = 6$ 5. $8^{-1/3} = \dfrac{1}{2}$ 6. $\left(\dfrac{3}{5}\right)^2 = \dfrac{9}{25}$

II   Change the following to exponential form.

7. $\log_2 64 = 6$ 8. $\log_{13} 169 = 2$ 9. $\log_{16} 2 = \dfrac{1}{4}$

10. $\log_5\left(\dfrac{1}{125}\right) = -3$ 11. $\log_{10} 0.0001 = -4$ 12. $\log_{1/2} 32 = -5$

III   Indicate which of the following expressions **are** defined. For those which are **not** defined, state the reason.

13. $\log_{-3} x = 2$ 14. $\log_2 (-5) = x$ 15. $\log_7\left(\dfrac{1}{49}\right) = x$

16. $\log_0 9 = x$ 17. $\log_{0.8} x = 2$ 18. $\log_1 25 = x$

19. $\log_{12} 0 = x$

DRILL EXERCISES (continued)

IV   Evaluate the following.

20. $7 \log_2 2$ 21. $\log_3 3^4$ 22. $\frac{1}{2} \ln e$

23. $4 \log 10$ 24. $\log 10^{2x-1} = 5$ 25. $\ln e^{y+3} = 4$
                                                  Solve for x.   Solve for y.

V   Use the laws of logarithms to write the following as sums and/or differences of logarithms.

26. $\log_4 (2x^5)$ 27. $\ln\left(\dfrac{ab}{10}\right)$ 28. $\log_3\left(\dfrac{m-n}{nt}\right)$

29. $\ln\left(\dfrac{x+y}{a}\right)^{1/2}$ 30. $\log_2 (abc)$ 31. $\log\left(\dfrac{m^{1/2} n^4}{t^2}\right)$

32. $\log_5\left(\dfrac{\sqrt{v}}{b^3}\right)$ 33. $\log_7 \sqrt[5]{xy}$ 34. $\log \sqrt[3]{\dfrac{5x^2 y}{2z}}$

VI   Write the following as single logarithmic expressions.

35. $\log_2 4 + \log_2 8$ 36. $\log 5 - 2 \log x$ 37. $\frac{1}{2}[\log_2 a + \log_2 b]$

38. $\frac{1}{2} \log 16 - \frac{2}{3} \log 64$ 39. $3 \log_5 a - 2 \log_5 a$

40. $\frac{1}{2} \log_3 (a - b) + \log_3 c$ 41. $\frac{2}{3}[\ln(x^2 - 1) + \ln a - \ln(x + 1)]$

VII   Use the appropriate laws of logarithms to solve the given equations as indicated.

42. Solve for log m: $\log \sqrt{m} = 5$.

43. Solve for log b: $\log\left(\dfrac{a}{b}\right) = \log c$.

44. Solve for ln t: $\ln(Kt) = \ln 116$.

45. Solve for log x: $\log \sqrt[5]{2x^2} = 9$.

DRILL EXERCISES (continued)

46. Solve for ln x: $2 \ln(xy) = \ln x - c$.

47. Solve for log a: $\log \sqrt{a} - \log b = -\frac{1}{2} \log a$.

48. Solve for ln a: $\ln(\pi a^3) + \ln b = \ln(Ka^2)$.

49. Solve for log x: $2 \log x + 5y - 1 = \log(xy)$.

50. Solve for log V: $10 = \log \sqrt[5]{\dfrac{Ra}{V}}$.

VIII   Change to following to exponential form.

51. $x = \log(mt)$

52. $m = \ln(x + b) - \ln 5$

53. $\ln y = \ln 0.6 - Kt$

54. $x - 5 = \log(y + \pi)$

55. $10 \log(1 + i) = \log 2$

56. $\ln 250 - 0.4t = \ln 125$

57. $\frac{2}{3} \log R + 0.05 = \log P$
      Solve for P.

58. $y = \dfrac{2 \ln(\frac{x}{a})}{t}$
      Solve for x.

IX   Change the following to logarithmic form.

59. $259 = 10^x$

60. $\pi e^{-0.5t} = e^{0.09x}$ Solve for x.

61. $e^x \cdot e^y = Kt$

62. $a + b = 10^{(t-2)}$

63. $i = Ie^{-Kt}$

64. $7R = K\left(1 - e^{-1/2t}\right)$ Solve for t.

65. $\dfrac{10^x}{10^{y-1}} = 2a + 1$ Solve for x.

ASSIGNMENT EXERCISES

Note: All variable expressions are restricted to positive values since the logarithms are not defined for negative numbers.

I   Change the following to logarithmic form.

1. $5^2 = 25$

2. $4^{-3} = \dfrac{1}{64}$

3. $27^{1/3} = 3$

4. $\left(\dfrac{2}{7}\right)^2 = \dfrac{4}{49}$

5. $64^{-1/2} = \dfrac{1}{8}$

6. $2^8 = 256$

ASSIGNMENT EXERCISES (continued)

II Change the following to exponential form.

7. $\log_3 243 = 5$ 8. $\log_{15} 225 = 2$ 9. $\log_{10}(0.001) = -3$

10. $\log_{1/3} 81 = -4$ 11. $\log_4\left(\dfrac{1}{64}\right) = -3$ 12. $\log_{36} 6 = \dfrac{1}{2}$

III Indicate which of the following expressions **are** defined. For those which are **not** defined, state the reason.

13. $\log_{-2} 32 = x$ 14. $\log_1 7 = x$ 15. $\log_3(-81) = x$

16. $\log_{1/4} 1 = x$ 17. $\log_{0.7} x = 2$ 18. $\log_5 0 = x$

19. $\log_0 2 = x$

IV Evaluate the following.

20. $5 \ln e$ 21. $8 \log_4 4$ 22. $\log_{16} 16^7$

23. $\log_3 3^{1.2}$ 24. $\ln e^{2x+2} = -4$ 25. $\log 10^{y-7} = 2$
                                       Solve for x.   Solve for y.

V Use the laws of logarithms to write the following as sums and/or differences of logarithms.

26. $\ln\left(\dfrac{mn}{11}\right)$ 27. $\log_7(xyz)$ 28. $\log_3(9\sqrt[6]{y})$

29. $\log_5(3x^4)$ 30. $\ln\left(\dfrac{r+s}{t}\right)^{1/3}$ 31. $\log\left(\dfrac{a+b}{2c}\right)$

32. $\log\left(\dfrac{x^4 y^{1/2}}{z^2}\right)$ 33. $\log_2\left(\dfrac{a^2}{\sqrt{b}}\right)$ 34. $\log\sqrt[5]{\dfrac{4xy^4}{7z}}$

VI Write the following as single logarithmic expressions.

35. $5 \log y - \log 7$ 36. $\log_3 17 + \log_3 4$

37. $\dfrac{1}{3} \log 8 - \dfrac{3}{4} \log 81$ 38. $\dfrac{1}{6}[\ln m + \ln n]$

39. $\dfrac{2}{3}[\log a - 5 \log y + 2 \log 7]$ 40. $\dfrac{1}{3} \log_2(x+y) + \log_2 z$

41. $\dfrac{1}{5}[\ln(x^2 - 3x + 2) - \ln(x-2) + \ln a]$

ASSIGNMENT EXERCISES (continued)

VII Use the appropriate laws of logarithms to solve the given equations as indicated.

42. Solve for log a: $\log \sqrt[3]{a} = 7$.

43. Solve for log y: $\log(xy^3) = \log z$.

44. Solve for log P_1: $M = 10 \log\left(\dfrac{P_2}{P_1}\right)$.

45. Solve for ln c: $3 \ln(abc) = \ln c + d$.

46. Solve for log b: $\log \sqrt[4]{\dfrac{3}{b^2}} = 12$.

47. Solve for log t: $\dfrac{1}{4} \log t - \log x = \log \sqrt{2t}$.

48. Solve for ln p: $\ln \sqrt{m} = \ln(4p^2)$.

49. Solve for ln x: $\ln(kx^5) - \ln y = \ln(\pi x^2)$.

50. Solve for log y: $\log(y - 2) - \log(y^2 - 2y) = 5 \log b$.

VIII Change to following to exponential form.

51. $p = \dfrac{1}{4} \log\left(\dfrac{m}{n}\right)$

52. $\pi - x = \log(y + 4)$

53. $7 + \ln t = 3 \ln a$

54. $\ln \pi + b \ln x = -0.25k$

55. $\log 3 = 5 \log(1 - r)$

56. $\ln 82 = \ln 246 - 0.2t$

57. $\log 0.04 + 2 \log y = \log x$
    Solve for x.

58. $x = \dfrac{1/2 \, \ln(by)}{z}$
    Solve for y.

IX Change the following to logarithmic form.

59. $R = ke^{-cw}$

60. $10^{\pi x - 2} = y + 5$

61. $e^y = 145$

62. $3e^{ky} = e^{-0.2x}$ Solve for y.

63. $A + B = 10^{C-2}$ Solve for c.

64. $e^x e^{-y} = ct$

65. $a(1 + e^{0.5t}) = 6b$
    Solve for t.

66. $\dfrac{10^{m+1}}{10^n} = \pi + x$ Solve for m.

UNIT 21

EVALUATING EXPONENTIAL AND LOGARITHMIC EXPRESSIONS, CHANGE OF BASE

Objectives: After having worked through this unit, the student should be capable of:

1.   using the calculator to find common and natural logarithms;
2.   checking answers obtained by use of a calculator;
3.   using the calculator to find antilogarithms;
4.   evaluating logarithms to any base by "changing the base";
5.   solving certain exponential and logarithmic equations.

1

FINDING COMMON LOGARITHMS

We have seen in the previous unit that logarithms to the base 10 are called common logarithms and when we write "log N", we mean "$\log_{10} N$".

Remember:

log 100 = 2   means 10^2 = 100.

The common logarithm of 100 is the exponent to which we have to raise 10 to obtain 100.

Since 10^4 = 10 000, we have   log 10 000 = 4.

Since 10^3 = 1000,   we have   log 1000   = 3.

Since 10^2 = 100,   we have   log 100   = 2.

Since 10^1 = 10,   we have   log 10   = 1.

Since 10^0 = 1,   we have   log 1   = 0.

Since 10^{-1} = 0.1,   we have   log 0.1   = –1.

Since 10^{-2} = 0.01,   we have   log 0.01   = –2.

Since 10^{-3} = 0.001,   we have   log 0.001 = –3, etc.

We can see that for a power of 10 the common logarithm is an integer.

2

The common logarithm of a number between 1 and 10 is a decimal between 0 and 1.

Examples:

1.   log 6 is a decimal between 0 and 1, since 6 is between 10^0 = 1   and 10^1 = 10.

2.   log 3.56 is a decimal between 0 and 1, since 3.56 is between 1 and 10.

3.   log 9.01 is a decimal between 0 and 1, since 9.01 is between 1 and 10.

3

USING THE CALCULATOR TO FIND COMMON LOGARITHMS

Most scientific calculators have the following two keys:

[log]   refers to common logarithms.   log y = x.

[ln]   refers to natural logarithms.   ln y = x.

Your calculator may operate differently from the way shown in the following examples. If this is the case, consult the manual for your calculator, then work through the following examples.

Example 1:

Using the calculator, find log 6.

Solution:	Operations	Display
Clear calculator:	Press [C]	0
	Enter 6	6
	Press [log]	0.77815125

Hence, log 6 = 0.7782 (rounded to four decimal places).

4

Example 2:
Using the calculator, find log 3.56.

Solution:

Operations	Display
Press [C]	0
Enter 3.56	3.56
Press [log]	0.551449998

Hence, log 3.56 = 0.5514 (rounded to four decimal places).

5

CHECK POINT 1
Use your calculator to find the following common logarithms.
Round to six decimal places.

x	log x
1.00	
1.589	
2.00	
2.36	
2.8412	
3.00	
3.756	
4.00	
4.4531	
4.98	

x	log x
5.00	
5.645	
6.00	
6.5983	
7.00	
7.814	
8.00	
8.32092	
9.00	
9.75	

6

We observe that the common logarithms of the numbers between 1 and 10 are:

(a)   decimals between 0 and 1, since $1 = 10^0$ and $10 = 10^1$;

(b)   except for numbers close to 1, the first digit of the logarithm is larger or equal to the first digit of the number.

Examples:

1.   log 2.41 = 0.382...

2.   log 8.32092 = 0.920...

3.   log 5.645 = 0.751...

In what follows, we will use the facts (a) and (b) above and the concept of scientific notation to check answers.

REVIEW: SCIENTIFIC NOTATION

We say that a number is in scientific notation when it is written as a product of a number between 1 and 10 and the appropriate power of ten.

Examples:

Number		Number in Scientific Notation
356	=	3.56×10^2
0.901	=	9.01×10^{-1}
64080	=	6.408×10^4
7.19	=	7.19×10^0

ESTIMATING COMMON LOGARITHMS

When using a calculator to find the common logarithm of numbers, we should check the answer to avoid mistakes by inadvertently pressing a wrong button.

Visualizing mentally the scientific notation of the number will provide us with a rough estimate of its common logarithm.

Example 1:

Estimate log 356.

Solution:

Visualize: $356 = 3.56 \times 10^2$

$$\log 356 = \log 3.56 + \log 10^2$$
$$= \begin{pmatrix} \text{larger than} \\ 0.3 \end{pmatrix} + 2$$

Hence, a rough estimate of log 356 would be 2.4

Example 2:

Estimate log 0.00849.

Solution:

Visualize: $0.00849 = 8.49 \times 10^{-3}$

$$\log 0.00849 = \log 8.49 + \log 10^{-3}$$
$$= \begin{pmatrix} \text{larger than} \\ 0.8 \end{pmatrix} + (-3)$$

Hence, a rough estimate of log 0.00849 is $-3 + 0.9 = -2.1$.

CHECK POINT 2

Change the numbers to scientific notation to estimate the following common logarithms.

1.   log 2503   2.   log 0.094   3.   log 194.56   4.   log 0.00084

11

FINDING COMMON LOGARITHMS AND CHECKING THE ANSWERS

Note:

Your calculator may operate differently from the way shown in the following examples. If this is the case, consult the manual for your calculator, then work through the following examples.

Use the calculator to find the following common logarithms. Round the answers to four decimal places.

Examples:	Operations	Display	Answer (rounded)	Check
log 6905 = ?	Press [C] Enter 6905 Press [log]	0 6905 3.839163683	log 6905 = 3.8395	6905 = 6.905×10^3 log 6905 $\doteq$ 3 + 0.7 = 3.7
log 0.0278 = ?	Press [C] Enter 0.0278 Press [log]	0 0.0278 -1.555955204	log 0.0278 = -1.5560	0.0278 = 2.78×10^{-2} log 0.0278 $\doteq$ -2 + 0.3 = -1.7 Note: The log of 2.78 is positive
log 184269.51 = ?	Press [C] Enter 184269.51 Press [log]	0 184269.51 5.265453481	log 184269.51 = 5.2655	184269.51 =1.8426951×10^5 log 184269.51 $\doteq$ 5 + 0.2 = 5.2
log 0.0007 = ?	Press [C] Enter 0.0007 Press [log]	0 0.0007 -3.15490196	log 0.0007 = -3.1549	0.0007 = 7.0×10^{-4} log 0.0007 $\doteq$ -4 + 0.8 = -3.2

12

CHECK POINT 3

Use your calculator to find the common logarithms of the given numbers. Round answers to six decimal places. Check your answers by estimation.

x	log x	Estimate
17.52		
0.984		
640317.8		
849.726		
5000		
0.00293		
16428		
0.000075		

USING THE CALCULATOR TO FIND NATURAL LOGARITHMS

Example 1:
Using the calculator, find ln 27.

Solution:

Operation	Display
Press [C]	0
Enter 27	27
Press [ln]	3.295836866

Since ln x means $\log_e x$ and e = 2.718..., we cannot use scientific notation to check answers.

If we approximate e by 3, we see that 3^3 = 27.

Hence, ln 27 = 3.2958 (rounded) is a reasonable answer.

Checking Answers
Approximating e by 3 to check answers is in most cases too inaccurate and cumbersome. Instead, we can use the fact that

$$\ln 27 = 3.295836866 \text{ means } e^{3.295836866} = 27.$$

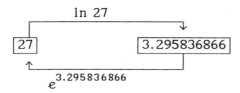

Hence, if 3.295836866 is the correct answer to ln 27, then $e^{3.295836866}$ should equal 27.

When evaluating natural logarithms using the calculator, proceed as shown in the following examples.

Example 1:
Find ln 27.

Solution:

Operation	Display	
Press [C]	0	Clear calculator
Enter 27	27	
Press [ln]	3.295836866	Write answer down, rounded to any desired decimal place.

Check:
To check, leave answer in calculator and

Press [2nd F][e^x] 27   Check.

Note:

1. If your calculator operates differently, consult the manual of how to find e^x.

2. Checking logarithms by saying

$$\text{if } \ln 27 = 3.295836866 \text{ then } e^{3.295836866} = 27,$$

is similar to checking division by saying

$$\text{if } 12 \div 3 = 4, \text{ then } 4 \times 3 = 12.$$

16

More Examples:

Round to four decimal places.

Examples:	Operations	Display	Answer (rounded)	Check
ln 849.5 = ?	Press [C] Enter 849.5 Press [ln]	0 849.5 6.744647941	ln 849.5 = 6.7446	Press [2nd F][e^x] Display 849.5
ln 0.0741 = ?	Press [C] Enter 0.0741 Press [ln]	0 0.0741 -2.602339747	ln 0.0741=-2.6023	Press [2nd F][e^x] Display 0.0741

17

CHECK POINT 4

Use your calculator to find the natural logarithms of the given numbers. Round answers to four decimal places. Check your answers.

x	ln x = y	Check: $e^y = x$
29.54		
18075		
0.00243		
519.6		
0.82		

ANTILOGARITHM

In many practical problems, we are required to deal with problems such as solve log x = 2.75 for x.

In this case, we are given the logarithm and the base and we are to find the corresponding number.

log x = 2.75 can be rewritten as $10^{2.75}$ = x.

We refer to finding x as finding the **antilogarithm** of 2.75.

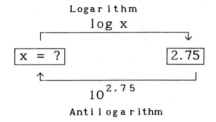

Finding the antilogarithm is the reverse process of finding the logarithm.

The problem, solve log x = 2.75 for x, is sometimes stated as "find the antilog 2.75".

Example 1:

Solve log x = 2.75 for x.

Solution:

Since log x = 2.75 means $10^{2.75}$ = x, we use the calculator to find $10^{2.75}$.

Operation	Display	
Press [C]	0	Clear calculator.
Enter 2.75	2.75	
Press [2nd F][10^x]	562.3413252	Write answer down to any desired decimal place.
Press [log]	2.75	Check.

Hence, x = 562.34 (rounded to two decimal places).

Example 2:
Solve ln x= −3.5408.

Solution:
Since ln x = −3.5408 means $e^{-3.5408} = x$,

we use the calculator to find $e^{-3.5408}$.

Operation	Display	
Press [C]	0	Clear calculator.
Enter 3.5408	3.5408	
Press [+/−]	−3.5408	
Press [2nd F][e^x]	0.028990125	Write answer down to any desired decimal place.
Press [ln]	−3.540800024	Check.

Hence, x = 0.02899 (rounded to five decimal places).

Note:
Since the calculator can display only a certain number of digits, rounding occurs. Hence, in the reverse process, digits such as 2 and 4 may appear in the eighth and ninth decimal place.

CHECK POINT 5

Find the antilogarithms of the following.

log x	x
4.7521	
0.2518	
−3.496	
−1.28	
0.5	

ln x	x
2.80	
−3.8675	
0.75	
−0.5	
6.29248	

EVALUATING LOGARITHMS TO ANY BASE

To show how to evaluate a logarithm to any base a, we make use of the following properties.

(21.1) $\boxed{\text{If } \; x = y, \text{ then } \; \log_a x = \log_a y, \text{ where } \; x > 0 \; \text{ and } \; y > 0.}$

(21.2) $\boxed{\text{If } \log_a x = \log_a y, \text{ then } \; x = y, \text{ where } \; x > 0 \; \text{ and } \; y > 0.}$

Using (21.1) we can show that $a^{\log_a x} = x$.

Clearly, $\qquad \log_a x = \log_a x$

∴ $\qquad (\log_a x)(\log_a a) = \log_a x$ **Remember**: $\log_a a = 1$

$\qquad \log_a a^{\log_a x} = \log_a x$ (20.3)

$\qquad a^{\log_a x} = x$ (21.2)

Hence,

(21.3) $\boxed{a^{\log_a x} = x}$

We can use (21.3) to simplify expressions.

Examples:

1. $10^{\log x} = x$

2. $5^{\log_5 (R+1)} = R + 1$

3. $e^{2 \ln y} = e^{\ln y^2} = y^2$

4. $10^{1/2 \log(x-1)} = 10^{\log(x-1)^{1/2}} = \sqrt{x-1}$

5. $2^{2/3 \log_2 z} = 2^{\log_2 z^{2/3}} = \sqrt[3]{z^2}$

CHECK POINT 6

Simplify the following.

1. $2^{\log_2 10}$

2. $10^{\log(v-1)}$

3. $e^{2 \ln x}$

4. $8^{1/2 \log_8 (R+3)}$

5. $10^{2/3 \log(2y+1)}$

CHANGE OF BASE

Problems involving the evaluation of a logarithm to a base other than 10 or e can be solved by a change of base.

For example, to evaluate $\log_5 39$, we change the problem to one which involves the evaluation of common or natural logarithms.

Example:

Evaluate $\log_5 39 = x$.

Solution:

Step 1. Change $\log_5 39 = x$ to exponential form.

$5^x = 39$

Step 2. Take the common logarithm of both sides.

$\log 5^x = \log 39$

Step 3. Solve for x using law 20.3.

$x \log 5 = \log 39$

$$x = \frac{\log 39}{\log 5}$$

$$= \frac{1.5911}{0.6990} \quad \text{(rounded to four decimal places)}$$

$$= 2.2763$$

Hence, $\log_5 39 = 2.2763$.

Similar to the solution of the previous equation, we can derive a general change-of-base property.

If $\log_a y = x$, then $a^x = y$.

Taking the logarithm to base b of both sides of $a^x = y$, we have

$$\log_b a^x = \log_b y$$

$$x \log_b a = \log_b y \qquad \text{Law 20.3}$$

$$x = \frac{\log_b y}{\log_b a} \qquad \text{Solve for x.}$$

Since $x = \log_a y$, we have

(21.4)
$$\boxed{\log_a y = \frac{\log_b y}{\log_b a}}$$

Example 1:
Find $\log_3 258$.

Solution:
We can evaluate $\log_3 258$ by changing it to a common or natural logarithm, using the above relationship (21.4)

(a) Changing to base 10 (base of common logarithm), we have $a = 3.$, $b = 10$ and $y = 258$.

$$\therefore \quad \log_3 258 = \frac{\log_{10} 258}{\log_{10} 3} = \frac{\log 258}{\log 3} = \frac{2.41162}{0.47712} = 5.0545$$

Hence, $\log_3 258 = 5.0545$.

(b) Changing to base e (base of natural logarithm), we have $a = 3$, $b = e$ and $y = 258$.

$$\therefore \quad \log_3 258 = \frac{\log_e 258}{\log_e 3} = \frac{\ln 258}{\ln 3} = \frac{5.55296}{1.09861} = 5.0545$$

Hence, $\log_3 258 = 5.0545$.

CHANGING COMMON TO NATURAL LOGARITHMS

To change common to natural logarithms, we use the relationship (21.4).

From $\quad \log_a y = \dfrac{\log_b y}{\log_b a}$, we see that

$$\log_e y = \frac{\log_{10} y}{\log_{10} e}$$

or $\quad \ln y = \dfrac{\log y}{\log e}$ $\qquad \dfrac{1}{\log e} = 2.3026$, rounded

Hence,

(21.5) $\quad \boxed{\ln y = 2.3026 \log y}$

CHANGING NATURAL TO COMMON LOGARITHMS

From $\quad \log_a y = \dfrac{\log_b y}{\log_b a}$, we see that

$$\log_{10} y = \frac{\log_e y}{\log_e 10}$$

or $\quad \log y = \dfrac{\ln y}{\ln 10}$ $\qquad \dfrac{1}{\ln 10} = 0.4343$, rounded

Hence,

(21.6) $\quad \boxed{\log y = 0.4343 \ln y}$

Examples:

1. If $\ln 250 = 5.5215$, then $\log 250 = 0.4343 \ln 250$

    $= (0.4343)(5.5215)$

    $= 2.3980$

2. If $\log 0.05 = -1.3010$, then $\ln 0.05 = 2.3026 \log 0.05$

    $= (2.3026)(-1.3010)$

    $= -2.9957$

CHECK POINT 7

Evaluate the following logarithms by changing to base 10.
(Round to four decmal places.)

1. $\log_5 16.9$ 2. $\log_{12} 0.598$ 3. $\log_8 7.5$

4. $\ln 0.06$ 5. $\log_2 45$ 6. $\ln 19.8$

Evaluate the following logarithms by changing to base e.
(Round to four decimal places.)

7. $\log_4 0.012$ 8. $\log_{16} 11.5$ 9. $\log 100$

10. $\log_8 594.6$ 11. $\log 0.008$ 12. $\log_7 7269$

SOLVING EXPONENTIAL EQUATIONS

Solving exponential equations involves finding an unknown base or an unknown exponent.

There is no general method of solving these equations as is the case, for example, with quadratic equations.

However, to solve exponential equations such as

$$10^{x-4} = 811 \quad \text{and} \quad 74.5 = 0.82e^{-0.5x}$$

we change to logarithmic form.

If the equation involves exponential expressions with different bases or unknown bases

$$3.5(8^{x+3}) = 15^{x-1} \quad \text{and} \quad V^{2.5} = 1.6V^{0.8}$$

we take logarithms of both sides of the equation and solve for the variable.

Example 1:

Solve $10^{x-4} = 811$.

Solution:

$$10^{x-4} = 811 \quad \text{Change to logarithmic form.}$$

$$x - 4 = \log 811$$

$$x - 4 = 2.909$$

$$x = 6.909$$

Example 2:

Solve $3.5(8^{x+3}) = 15^{x-1}$.

Solution:

Since we have two different bases, we take the logarithm of both sides. In this case, we can either take the common or the natural logarithm. Let us take the common logarithm of both sides of

$3.5(8^{x+3}) = 15^{x-1}$

$\log [3.5(8^{x+3})] = \log 15^{x-1}$

$\log 3.5 + (x + 3) \log 8 = (x - 1) \log 15$ Laws 20.1 and 20.3

$\log 3.5 + x \log 8 + 3 \log 8 = x \log 15 - \log 15$ Collect all terms with x on the left side of the equation.

$x \log 8 - x \log 15 = -\log 3.5 - 3 \log 8 - \log 15$

$(\log 8 - \log 15) x = -(\log 3.5 + 3 \log 8 + \log 15)$ Factor out x.

$x = \dfrac{-(\log 3.5 + 3 \log 8 + \log 15)}{(\log 8 - \log 15)}$ Divide both sides by the coefficient of x.

$ = \dfrac{-4.429429}{-0.273001} = 16.2250$ (rounded)

Example 3:

Solve $74.5 = 0.82e^{-0.5x}$

Solution:

$74.5 = 0.82e^{-0.5x}$ Divide both sides by 0.82.

$\dfrac{74.5}{0.82} = \dfrac{0.82}{0.82} e^{-0.5x}$

$90.8537 = e^{-0.5x}$ Change to logarithmic form

$\ln 90.8537 = -0.5x$

$4.5093 = -0.5x$ Solve for x.

$x = -9.0185$

36

Example 4:

Solve $V^{2.5} = 1.6V^{0.8}$.

Solution:

$$V^{2.5} = 1.6V^{0.8}$$

Since we have an unknown base, take the log of both sides.

$$\log V^{2.5} = \log (1.6V^{0.8})$$

$$2.5 \log V = \log 1.6 + 0.8 \log V$$

Laws 20.2 and 20.3.

$$2.5 \log V - 0.8 \log V = \log 1.6$$

Collect log V on the left side of the equation and combine the terms.

$$1.7 \log V = 0.2041$$

$$\log V = 0.12$$

Change to exponential form.

$$V = 10^{0.12} = 1.318 \text{ (rounded)}$$

37

CHECK POINT 8

Solve the following equations.

1. $10^{x+3} = 21$ 2. $4(3^P) = 5$ 3. $12e^{1.5y} = 612$

4. $2\pi e^{-1.5} = e^{0.2x}$

38

SOLVING LOGARITHMIC EQUATIONS

When solving logarithmic equations, we should be constantly aware of the basic properties. These properties often help us to rewrite the equation into a solvable form.

Sometimes it may also be necessary to change logarithmic expressions to their exponential form.

In addition, we may have to use the following property.

(21.7) | If $m = n$, then $a^m = a^n$ where $a > 0$ and $a \neq 1$. |

39

Example 1:

Solve $1.08 = 0.9 \log(5x)$.

Solution:

$$1.08 = 0.9 \log(5x)$$

$$\log(5x) = \frac{1.08}{0.9}$$

$$\log 5 + \log x = 1.20 \qquad (20.1)$$

$$\log x = 1.20 - \log 5$$

$$\log x = 0.5010$$

$$x = 10^{0.5010}$$

$$x = 3.17$$

Example 2:

Solve $2.15 = \ln \sqrt{\dfrac{43.5}{R}}$

Solution:

$2.15 = \ln \left(\dfrac{43.5}{R}\right)^{1/2}$

$2.15 = \dfrac{1}{2} (\ln 43.5 - \ln R)$ Laws 20.3 and 20.2.

$2.15 = \dfrac{1}{2} \ln 43.5 - \dfrac{1}{2} \ln R$ Solve for ln R.

$\dfrac{1}{2} \ln R = \dfrac{1}{2} \ln 43.5 - 2.15$

$\dfrac{1}{2} \ln R = 1.89 - 2.15$

$\ln R = -0.52$ Change to exponential form.

$R = e^{-0.52}$

$R = 0.5945$

Example 3:

Solve $\log 2 + \dfrac{1}{2} \log (x + 1) = 1$

Solution:

$\log 2 + \dfrac{1}{2} \log (x + 1) = 1$

$\log 2 + \log (x + 1)^{1/2} = 1$ (20.3)

$\log 2(x + 1)^{1/2} = 1$ (20.1)

$2(x + 1)^{1/2} = 10^1$ Change to exponential form.

$(x + 1)^{1/2} = 5$

$\left((x + 1)^{1/2}\right)^2 = 5^2$ Square both sides.

$x + 1 = 25$

$x = 24$

42

Example 4:

Solve $\frac{1}{2} \log(x + 2) - \log x = 0$.

Solution:

$$\frac{1}{2} \log(x + 2) - \log x = 0 \qquad \text{Transpose log x.}$$

$$\log (x + 2)^{1/2} = \log x \qquad \text{If } \log a = \log b, \text{ then } a = b.$$

$$(x + 2)^{1/2} = x \qquad \text{Square both sides.}$$

$$x + 2 = x^2$$

$$-x^2 + x + 2 = 0$$

$$x^2 - x - 2 = 0$$

$$(x - 2)(x + 1) = 0$$

$$\therefore \quad x = 2 \quad \text{or} \quad x = -1$$

Since the logarithms of negative numbers are not defined, we have x = 2 as the only solution.

43

CHECK POINT 9

Solve the following equations.

1. $0.5 \log(2a) = 3.75$

2. $1 = \frac{1}{2} \log (3m + 10)$

3. $\ln \sqrt{\frac{5}{V}} = 1.4$

4. $\ln(x - 2) = \frac{1}{2} \ln x$

DRILL EXERCISES

Use a calculator to find the logarithms of the given numbers.
Round answers to five decimal places.
Check each answer by finding the antilog of your answer.

	x	$\log x = y$	Check: $10^y = x$		x	$\ln x = y$	Check: $e^y = x$
1.	0.058			8.	6.0942		
2.	1264.9			9.	0.175		
3.	87.09			10.	5000		
4.	0.0005			11.	12.59		
5.	642.75			12.	782.04		
6.	0.25			13.	94.081		
7.	568147			14.	0.0015		

DRILL EXERCISES (continued)

Find the antilogarithms of the following.

log x	x
15. 3.0125	
16. 0.7234	
17. -2.6109	
18. 1.1792	
19. -0.5	
20. 0.75	

ln x	x
21. 3.60	
22. -1.8965	
23. 0.5	
24. 2.75	
25. -0.09	
26. 5.3065	

Change the base to common logarithms to evaluate the following.

27. $\log_3 21.5$

28. $\log_7 0.16$

29. $\log_{12} 10.9$

30. $\ln 25$

31. $\log_2 25$

32. $\log_5 0.07$

Change the base to natural logarithms to evaluate the following.

33. $\log_6 351$

34. $\log_8 0.125$

35. $\log 1000$

36. $\log_3 88.7$

37. $\log_4 163.82$

38. $\log_2 36$

Solve the following equations.

39. $10^{y-1} = 106$

40. $0.5^x = 0.27$

41. $3^{x+1} = 16$

42. $7^{2x+1} = 5^x$

43. $10e^{-0.75R} = 516$

44. $e^{-2.5} = \pi e^{0.15t}$

45. $6.4 = 1.6 \ln x$

46. $20 \log(3y) = 5.6$

47. $\ln\sqrt{\dfrac{1.46}{L}} = 0.5$

48. $3 \log(2x - 3) = 0$

49. $\dfrac{1}{2} \ln(a + 1) + \ln 5 = 1$

50. $\log(x + 1) + \log(x - 2) = 1$

ASSIGNMENT EXERCISES

Use a calculator to find the logarithms of the given numbers.
Round answers to five decimal places.
Check each answer by finding the antilog of your answer.

	x	log x = y	Check: 10^y = x
1.	470		
2.	2.48		
3.	0.072		
4.	9651.2		
5.	0.98		
6.	61.7		
7.	7536.5		

	x	ln x = y	Check: e^y = x
8.	5.01		
9.	827.401		
10.	0.927		
11.	46.18		
12.	1230		
13.	0.0059		
14.	67.4		

Find the antilogarithms of the following.

log x	x
15. -2.7043	
16. 0.06519	
17. 1.6824	
18. 0.0035	
19. -1.9	
20. 0.67	

ln x	x
21. -1.6990	
22. 2.15	
23. 3.067	
24. -0.25	
25. 4.739	
26. 0.2701	

Change the base to common logarithms to evaluate the following.

27. $\log_3 8.61$

28. $\log_4 104$

29. ln 912

30. $\log_2 33.6$

31. $\log_7 (0.05)$

32. $\log_5 2.27$

Change the base to natural logarithms to evaluate the following.

33. $\log_9 182$

34. $\log_3 243$

35. $\log_5 9.72$

36. $\log_2 0.356$

37. log 0.1

38. $\log_6 316.9$

ASSIGNMENT EXERCISES (continued)

Solve the following equations.

39. $3^{y+1} = 7$

40. $0.2^{x-1} = 0.5^{2x+1}$

41. $15^{2x-1} = 3^{x-2}$

42. $5^{-x} = 0.082$

43. $4\pi e^{2.1t} = e^{-0.7}$

44. $2e^{3x+5} = 47.6$

45. $\log(8x)^{0.4} = 1$

46. $12 = 4.8 \ln P$

47. $0 = \log(3m - 9)^2$

48. $\ln\sqrt{\dfrac{2.83}{k}} = 4.1$

49. $\ln(2x - 1) + \ln x = 0$

50. $\log(6y + 5) - \log 2 = \log 3 - \log y$

GRAPHS OF FUNCTIONS, EXPONENTIAL GROWTH AND DECAY

Objectives: After having worked through this unit, the student should be capable of:

1. finding the x- and y-intercepts of a linear function and of a quadratic function;
2. finding the coordinates of the vertex of the graph of a quadratic function;
3. identifying the zeros of functions;
4. identifying and graphing a constant function and the identity function;
5. sketching the graphs of linear, quadratic, exponential and logarithmic functions;
6. solving applied problems involving exponential growth and decay, using laws of exponents and logarithms including change of base.

INTRODUCTION

In this unit, we shall discuss a method of pictorially representing the relationship between two variables of an equation. If for example one variable "A" represents the area of a square and the other variable "s" represents one side of the same square, the formula

$$A = s^2$$

defines a **relationship** between the pair of variables **A** and **s**.

If we know the length of the side "s",
we can find the area of the square "A".

For example, if s = 1, then A = 1^2 = 1;

if s = 3, then A = 3^2 = 9;

if s = 5, then A = 5^2 = 25.

INDEPENDENT AND DEPENDENT VARIABLES

From the above illustrations, we see that the value of the variable "A" depends on the value we choose for the variable "s".

Hence, A is called the **dependent** variable and
s is called the **independent** variable.

A SOLUTION OF AN EQUATION IN TWO VARIABLES

A solution of an equation in two variables is any ordered pair of numbers (a,b) which makes the equation true. The first number "a" represents the value of the independent variable. The second number "b" represents the value of the dependent variable.

Example:
1. (2,7) is a solution of   y = x + 5, since   7 = 2 + 5 is true.
2. (-9,-4) is a solution of   y = x + 5, since   -4 = -9 + 5 is true.

Since for any arbitrarily chosen value of x we can calculate a value of y, the equation y = x + 5 has an infinite number of ordered pairs as solutions.

THE GRAPH OF AN EQUATION

The graph of an equation consists of all points in the coordinate plane whose ordered pairs (x,y) are solutions of the equation.

Using a table of values, we can normally sketch a good approximation to the graph of simple equations. In constructing a table of values, we can choose any value we want for the independent variable. However, it is generally advisable to choose:

    (a)   numbers close to the origin, and
    (b)   numbers which will make the evaluation of the dependent variable as simple as possible.

METHOD OF GRAPHING EQUATIONS

Step 1. Construct a table of values.

Step 2. Plot the ordered pairs.

Step 3. Join the plotted points by a smooth curve and label the graph with its equation.

Example 1:

Graph the equation $A = \pi r^2$.

Solution:

Step 1. Construct a table of values.

> **Note:** Since r represents the radius of a circle, we will not choose negative values for r. ($\pi = 3.14$ used)

$A = \pi r^2$.

r	0	1	2	3	4
A	0	3.14	12.56	28.26	50.24

Step 2. Plot the ordered pairs (0,0), (1, 3.14), (2, 12.56), (3, 28.26) and (4, 50.24).

> **Note:** (a) The axes need not be scaled the same. Here, we scale each axis to fit the points of the problem.
> (b) To plot a point such as (2, 12.56) on a scale like the following, we estimate the approximate location.

Step 3. Join the points by a smooth curve and label it with the equation.

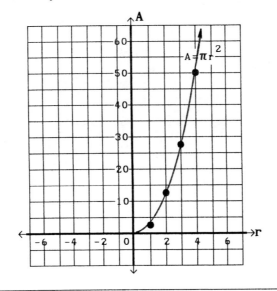

Example 2:
Graph the equation $P = \dfrac{1000}{V}$ for positive values of V.

Solution:

Step 1. Construct a table of values. $\dfrac{1000}{V}$ is not defined when V = 0.

$$P = \frac{1000}{V}$$

V	50	100	150	200	250	300	350
P	20	10	6.7	5	4	3.3	2.9

Step 2 and 3. Plot the ordered pairs and join the points by a smooth curve.

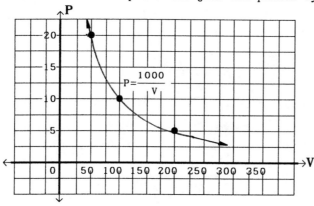

7

CHECK POINT 1

Graph the following equations.

1. $y = 2x - 1$ 2. $A = s^2$

3. $L = \frac{600}{W}$ (Use positive values for W.)   4. $y = -x^2$

8

We shall now briefly discuss the important concept of a function. Functions are often found when we apply mathematics to such areas as business, technology and the physical and social sciences. To discuss a function we need to know what we mean by a set.

9

SET

A set is a well defined collection of objects. "Well defined" means that the membership of a set must be clearly described. One way of describing precisely the membership of a set is to clearly state the conditions which an object must satisfy to be a member of the set. To do so, we use the notation referred to as "**set builder notation**".

Example:

The notation $f = \{(x,y)|y = 2x + 1\}$ represents the set of all ordered pairs (x,y) satisfying the equation $y = 2x + 1$.

The symbol | inside the brackets is read as "such that".

10

FUNCTION:

A function is a set of ordered pairs with the special condition that no two distinct ordered pairs have the same first element.

Examples:

1. The formula $A = s^2$ relating the area of a square to one of its equal

    sides, defines the function $f = \{(s,A) \mid s > 0 \text{ and } A = s^2\}$.

    The values of s and A in the ordered pair (s,A) are determined by the
    equation $A = s^2$ which will give us only one value of A for any
    given value of s.

2. If a car travels at 100 km/h, the formula $d = 100t$ defines a function
    relating the distance travelled to the time travelled.
    We see that the distance (d) depends on the time (t) travelled.

    The function is the set of ordered pairs $\{(t,d) \mid t > 0 \text{ and } d = 100t\}$.
    $d = 100t$ will give only one value of d for any given value of t.

3. Given a constant temperature, the formula $V = \dfrac{k}{p}$ where k is a constant,

    defines a function relating volume and pressure in such a way that the
    volume (V) of a gas depends on the pressure (p). We note that the
    volume increases as the pressure decreases. Again the equation will
    give only one value of V for any given value of p.

GRAPHS OF FUNCTIONS

Since functions are sets of ordered pairs, we shall be able to picture
functions using graphs.

The graph of a function f consists of all points in the plane whose
coordinates are ordered pairs belonging to the set f.

When discussing the graph of functions defined by an equation, we often
refer to this graph as the graph of the equation.

f(x) NOTATION

When discussing functions and their graphs, it is convenient to use the
notation $f(x)$ instead of the variable y.

We read $f(x)$ as "f of x".

For example, instead of $\quad y = 3x^2 - 5x - 10$,

$$\text{we write} \quad f(x) = 3x^2 - 5x - 10.$$

The value of $f(x)$ when $x = 2$ is given by

$$f(2) = 3(2)^2 - 5(2) - 10 = -8.$$

Similarly, $\quad f(0) = 3(0)^2 - 5(0) - 10 = -10$

$$f(-3) = 3(-3)^2 - 5(-3) - 10 = 32$$

Note:

1. We could have used y instead of $f(x)$ and written $y = 3x^2 - 5x - 10$,
    but $y = 32$ does not tell us that $x = -3$ was used to evaluate y,
    while $f(-3) = 32$ does.

2. It is conventional in most situations to represent sets which are
    functions by lower case letters such as f, g, h, etc.

14

CHECK POINT 2

1. If f(x) = 5x - 3, find f(0), f(-4), f(3).

2. If g(t) = 2t^2 + 4t - 3, find g(-2), g(0), g(4).

3. If h(v) = 3v + 7, find h(-1), h(0), h(3).

15

Graphs of functions which belong to the same type have normally the same general shape. If we have an idea of the general shape of the graph of a function, we need the coordinates of only a few key points to sketch the graph.

16

GRAPHS OF LINEAR FUNCTIONS

Functions defined by equations of the form f(x) = ax + b or y = ax + b where a, b are real numbers are called **linear functions**.

Examples:

f(x) = 2x - 3   defines a linear function. (a = 2, b = -3)

$y = -\dfrac{x}{5}$ defines a linear function. (a = $-\dfrac{1}{5}$, b = 0)

g(x) = 7   defines a linear function. (a = 0, b = 7)

The graph of a linear function is a straight line.

17

REVIEW: X- AND Y-INTERCEPTS OF A GRAPH

The x-intercept is the point where the graph crosses the x-axis.
Its coordinates are (x,0) since the y-coordinate of any point on the x-axis is zero.

The y-intercept is the point where the graph crosses the y-axis.
Its coordinates are (0,y) since the x-coordinate of any point on the y-axis is zero.

A line which is not parallel to either the x- or y-axis will cross (intercept) both axes.

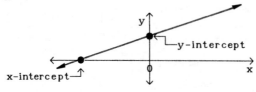

18

SKETCHING THE GRAPH OF A LINEAR FUNCTION

To sketch the graph of a linear function, we need only two points.
Since the coordinates of the x- and y-intercepts are easy to find, we normally use these points to sketch the graph of a linear function. As a check, we may find the coordinates of a third point. If the three points do not lie on the same straight line, the coordinates of at least one point are incorrect and we check our calculations.

REVIEW

When we discussed the graph of the linear equation **y = mx + b** where m and b are real numbers, we said that m is the slope (incline) and b is the y-intercept of the line graph.

Since the equation f(x) = ax + b has the same form as y = mx + b, the constant a is the slope and b is the y-intercept of the graph of the linear function.

> If a > 0, the functional values f(x) <u>increase</u> as the values of x increase;
>
> If a < 0, the functional values f(x) <u>decrease</u> as the values of x increase.

Example 1:

Draw the graph of the function defined by $f(x) = \frac{2}{3}x$.

Solution:

Since b = 0, the graph intercepts the x- and y-axis at the origin. Hence, to make up a table of values we choose the coordinates of the origin and an x value on either side of the origin.

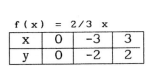

f(x) = 2/3 x

x	0	-3	3
y	0	-2	2

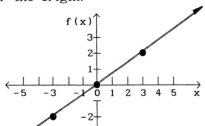

The slope of the graph is $a = \frac{2}{3}$,

that is, f(x) **increases** by $\frac{2}{3}$ as x increases by 1.

Example 2:

Draw the graph of the function defined by $y = -2x + \frac{3}{4}$.

Solution:

Step 1. Make up a table of values.

y = -2x+3/4

x	0	3/8	1
y	3/4	0	-5/4

Step 2. Plot the points and draw the graph.

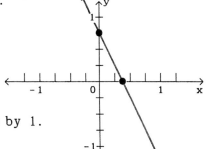

The slope of the graph is a = -2, that is, f(x) **decreases** by 2 as x increases by 1.

THE CONSTANT FUNCTION

If $a = 0$, we have that $f(x) = 0x + b$. Hence, for any value of x, $f(x) = b$.

This special linear function is called the **constant function**.

Examples:

Each of the following equations defines a constant function.

$$f(x) = 4, \qquad g(x) = -\frac{2}{3}, \qquad y = 8.67, \qquad h(x) = \pi.$$

The graph of a constant function defined by $f(x) = b$, is a line parallel to the x-axis such that b is the y-intercept.

Examples:

The following are graphs of constant functions.

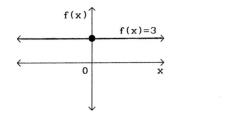 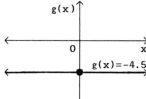

The slope of the constant function is $a = 0$.

THE IDENTITY FUNCTION

If $a = 1$ and $b = 0$, we have that $f(x) = x + 0$.
Hence, for any value of x that we substitute into $f(x) = x$, we obtain x as the functional value.

Examples:

$f(2) = 2$ and the corresponding ordered pair is (2,2).
$f(-6) = -6$ and the corresponding ordered pair is (-6,-6).

This special linear function whose elements are the ordered pairs (x,x) is called the **identity function**.

The graph of the identity function defined by $f(x) = x$ is a line through

the origin whose slope is $a = 1$.

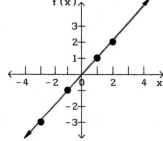

25

CHECK POINT 3

Draw the graphs of the functions defined by the given equations. Indicate the slope and the x- and y-intercepts in each case.

1. $f(x) = 4x - 5$ 2. $g(x) = -3x$

3. $y = \dfrac{3}{2}$ 4. $h(x) = -\dfrac{x}{3} + 1$

5. $y = x$ 6. $f(x) = -3$

26

GRAPHS OF QUADRATIC FUNCTIONS
Functions defined by the equation of the form

$$f(x) = ax^2 + bx + c \qquad or \qquad y = ax^2 + bx + c$$

where a, b, c are real numbers and a $\neq$ 0, are called **quadratic functions.**

Examples:
Each of the following equations defines a quadratic function.

$y = 3x^2 - 4x + 1$ $g(x) = x^2 + 9$

$f(x) = -5x^2 - x$ $h(x) = -x^2$

27

THE GENERAL SHAPE OF THE GRAPH OF QUADRATIC EQUATIONS
The following are graphs of quadratic functions defined by the equations

$f(x) = x^2 - 4$ and $g(x) = -x^2 + 9$.

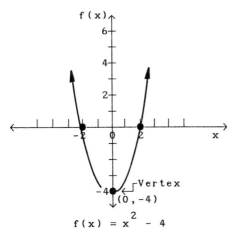

$$f(x) = x^2 - 4$$

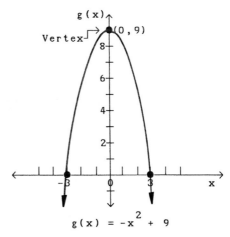

$$g(x) = -x^2 + 9$$

Any graph of a quadratic function has the **same general shape of a parabola** as shown in these two examples.

28

VERTEX OR EXTREME POINT
The graph of $f(x) = x^2 - 4$ has a **minimum point** at (0,-4).
The graph of $g(x) = -x^2 + 9$ has a **maximum point** at (0,9).

We call the minimum or maximum point on the graph of a quadratic function **the vertex** or **extreme point.**

29

Comparing $f(x) = x^2 - 4$ and $g(x) = -x^2 + 9$

to the general equation $f(x) = ax^2 + bx + c$, we see that for

(a) $f(x) = x^2 - 4$, $a = 1$ is positive and the graph opens upward.

(b) $g(x) = -x^2 + 9$, $a = -1$ is negative and the graph opens downward.

This is true in general for functions defined by $f(x) = ax^2 + bx + c$, that is,

> **when   a > 0, the graph opens upward**
>
> **when   a < 0, the graph opens downward.**

30

If we know the coordinates of the vertex, we can select values of   x   to the right and left of it to make up a table of values.
To sketch the graph we plot these points and draw a smooth curve to resemble a parabola through them.

We find the x coordinate of the vertex of the graph for any quadratic equation by using the following equation where a is the coefficient of x^2 and b is the coefficient of x in $f(x) = ax^2 + bx + c$

> $x = -\dfrac{b}{2a}$ **is the x coordinate of the vertex**

of the graph of any quadratic function.

To find the y-coordinate of the vertex, we substitute the value
$x = -\dfrac{b}{2a}$ into $f(x) = ax^2 + bx + c$ and evaluate.

31

Example 1:
Given $f(x) = x^2 - 4x - 5$ find the coordinates of the vertex.

Solution:
Step 1.
Find the x-coordinate using $x = -\dfrac{b}{2a}$.

$a = 1$, $b = -4$, $\therefore$ $x = -\dfrac{(-4)}{2(1)} = 2$.

Step 2.
Find the y-coordinate if $x = 2$.

$f(x) = x^2 - 4x - 5$

$f(2) = 2^2 - 4(2) - 5$

$= -9$

Hence, the   coordinates of the vertex are $(2, -9)$.
Note:
Since $a = 1$ is positive the graph opens upward.

Example 2:
Given $f(x) = -2x^2 + 5x + 9$, find the coordinates of the vertex.

Solution:
Step 1. $a = -2$, $b = 5$ and $x = -\dfrac{5}{2(-2)} = \dfrac{5}{4}$

Step 2. $f\left(\dfrac{5}{4}\right) = -2\left(\dfrac{5}{4}\right)^2 + 5\left(\dfrac{5}{4}\right) + 9 = -\dfrac{50}{16} + \dfrac{25}{4} + 9 = \dfrac{97}{8} = 12\dfrac{1}{8}$

Hence, the coordinates of the vertex are $\left(\dfrac{5}{4},\ 12\dfrac{1}{8}\right)$.

Note:
Since $a = -2$ is negative the graph opens downward.

CHECK POINT 4

For each of the following find the coordinates of the vertex and state whether the graph opens upward or downward.

1. $f(x) = 2x^2 - 3x + 1$ 2. $y = -x^2 + 2$

3. $g(x) = 3x^2 - x$ 4. $h(x) = -9x^2 - 1$

5. $y = x^2$ 6. $f(x) = -\dfrac{x^2}{2} - x + 5$

7. $y = -4x^2 + 12x - 9$

ZEROS OF A FUNCTION
The zeros of any function defined by an equation are the values of x for which the functional value $f(x) = 0$.

Hence, to find the zeros of any function, we solve the equation $f(x) = 0$ for values of x.

Examples:
1.   For the linear function defined by $f(x) = 2x - 1$, we solve $0 = 2x - 1$ to find that $x = \dfrac{1}{2}$ is a zero of this linear function.

2.   For the quadratic function defined by $f(x) = x^2 - 3x - 4$, we solve $0 = x^2 - 3x - 4$ or $x^2 - 3x - 4 = 0$ to find that $(x - 4)(x + 1) = 0$, that is, $x = 4$ and $x = -1$ are zeros of this quadratic function.

FINDING THE ZEROS OF A QUADRATIC FUNCTION
In general, to find the zeros of a quadratic function, we solve the equation
$$ax^2 + bx + c = 0$$

by factoring or by using the quadratic formula
$$x = \frac{-b \pm \sqrt{b^2 - 4ac}}{2a}.$$

Note:

You may want to review at this time the unit on solving quadratic equations by factoring and the unit on the quadratic formula. (Pre-Calculus Part 1)

36

CHECK POINT 5

Find the zeros for the function defined by the following equations.
Write irrational solutions as decimals rounded to two decimal places.

1. $f(x) = x^2$

2. $y = -x^2 + 4$

3. $g(x) = x^2 - 4x + 3$

4. $h(x) = -3x^2 + 10x - 4$

5. $y = 4x^2 - 12x + 9$

6. $f(x) = 2x^2 + 8x - 5$

37

Note:

When solving equations of the form $ax^2 + bx + c = 0$, we may obtain values of x which are complex numbers. In this case, the graph of the quadratic function does not cross the x-axis, that is, the graph does not have x-intercepts as shown below.

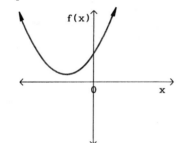 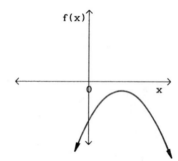

Example:

To find the zeros of $f(x) = x^2 + 9$, we solve $x^2 + 9 = 0$ and obtain $x = \pm \sqrt{-9} = \pm 3j$ which is not a real number. Hence, the graph of the function defined by $f(x) = x^2 + 9$ does not cross the x-axis.

38

SKETCHING THE GRAPH OF A QUADRATIC FUNCTION

Knowing the general shape of the graph of a quadratic function, we can quickly sketch the graph of most quadratic functions using the following method.

Step 1.
    Find the coordinates of the vertex $\left(x_v = \dfrac{-b}{2a}\right)$.

Step 2.
    Check whether $a > 0$, that is, the graph opens upward,
             or $a < 0$, that is, the graph opens downward.

Step 3.
    If possible find the real zeros of $f(x)$ or make a table of values using one or two values of x to find $f(x)$ on both sides of the vertex point.

Step 4.
    Plot the point and sketch the graph.

Example 1:
   Sketch the graph of the function defined by $f(x) = x^2 - 2x - 8$.

Solution:
   Step 1.
      Find the coordinates of the vertex $a = 1$, $b = -2$ and $x_v = \dfrac{-(-2)}{2(1)} = 1$

      $f(1) = 1^2 - 2(1) - 8 = -9$.

      Hence, the coordinates of the vertex are $(1,-9)$.

   Step 2.
      Since $a = 1$ is positive, the graph opens upward and will cross the x-axis since the vertex is below the x-axis.
   Step 3.
      Find the zeros of $f(x)$ $x^2 - 2x - 8 = 0$

      $(x - 4)(x + 2) = 0$

      Hence, $x = 4$ and $x = -2$ are zeros of the function.

   Step 4.
      Plot the coordinates of the vertex and the zeros and then sketch the graph.

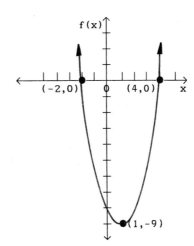

   Note:
      If we want a more accurate sketch, we make up a table of values choosing x-values on either side of the vertex.

Example 2:
   Sketch the graph of the function defined by $y = -x^2 + 2x - 1$.

Solution:
   Step 1.

      Find the coordinates of the vertex. $a = -1$, $b = 2$ and $x_v = \dfrac{-2}{2(-1)} = 1$

      $f(1) = -1^2 + 2(1) - 1 = 0$
      Hence, the coordinates of the vertex are $(1,0)$, that is, the vertex is on the x-axis.

   Step 2.
      Since $a = -1$ is negative, the graph opens downward.

   Step 3.
      Since the vertex is on the x-axis, no other zeros exist.
      Hence, we make up a table of values choosing x-values to the right and left of the vertex.

$y = -x^2 + 2x - 1$				
x	0	2	-1	3
y	-1	-1	-4	-4

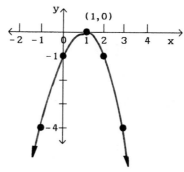

Note:

We see from the table of values and the graph that for x-values which are the same distance from the vertex, the functional values f(x) are equal and we say that the graph of f is **symmetric about the line x = 1.**

41

Example 3:

Sketch the graph of the function defined by $f(x) = x^2 - 4x + 10$

Solution:

Step 1.

Find the coordinates of the vertex. $x_v = \dfrac{-b}{2a} = \dfrac{-(-4)}{2(1)} = 2$

$f(2) = 2^2 - 4(2) + 10 = 6$

Hence, the coordinates of the vertex are (2,6).

Step 2.

Since a = 1 is positive the graph opens upward. The vertex is above the x-axis and the graph opens upward. Hence, the graph does not cross the x-axis, that is, the function has no zeros that are real numbers.

Step 3.

Since no real zeros exist, we make up a table of values.

$f(x) = x^2 - 4x + 10$				
x	0	1	3	4
f(x)	10	7	7	10

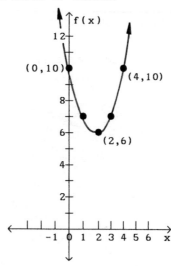

Step 4.

Plot the points and sketch the graph.

42

CHECK POINT 6

Sketch the graphs of the functions defined by the given equations.

1. $f(x) = -x^2$

2. $y = x^2 - 4$

3. $g(x) = x^2 - 6x + 2$

4. $f(x) = -2x^2 + x - 3$

5. $y = 6x - x^2$

6. $f(x) = -2x^2 - 4x - 5$

43

EXPONENTIAL AND LOGARITHMIC GRAPHS

In what follows, we shall discuss two important and very useful functions called the exponential function and the logarithmic function.

As their names suggest, these functions are defined by exponential and logarithmic expressions which we have already discussed.
We shall start our discussion of these functions by introducing some new notation and the concept of an asymptote.

44

Notation

The notation

$x \to -\infty$ is used to indicate that the values of x decrease, that is, the x values are getting smaller and smaller, without a lower bound.

For example, x = -1, -10, -100, -1000, -10000, ...

$x \to +\infty$ or $x \to \infty$ is used to indicate that the values of x increase, that is, the x values are getting larger and larger, without an upper bound.

For example, x = 1, 10, 100, 1000, 10000, 100000,...

$x \to a^+$ is used to indicate that the values of x approach the value of x = a from the right side of a.

For example, if x = 0, the values x = 2, 1, 0.5, 0.01, 0.001, 0.0001 are getting closer and closer to zero from the right side of zero.

Note:

An important part of this idea is that for **any** chosen number k, no matter how close to a, x will eventually take values closer to a than k.

$x \to a^-$ is used to indicate that the values of x approach the value x = a from the left side of a.

For example, if a = 0, the values x = -2, -1, -0.5, -0.01, -0.001, -0.0001 are getting closer and closer to zero from the left side of zero.

$x \to a$ is used to indicate that the values of x approach x = a from both sides of a.

Asymptote

An asymptote is a straight line to which a graph approaches arbitrarily close as points on the graph are chosen arbitrarily distant from the origin. If the asymptote is a horizontal straight line y = c, we call it a horizontal asymptote. (See Figure 1.)

If the asymptote is a vertical straight line x = c, we call it a vertical asymptote. (See Figure 2.)

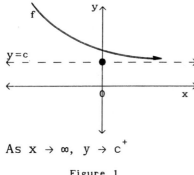

As $x \to \infty$, $y \to c^+$

Figure 1

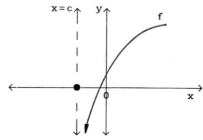

As $x \to c^+$, $y \to -\infty$

Figure 2

EXPONENTIAL FUNCTION

The function f defined by

$$y = a^x \quad (a > 0)$$

is called the exponential function with base a.

The Exponential Function defined by $y = 2^x$

To discuss the general shape of graphs of exponential functions, we shall use the function with base a = 2, that is, $y = 2^x$.

Since $y = 2^x$ is defined for all real numbers, we shall make up a table of values choosing values of x to either side of zero.

$y = 2^x$

x	-10	-5	-3	-1	0	1	3	5	10
y	0.001	0.031	0.125	0.5	1	2	8	32	1024

Note:

From the table of values we see that
1. As x is getting larger, y is getting rapidly larger, that is, as $x \to \infty$ we have that $y \to \infty$.
2. As x is getting smaller, y is getting smaller. For any negative value x = -a, we have that $y = 2^{-a} = \dfrac{1}{2^a}$ is a positive fraction.

Hence, y can never be negative or zero but as x is getting smaller, y is getting closer to zero, that is, as $x \to -\infty$, $y \to 0^+$.

Therefore, the line **y = 0 is a horizontal asymptote.**

Using this information and the table of values, we can sketch the graph of $y = 2^x$.

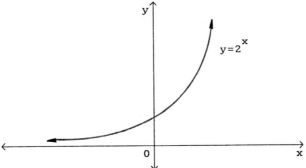

We refer to graphs of exponential functions as exponential graphs.

The General Shape of Exponential Graphs

The general shape of graphs of exponential functions depends on the size of the base a in $y = a^x$.

We can illustrate this by sketching the graphs of the function defined by $y = \left(\frac{1}{2}\right)^x$.

$y = (1/2)^x$

x	-3	-2	-1	0	1	2	3
y	8	4	2	1	1/2	1/4	1/8

$y = (1/2)^x$

The general shape of any exponential graph is similar to either the graph of $y = 2^x$ or $y = \left(\frac{1}{2}\right)^x$, that is, the graphs of exponential functions defined by $y = a^x$ behave as follows:

If $a > 1$, the graph is strictly monotone **increasing** and has the same general shape as the graph of $y = 2^x$.

If $0 < a < 1$, the graph is strictly monotone **decreasing** and has the same general shape as the graph of $y = \left(\frac{1}{2}\right)^x$.

Note:

$$y = 2^{-x} = \frac{1}{2^x} = \left(\frac{1}{2}\right)^x \qquad \text{and} \qquad y = \left(\frac{1}{2}\right)^{-x} = \frac{1}{2^{-x}} = 2^x.$$

Examples:

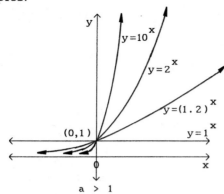

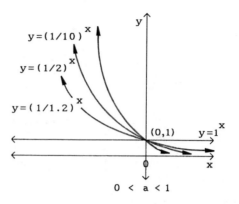

Note:

1. Since $a^0 = 1$ for any value $a \neq 0$, all graphs of exponential functions defined by $y = a^x$ $(a > 0)$ intercept the y-axis at $(0,1)$.

2. The line $y = 0$, that is, the x-axis is a horizontal asymptote for the graphs of $y = a^x$ $(a > 0$ and $a \neq 1)$.

Check Point 7

Sketch the exponential graphs of the functions defined by the given equations. Use the key $[y^x]$ on your scientific calculator to obtain the functional values.

1. $y = (2.5)^x$

2. $g(x) = (1.5)^x$

3. $f(x) = 3^{-x}$

4. $y = \left(\dfrac{1}{3}\right)^{-x}$

The Graph of $f(x) = e^x$

In science and technology the most frequently used exponential function is defined by

$$f(x) = e^x$$

where e is the irrational number $e = 2.71828183\ldots$

The graph of this function has the general shape of an exponential graph. To obtain the functional value e^x, we use the key $[e^x]$ which is on any scientific calculator.

Example:

Sketch the graphs of $f(x) = e^{1.2x}$ and $g(x) = e^{-0.5x} = \left(\dfrac{1}{e}\right)^{0.5x}$.

Solution:

Make up a table of values as follows.

$f(x) = e^{1.2x}$

x	-2	-1	0	1	2
1.2x	-2.4	-1.2	0	1.2	2.4
f(x)	0.1	0.3	1	3.3	11.0

$g(x) = e^{-0.5x} = (1/e)^{0.5x}$

x	-2	-1	0	1	2
-0.5x	1	0.5	0	-0.5	-1.0
g(x)	2.7	1.6	1	0.6	0.4

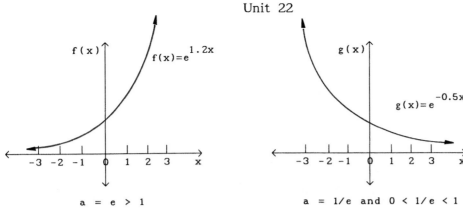

$a = e > 1$ $a = 1/e$ and $0 < 1/e < 1$

Note:

We say that a function f is **one-to-one** if for any two values x_1 and x_2 such that $x_1 \neq x_2$, we have that $f(x_1) \neq f(x_2)$.

Given the one-to-one function f, the **inverse function** g is the set of ordered pairs obtained from f by interchanging in each ordered pair of f the first and second elements.

The exponential function is a one-to-one function and has an inverse.

―― 52

Check Point 8

Sketch the graph of the functions defined by $f(x) = e^x$ and $g(x) = e^{-x}$.

―― 53

THE LOGARITHMIC FUNCTION

The inverse of the exponential function defined by $y = a$ ($a > 0$ and $a \neq 1$) is a new function defined by

$$y = \log_a x \quad \text{and} \quad x > 0$$

and is called the logarithm of x to the base a.

You may recall that logarithms are exponents and that $y = \log_a x$ means

> **y is the exponent to which we have to raise the base a to obtain the number x, that is, $a^y = x$.**

Conversely, $y = a^x$ can be written as $x = \log_a y$ since x is the exponent to which we have to raise the base a to obtain the number y.

Since $x = \log_a y$ defines the inverse of $y = a^x$, the variable x in $x = \log_a y$ is the dependent variable and the variable y is the independent variable. Hence, we interchange x and y in $x = \log_a y$ and obtain

$y = \log x.$

―― 54

The General Shape of Logarithmic Graphs

Since any logarithmic function is the inverse of some exponential function, all logarithmic graphs are the reflections about $y = x$ of exponential graphs. Hence, the shape of a logarithmic graph depends on the base of the logarithm.

Since most scientific calculators have only the keys [log x] and [ℓn x], we shall graph functions with bases other than 10 and e as inverses of corresponding exponential functions. We do this by interchanging the x and y values in the tables of values. (log x means log to the base 10)

Example:

Sketch the graphs of $y = \log_2 x$ and $y = \log_{(1/2)} x$.

Solution:

We make up a table of values using the corresponding exponential functions $y = 2^x$ and $y = (1/2)^x$.

Since $y = \log_2 x$ is the inverse function of $y = 2^x$, we interchange the x and y values in the table for $y = 2^x$ to obtain a table of values for $y = \log_2 x$. We do the same with $y = (1/2)^x$ and its inverse $y = \log_{(1/2)} x$

Step 1.

Make up tables of values for $y = 2^x$ and $y = (1/2)^x$.

$y = 2^x$

x	-2	-1	0	1	2
y	1/4	1/2	1	2	4

$y = (1/2)^x$

x	-2	-1	0	1	2
y	4	2	1	1/2	1/4

Step 2.

Interchange the x and y values in the tables for $y = 2^x$ and $y = (1/2)^x$ to obtain tables for $y = \log_2 x$ and $y = \log_{(1/2)} x$, respectively.

$y = \log_2 x$

x	1/4	1/2	1	2	4
y	-2	-1	0	1	2

$y = \log_{(1/2)} x$

x	4	2	1	1/2	1/4
y	-2	-1	0	1	2

Plot the points and sketch the graphs.

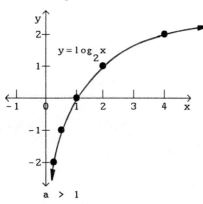

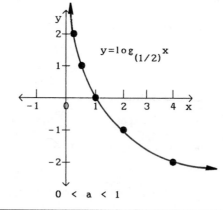

In general the shape of any logarithmic graph is similar to either the graph of $y = \log_2 x$ or $y = \log_{(1/2)} x$, that is, the graphs of logarithmic functions defined by $y = \log_a x$ behave as follows.

If $a > 1$, the graph is strictly monotone **increasing** and has the general shape of $y = \log_2 x$.

If $0 < a < 1$, the graph is strictly monotone **decreasing** and has the general shape of $y = \log_{(1/2)} x$.

Examples

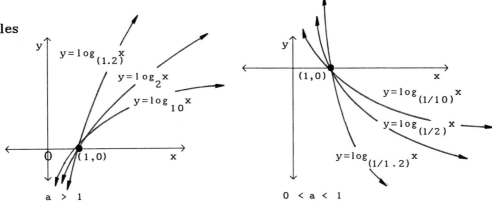

a > 1   0 < a < 1

Note:

1. Since $\log_a 1 = 0$ for any base a (a > 0 and a ≠ 1), all graphs of logarithmic functions defined by $y = \log_a x$ (x > 0) intercept the x-axis at (1,0).

2. The line x = 0, that is the y-axis is a vertical asymptote for the graphs of $y = \log_a x$ (a > 0 and a ≠ 1).

56

The Graph of $f(x) = \log_e x$

As was the case with the exponential function $f(x) = e^x$, the logarithmic function with base e is an important function in science and technology. We give this function a special name and call it the **natural logarithmic function**. We denote $\log_e x$ by *ln* x. The graph of the natural logarithmic function has the general shape of a logarithmic graph. To make up a table of values, we can use the key [*ln*x] given on any scientific calculator.

Example:

Sketch the graph of $f(x) = ln(x/2)$.

Solution:

Make up a table of values using a scientific calculator to find the values $ln(x/2)$. Since $f(x) = ln(x/2)$ is defined only for x > 0, we start choosing values of x close to zero.

$f(x) = ln(x/2)$

x	0.25	0.5	1	2	4	6	8
x/2	0.125	0.25	0.5	1	2	3	4
f(x)	-2.1	-1.4	-0.7	0	0.7	1.1	1.4

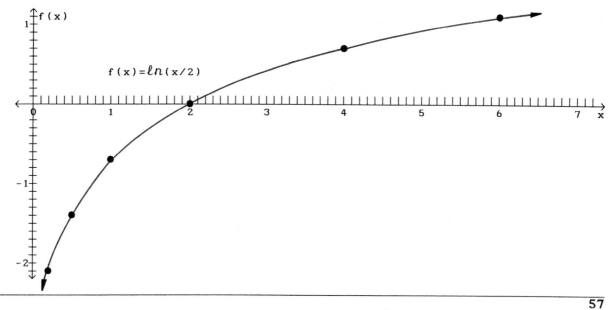

$$f(x) = \ell n(x/2)$$

57

Check Point 9

1. Sketch the graphs of $y = \log_3 x$ and $y = \log_{(1/3)} x$ by first making up a table of values for $y = 3^x$ and $y = (1/3)^x$.

2. Sketch the graph of $y = \ell n(2x)$ by making up a table of values using a scientific calculator to find values of $\ell n(2x)$.

58

EXPONENTIAL GROWTH AND DECAY

Many natural phenomena involving growth and decay follow the general law

$$A(t) = A_0 e^{Kt}$$

where K is a nonzero constant, A_0 is the original amount and $A(t)$ is the amount present after time t.

59

Exponential Growth

Under certain conditions the growth of a colony of bacteria or insect population may be a function of time defined by

$$\boxed{P(t) = P_0 e^{Kt}}$$

where $K > 0$ is a constant of proportionality and P_0 is the population at time $t = 0$.

Since the exponent of $P(t) = P_0 e^{Kt}$ is positive and the time $t \geq 0$, the graph of an exponential growth function has the following general shape.

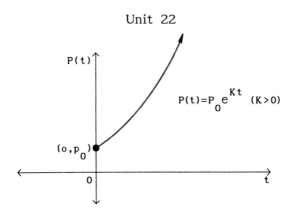

Example:

If $P(t) = 100e^{0.25t}$ gives the size of an insect population at time t in days, find P(t) for t = 0, 1, 2, 3, 4, 5 and sketch a graph showing the growth of the population during the first five days.

Solution:

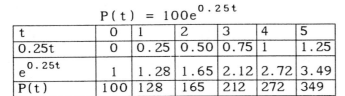

t	0	1	2	3	4	5
$0.25t$	0	0.25	0.50	0.75	1	1.25
$e^{0.25t}$	1	1.28	1.65	2.12	2.72	3.49
$P(t)$	100	128	165	212	272	349

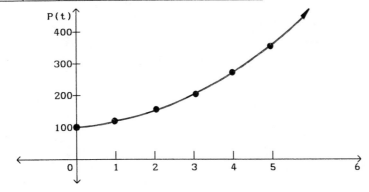

Exponential Decay

The decay of radioactive materials is a function of time defined by

$$D(t) = D_0 e^{-Kt}$$

where K > 0 is a constant of proportionality and D_0 is the amount present at time t = 0.

Since the exponent of $D(t) = D_0 e^{-Kt}$ is negative and the time $t \geq 0$, the graph of an exponential decay function has the following general shape.

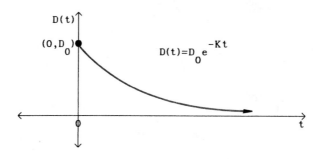

Example:

If $A(t) = A_0 e^{-0.087t}$ is the function according to which the radioactive

material iodine –131 decays in t days, find how much material is left after

t = 1, 2, 3, 4, 5 days if the initial amount A_0 = 500 g. Use the data to

sketch a graph and find t as a function of the amount present, A(t).

Solution:

Since A_0 = 500, we evaluate $A(t) = 500e^{-0.087t}$ for t = 1, 2, 3, 4, 5.

Step 1.

$$A(t) = 500e^{-0.087t}$$

t	1	2	3	4	5
$-0.087t$	-0.087	-0.174	-0.261	-0.348	-0.435
$e^{-0.087t}$	0.9167	0.8403	0.7703	0.7061	0.6473
A(t)	458	420	385	353	324

Step 2.

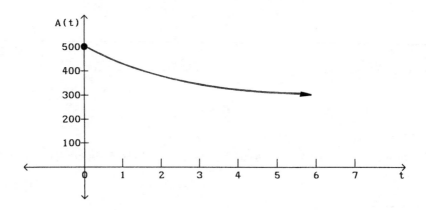

Step 3.

To find t as a function of A(t), we change

$$A(t) = A_0 e^{-Kt}$$

to its logarithmic form.

$$\frac{A(t)}{A_0} = e^{-Kt}$$

Remember: –Kt is the exponent ($\log_e$ = ℓn) to

which we have to raise e to get $\dfrac{A(t)}{A_0}$

$$\therefore \quad -Kt = \ln\left(\frac{A(t)}{A_0}\right)$$

$$t = -\frac{1}{K} \ln\left(\frac{A(t)}{A_0}\right)$$

For example, if we want to know how many days it takes for an amount of 500 g to decay to 1 g, we use

$$t = -\frac{1}{K} \ln\left(\frac{A(t)}{A_0}\right)$$

with $K = 0.087$, $A_0 = 500$ and $A(t) = 1$ we have
$$t = -\frac{1}{0.087} \ln\left(\frac{1}{500}\right)$$

$$= 71.4$$

Hence, 500 g of iodine-131 will decay to 1 g in about 71 days.

63

Half-life of radioactive decay

The time it takes for half of the original amount of radioactive substance to decay is called the half-life of the substance.

For example, it takes a quantity of 1000mg of radium 226, 1620 years to decay to 500mg. Hence, the half-life of radium 226 is 1620 years. The radioactive element polonium has a half-life of 138 days.

Carbon Dating

An application of radioactive decay is found in the dating of archaeological and geological specimens.

The most publicized application is carbon dating which measures the amount of carbon 14 a radioactive substance present in the specimens.
Living organisms stop taking in carbon 14 when they die and the carbon 14 present starts to decay with a half-life of 5568 years.

Knowing the half-life of a radio-active substance allows us to calculate the decay constant K in
$$A(t) = A_0 e^{-Kt}$$

Example 1:

Find the decay constant K for carbon 14 which has a half-life of 5568 years.

Solution:

Let the original amount $A_0 = 1$, then $A(t) = \frac{1}{2}$ when $t = 5568$.

Substituting into $A(t) = A_0 e^{-Kt}$, we have
$$0.5 = e^{-5568K}$$

and $\ln 0.5 = -5568K$

$$K = 0.00012 \text{ (rounded to 5 decimal places)}$$

Hence, the decay function for carbon 14 is given by

$$A(t) = A_0 e^{-0.00012t}$$

Example 2:

If it is found that 60% of the original carbon 14 is present in a bone, find the age of the bone.

Solution:

Since the original amount represents 100%, A_0 = 1 and $A(t)$ = 0.6.

Since the decay function of carbon 14 is

$$A(t) = e^{-0.00012t}$$

we have

$$0.6 = e^{-0.00012t}$$

and

$$ln(0.6) = -0.00012t$$

$$t = 4257 \text{ years.}$$

Hence, the bone is about 4000 years old.

CHECK POINT 10

1. Find the decay constant for a radioactive substance which has a half-life of 22 years. Find the time it takes for 1000 mg of this substance to decay to 5 mg.

2. If the decay constant of a radioactive substance is K = 0.0283 and t is in days, find the half-life of this substance.

DRILL EXERCISES

Draw the graph of the functions defined by the given equations. Indicate the slope and the x- and y-intercepts in each case.

1. $f(x) = -x + 1$ 2. $y = -3$ 3. $f(x) = x$

4. $y = 3x - 4$ 5. $g(x) = \dfrac{x}{2}$ 6. $h(x) = -\dfrac{4x}{3} + 2$

For each of the following find the coordinates of the vertex and the zeros of of the function. Indicate whether the graph opens upward or downward.

7. $f(x) = x^2 - 16$ 8. $g(x) = -x^2 - 2x - 1$

9. $y = 2x^2 + x - 1$ 10. $h(x) = -4x^2 - 8x$

11. $f(x) = 9x^2 + 12x + 4$ 12. $y = -x^2 + 6x - 9$

13. $f(x) = -3x^2 + 12x - 5$ 14. $g(x) = 2x^2 - 2x - 12$

Sketch the graph of each of the following functions defined by the given equations. Use the coordinates of the vertex and the zeros if they exist.

15. $y = -2x^2$ 16. $f(x) = x^2 + 3x - 4$

17. $g(x) = 4x^2 - 1$ 18. $h(x) = -x^2 + 2x + 8$

19. $y = x^2 + 4$ 20. $f(x) = 2x^2 - x - 3$

DRILL EXERCISES (continued)

21. $g(x) = -2x^2 - 4x - 5$ 22. $h(x) = 3x^2 - 5x - 2$

Note:

As has been explained before it is customary to write $\log_{10}x$ as $\log x$ and $\log_e x$ as $\ln x$.

In each of the following, use the laws of exponents and logarithms to reduce expressions to their simplest form. If necessary review the laws.

23. $\left(0.5e^{1.6x}\right)\left(-1.5e^{-0.8x}\right)$ 24. $\left(A_0 2^{ax}\right) \div \left(B_0 2^{\pi x}\right)$

25. $\left(-\dfrac{1}{2}\ln(x+1)\right) + \left(\dfrac{1}{2}\ln(x-1)\right)$ 26. $\left(A_0 e^{-ct}\right) \div \left(A_1 e^{kt}\right)$

27. $\left(p\,\ln(x^2+x)\right) - \left(p\,\ln(x+1)\right)$

Simplify the following.

28. $f(x) = 10^{\log(x-1)^{1/2}}$ 29. $f(x) = e^{-2\ln x}$

30. $f(x) = x^2\log 10$ 31. $f(x) = \ln e^{\pi x}$

As we have seen in a previous unit, to change the base we use the equation

$$\log_a y = \frac{\log_b y}{\log_b a}.$$

Sketch the graph of the following functions by changing the given base to base 10 or base e as indicated and then use the calculator to find the values $\log x$ or $\ln x$ for a table of values.

Examples:

1. $f(x) = \log_5 x = \dfrac{\log x}{\log 5}$

2. or $f(x) = \log_{(1.2)}x = \dfrac{\ln x}{\ln(1.2)}.$

32. $f(x) = \log_3 x,$ change to base 10.

33. $f(x) = \log_{(1/2)}x,$ change to base e.

34. $f(x) = \log_{(1.5)}x,$ change to base 10.

DRILL EXERCISES (continued)

35. $f(x) = \log x$,   change to base e.

36. The equation $N(t) = N_0 2^t$ gives the number of bacteria in a certain culture after t hours. If the initial count $N_0 = 100$, find the number of bacteria after 24 hours and how long it will take for the bacteria count to reach 1000.

37. The equation $A(t) = A_0 e^{-0.0248t}$ gives the law according to which radioactive strontium 90, which is used in nuclear reactors, decays. Find the half-life in years of this substance.
    Find how long it takes for 1000 mg of strontium 90 to decay to 1 mg.

    (t = time in years, A_0 = initial amount, $A(t)$ = Amount at time t).

38. If the radioactive substance radium 226 has a half-life of 1620 years, find its constant of decay K to two significant digits. Find the number of years it will take for 1000 mg to decay to 5 mg.

39. If $A(t) = A_0 e^{-0.00012t}$ gives the amount of carbon 14 present after time t in years. Find the age of various fossils having lost 75%, 80%, 90% and 95% of their carbon 14.

40. In an excavation a human skull was found. If measurements indicate that 3% of carbon 14 have been lost find how long ago this person died.

ASSIGNMENT EXERCISES

Draw the graph of the functions defined by the given equations. Indicate the slope and the x- and y-intercepts in each case.

1. $f(x) = \dfrac{x}{2} + 1$ 2. $g(x) = 3x$ 3. $h(x) = \dfrac{3}{4}$

4. $y = 2x - 5$ 5. $y = -x$ 6. $g(x) = 4 - \dfrac{2}{3}x$

Sketch the graph of each of the following functions defined by the given equations. Use the coordinates of the vertex and the zeros if they exist.

7. $y = x^2 + 1$ 8. $y = x^2 - 2x + 1$

9. $f(x) = 2 - x^2$ 10. $g(x) = x^2 - 4$

11. $h(x) = 6 - x - x^2$ 12. $f(x) = 2x^2 + 5x + 2$

13. $g(x) = 4x^2 - 11x - 20$ 14. $h(x) = 3 + 8x - 3x^2$

ASSIGNMENT EXERCISES (continued)

15. If the radioactive element polonium has a half-life of 138 days, find its constant of decay K, relative to days.

16. Find the number of days it will take 100 mg of polonium to decay to 1 mg. (Use the decay equation obtained in the previous question.)

17. Statistics indicate that the population growth of a certain country is 3% per year. Assuming that this trend will hold in the future, we can use $P(t) = P_0(1.03)^t$

    to obtain estimates of the size of the population at future dates. If the current population is 23 million estimate the number of years it will take the population to double. (Change the base to 10 or e). Calculate P(t) for t = 5, 10, 15, 20, 25 and sketch a graph showing the growth of the population.

Appendix

Solutions to Check Points
and
Answers to Drill Exercises

Fanshawe College

SOLUTIONS TO CHECK POINTS

Check Point 3, Frame 22

I
1. T Commutative property of addition.

2. F $\dfrac{0}{185} = 0.$

3. T Identity element of multiplication.

4. T Associative property of multiplication.

5. F Division is not commutative.

6. F $8\pi \div 0$ is undefined.

7. T Distributive property of multiplication with respect to addition.

8. F Subtraction is not commutative.

9. T Identity element of addition.

10. F $42 \times 0 = 0.$

II 11. $7(5 + 9) = 7(5) + 7(9)$
$= 35 + 63$
$= 98$

12. $8(9 + 10) = 8(9) + 8(10)$
$= 72 + 80$
$= 152$

Check Point 4, Frame 27

I

II 1. $0 > -9$ 2. $6 < 2\pi$

3. $-\sqrt{2} > -1.6$ 4. $\dfrac{1}{3} = \sqrt{\dfrac{1}{9}}$

SOLUTIONS TO CHECK POINTS

Check Point 1, Frame **5**

1.
```
2 | 18
3 | 9
3 | 3
  | 1
```
Hence, $18 = 2 \cdot 3^2$

2.
```
3 | 45
3 | 15
5 | 5
  | 1
```
Hence, $45 = 3^2 \cdot 5$

3.
```
2  | 610
5  | 305
61 | 61
   | 1
```
Hence, $610 = 2 \cdot 5 \cdot 61$

4.
```
2 | 324
2 | 162
3 | 81
3 | 27
3 | 9
3 | 3
  | 1
```
Hence, $324 = 2^2 \cdot 3^4$

Check Point 2, Frame **15**

$-\dfrac{1}{2}$ rational, real, complex

5 natural, integer, rational, real, complex

$\dfrac{12}{13}$ rational, real, complex

$\sqrt{7}\,j$ imaginary, complex

$-\pi$ irrational, real, complex

$\sqrt{\dfrac{9}{25}}$ rational, real, complex

$3 - 2j$ complex

0 integer, rational, real, complex

-37 integer, rational, real, complex

e irrational, real, complex

$\dfrac{5}{3}\,j$ imaginary, complex

1 natural, integer, rational, real, complex

$\sqrt{3} + \pi j$ complex

6.50 rational, real, complex

380

ANSWERS TO DRILL EXERCISES

1. Prime numbers: 2, 5, 13, 29, 37, 7, 11, 23 and 31

2. $8 = 2^3$

3. $26 = 2 \cdot 13$

4. $100 = 2^2 \cdot 5^2$

5. $94 = 2 \cdot 47$

6. $250 = 2 \cdot 5^3$

7. $115 = 5 \cdot 23$

8. $144 = 2^4 \cdot 3^2$

9. $730 = 2 \cdot 5 \cdot 73$

10. $1452 = 2^2 \cdot 3 \cdot 11^2$

11. $1925 = 5^2 \cdot 7 \cdot 11$

12.

	Integers	Natural	Rational	Irrational	Real	Imaginary	Complex
-6	x		x		x		x
$\sqrt{3}$				x	x		x
256	x	x	x		x		x
$\sqrt{\dfrac{1}{4}}$			x		x		x
-2π				x	x		x
$-\dfrac{11}{13}$			x		x		x
j						x	x
0	x		x		x		x
$1 - 2j$							x
$5 + 0j$	x	x	x		x		x
$0 - \sqrt{2}\,j$						x	x
-2670	x		x		x		x
1.75			x		x		x
$0.666\ldots$			x		x		x
$-\dfrac{1}{3}$			x		x		x

381

ANSWERS TO DRILL EXERCISES

13. F $\quad \dfrac{0}{2.75} = 0.$

14. T $\quad$ Distributive property of multiplication with respect to addition.

15. T $\quad$ Associative property of multiplication.

16. T $\quad$ Commutative property of addition.

17. F $\quad$ Subtraction is not commutative.

18. F $\quad 0 \times 1 = 0.$

19. F $\quad \dfrac{-2\pi}{0}$ is undefined.

20. T $\quad$ Identity element of addition.

21.

22. $\sqrt{2} > 1.2$

23. $\dfrac{-\pi}{2} > -1.75$

24. $e > 2.5$

25. $\sqrt{\dfrac{1}{16}} = 0.25$

26. $0 < 0.1$

27. $-6 < -0.75$

28. $0.06 > -1.5$

29. $\dfrac{1}{5} = 0.2$

30. $-150 < -2$

SOLUTIONS TO CHECK POINTS

Check Point 1, Frame 11

1. T since $|-6| = 6$ and $|6| = 6$

2. T since $|0| = 0$, $|-9| = 9$ and $0 < 9$

3. F since $|-7| = 7$

4. T since $|+2| = 2$, $|-1| = 1$, and $2 > 1$

5. $(+25) + (-10) = +15$

6. $(-5) - (+10) = (-5) + (-10) = -15$

7. $(-6) + (-12) = -18$

8. $(-8) - (-15) = (-8) + (+15) = +7$

Check Point 2, Frame 14

1. $(+16) - (-12) - (+4) + (-16) = (+16) + (+12) + (-4) + (-16)$
$$= 16 + 12 - 4 - 16$$
$$= 28 - 4 - 16$$
$$= 24 - 16$$
$$= 8$$

2. $(-10) + (-5) - (+2) + (+7) - (-9) = (-10) + (-5) + (-2) + (+7) + (+9)$
$$= -10 - 5 - 2 + 7 + 9$$
$$= -15 - 2 + 7 + 9$$
$$= -17 + 7 + 9$$
$$= -10 + 9$$
$$= -1$$

3. $(-3) - (-19) + (-25) - (+6) - (-8) = (-3) + (+19) + (-25) + (-6) + (+8)$
$$= -3 + 19 - 25 - 6 + 8$$
$$= 16 - 25 - 6 + 8$$
$$= -9 - 6 + 8$$
$$= -15 + 8$$
$$= -7$$

SOLUTIONS TO CHECK POINTS

Check Point 2, Frame 14

4. $16 - 2 + 3 - 19 - 5 - 3 + 10 = 14 + 3 - 19 - 5 - 3 + 10$
$$= 17 - 19 - 5 - 3 + 10$$
$$= -2 - 5 - 3 + 10$$
$$= -7 - 3 + 10$$
$$= -10 + 10$$
$$= 0$$

Check Point 3, Frame 21

1. $(-5)(+2)(+3)(-1) = (-10)(+3)(-1)$
$$= (-30)(-1)$$
$$= +30$$

2. $(-6)(-5)(-2) = (+30)(-2)$
$$= -60$$

3. $(+3)(-4)(-2)(+4)(-1) = (-12)(-8)(-1)$
$$= (+96)(-1)$$
$$= -96$$

4. $(+1)(-16)(-1)(+2)(-10)(-2) = (-16)(-2)(+20)$
$$= (+32)(+20)$$
$$= +640$$

Check Point 4, Frame 26

1. $(-3)^4 = (-3)(-3)(-3)(-3)$
$$= 81$$

2. $-6^2 = -(36)$
$$= -36$$

3. $(-2)^3 = (-2)(-2)(-2)$
$$= -8$$

4. $-1^6 = -(1)$
$$= -1$$

5. $(-1) \div (+1) = -1$

6. $(+10) \div (-5) = -2$

7. $(-6) \div (-3) = +2$

8. $\dfrac{-1}{-1} = +1$

9. $\dfrac{0}{-7} = 0$

10. $\dfrac{-45}{-9} = +5$

11. $\dfrac{-110}{0}$ is undefined

382

SOLUTIONS TO CHECK POINTS

Check Point 5, Frame 33

1. $-24 \div [(3)(2)] = -24 \div [6]$
$$= -4$$

2. $(3)\left\{5^2 + 3[4^2 - 2^2]\right\} = 3\left\{5^2 + 3[16 - 4]\right\}$
$$= (3)\left\{5^2 + 3[12]\right\}$$
$$= (3)\left\{5^2 + 36\right\}$$
$$= (3)\left\{25 + 36\right\}$$
$$= (3)\left\{61\right\}$$
$$= 183$$

3. $\left\{6 - 3^2(-4) + [2(5) - 2^3]\right\} \div (0 - 2)$
$$= \left\{6 - 9(-4) + [10 - 8]\right\} \div (-2)$$
$$= \left\{6 + 36 + 2\right\} \div (-2)$$
$$= 44 \div (-2)$$
$$= -22$$

4. $\left[(-3)(-4)(0)\right] \div 2^2 + 6 = 0 \div 4 + 6$
$$= 0 + 6$$
$$= 6$$

Check Point 6, Frame 45

1. $2x + 4 = 2(5) + 4$
$$= 10 + 4$$
$$= 14$$

2. $x - 3y = 5 - 3(-3)$
$$= 5 + 9$$
$$= 14$$

3. $xyz = (5)(-3)(2)$
$$= -30$$

4. $3xy + 5xz = 3(5)(-3) + 5(5)(2)$
$$= -45 + 50$$
$$= 5$$

5. $z(xz + 3y) + 6 = 2[(5)(2) + 3(-3)] + 6$
$$= 2[10 - 9] + 6$$
$$= 2[1] + 6$$
$$= 8$$

ANSWERS TO DRILL EXERCISES

1. -25	2. 47	3. 17
4. 5	5. -9	6. -15
7. 7	8. -7	9. -10
10. -19	11. $+6$	12. -20
13. 7	14. 2	15. 28
16. $+4$	17. -7	18. -24
19. -2	20. 2	21. 0
22. -4	23. -54	24. 0
25. 35	26. -32	27. 42
28. -33	29. -4	30. -42
31. 144	32. 0	33. 48
34. $+16$	35. 256	36. -9
37. -1	38. -25	39. -5
40. -17	41. 10	42. $\frac{-120}{0}$ is undefined
43. 12	44. -88	45. 0
46. 5	47. -20	48. 2
49. 192	50. 12	51. 56
52. 117		

383

SOLUTIONS TO CHECK POINTS

Check Point 1, Frame 11

1. $10^3 \cdot 10^{21} = 10^{3+21} = 10^{24}$
2. $\dfrac{8^{12}}{8^5} = 8^{12-5} = 8^7$
3. $x^4 \cdot y^6$ cannot be simplified.
4. $(5xy)^2 (5xy)^7 = (5xy)^{2+7} = (5xy)^9$
5. $a^7 \div a^5 = a^{7-5} = a^2$
6. $(y+2)^3 \div (y-1)^2$ cannot be simplified.
7. $(x+2)(x+2)^5 = (x+2)^{1+5} = (x+2)^6$
8. $a^2 + a^4$ cannot be simplified.
9. $(2a+3)^5 \div (2a+3)^4 = (2a+3)^{5-4} = 2a+3$
10. $x^8 - x^3 =$ cannot be simplified.
11. $(3x)^9 \div (3x)^4 = (3x)^{9-4} = (3x)^5$

Check Point 2, Frame 19

2. $(10^3)^4 = 10^{3 \cdot 4} = 10^{12}$
3. $(3x)^4 = 3^4 \cdot x^4$
4. $\left(\dfrac{a^2}{b^4}\right)^3 = \dfrac{a^{2 \cdot 3}}{b^{4 \cdot 3}} = \dfrac{a^6}{b^{12}}$
5. $\left(\dfrac{x^2}{3y}\right)^3 = \dfrac{x^{2 \cdot 3}}{3^3 y^3} = \dfrac{x^6}{27y^3}$
6. $\left(\dfrac{2n^3 m}{p^2 t^2}\right)^2 = \dfrac{2^2 n^6 m^2}{p^4 t^4} = \dfrac{4n^6 m^2}{p^4 t^4}$

Check Point 3, Frame 22

1. $10^0 = 1$
2. $7x^0 = 7 \cdot 1 = 7$
3. $(3x)^0 = 1$
4. $x^3(5x^2 + 3y^4 z)^0 = x^3 \cdot 1 = x^3$

SOLUTIONS TO CHECK POINTS

Check Point 4, Frame 27

1. $2^{-4} = \dfrac{1}{2^4} = \dfrac{1}{16}$
2. $10^{-3} = \dfrac{1}{10^3} = \dfrac{1}{1000}$
3. $\left(\dfrac{1}{2}\right)^{-2} = \dfrac{1^{-2}}{2^{-2}} = \dfrac{1}{2^{-2}} = \dfrac{2^4}{1} = 16$
4. $\left(\dfrac{2}{3}\right)^{-3} = \dfrac{2^{-3}}{3^{-3}} = \dfrac{1}{2^3} \cdot 3^3 = \dfrac{1}{8} \cdot 27 = \dfrac{27}{8}$

Check Point 5, Frame 29

1. $x^2 y^{-3} = x^2 \cdot \dfrac{1}{y^3} = \dfrac{x^2}{y^3}$
2. $\dfrac{5a^2}{b^{-4}} = 5a^2 \cdot \dfrac{1}{b^{-4}} = 5a^2 \cdot b^4 = 5a^2 b^4$
3. $2m^3 n^{-4} = 2 \cdot m^3 \cdot \dfrac{1}{n^4} = \dfrac{2m^3}{n^4}$
4. $\dfrac{a^{-5}}{b^6} = \dfrac{1}{a^5} \cdot \dfrac{1}{b^6} = \dfrac{1}{a^5 b^6}$
5. $7x^3 y^{-8} z^0 = 7 \cdot x^3 \cdot \dfrac{1}{y^8} \cdot 1 = \dfrac{7x^3}{y^8}$
6. $9m^0 n^{-6} t^{-2} = 9 \cdot 1 \cdot \dfrac{1}{n^6} \cdot \dfrac{1}{t^2} = \dfrac{9}{n^6 t^2}$

Check Point 6, Frame 36

1. $\dfrac{5x^{-3}}{yz} = \dfrac{5}{x^3 yz}$
2. $9(a-b)^{-4} = \dfrac{9}{(a-b)^4}$
3. $\dfrac{2^{-3} m^2 n^{-1}}{3^{-1} t^3} = \dfrac{3m^2}{2^3 n t^3} = \dfrac{3m^2}{8nt^3}$
4. $\left(\dfrac{-7x^2 y^{-3}}{2^{-3} z^4}\right)^2 = \dfrac{(-7)^2 x^4 y^{-6}}{2^{-6} z^8} = \dfrac{3136x^4}{y^6 z^8}$
5. $\left(\dfrac{3^{-1} r^2 t^{-3}}{-2sz^{-1}}\right)^{-3} = \dfrac{3^3 r^{-6} t^9}{(-2)^{-3} s^{-3} z^3} = \dfrac{27(-2)^3 s^3 t^9}{r^6 z^3} = \dfrac{-216s^3 t^9}{r^6 z^3}$

SOLUTIONS TO CHECK POINTS

Check Point 7, Frame 41

1. $(2a^7)(5a^{-10}) = 10a^{7+(-10)} = 10a^{-3} = \dfrac{10}{a^3}$

For $a = 3$, $\dfrac{10}{a^3} = \dfrac{10}{3^3} = \dfrac{10}{27}$

2. $x^0(x^{-2}y^3z)^2 = 1(x^{-4}y^6z^2) = \dfrac{y^6z^2}{x^4}$

For $x = 3$, $y = 2$, $z = 9$

$\dfrac{y^6z^2}{x^4} = \dfrac{2^6 \cdot 9^2}{3^4} = \dfrac{64 \cdot 81}{81} = 64$

3. $\left(\dfrac{m^{-1}n^{-3}}{2r^5}\right)^{-3} = \dfrac{m^3n^{-6}}{2^{-3}r^{-15}} = \dfrac{2^3m^3r^{15}}{n^6} = \dfrac{8m^3r^{15}}{n^6}$

For $m = 3$, $n = 2$, $r = 1$

$\dfrac{8m^3r^{15}}{n^6} = \dfrac{8(3)^3(1)^{15}}{2^6} = \dfrac{8(27)(1)}{64} = \dfrac{27}{8}$

4. $2x(x^{-3}y^2)^{-1}(3x^4y^3z^{-1})^0 = 2x(x^3y^{-2}) = \dfrac{2x^4}{y^2}$

For $x = 3$, $y = -5$, $z = 6$

$\dfrac{2x^4}{y^2} = \dfrac{2(3)^4}{(-5)^2} = \dfrac{(2)(81)}{25} = \dfrac{162}{25}$ or $6\dfrac{12}{25}$

5. $\left(\dfrac{x^{-2}y}{y^{-3}x^{-5}}\right)^{-2} = \dfrac{x^4y^{-2}}{y^6x^{10}} = \dfrac{1}{x^6y^8}$

For $x = -1$, $y = -2$

$\dfrac{1}{x^6y^8} = \dfrac{1}{(-1)^6(-2)^8} = \dfrac{1}{1(256)} = \dfrac{1}{256}$

6. $\left(\dfrac{3r^2t^{-1}}{r^{-3}t^4}\right)\left(\dfrac{4r^4}{9t}\right)^{-1} = \left(\dfrac{3r^5}{t^5}\right)\left(\dfrac{4^{-1}r^{-4}}{9^{-1}t^{-7}}\right) = \left(\dfrac{3r^5}{t^5}\right)\left(\dfrac{9t^7}{4r^4}\right) = \dfrac{27rt^2}{4}$

For $r = -7$, $t = 4$

$\dfrac{27rt^2}{4} = \dfrac{27(-7)(4)^2}{4} = 27(-7)(4) = -756$

SOLUTIONS TO CHECK POINTS

Check Point 8, Frame 43

1. $\dfrac{2.56}{10^3} = 2.56 \times 10^{-3}$ 2. $\dfrac{1}{R^2} = 1 \times R^{-2}$ 3. $\dfrac{g}{4\pi^2f^2} = g \times (4^{-1}\pi^{-2}f^{-2}) = 4^{-1}\pi^{-2}f^{-2}g$

Check Point 9, Frame 52

I 1. $801.43 = 8.0143 \times 10^2$ 2. $0.00091 = 9.1 \times 10^{-4}$

3. $7.21 = 7.21 \times 10^0$ 4. $81\ 400\ 000 = 8.14 \times 10^7$

II 5. $3.045 \times 10^{-2} = 0.030\ 45$ 6. $1.09 \times 10^4 = 10\ 900$

7. $5 \times 10^{-6} = 0.000\ 005$ 8. $7 \times 10^3 = 7000$

Check Point 10, Frame 60

1. $250\ 000 \times 30\ 000\ 000 = 2.5 \times 10^5 \times 3 \times 10^7 = 7.5 \times 10^{12}$

$= 7\ 500\ 000\ 000\ 000$

2. $\dfrac{0.0048}{0.000\ 016} = \dfrac{4.8 \times 10^{-3}}{1.6 \times 10^{-5}} = 3 \times 10^2 = 300$

3. $\dfrac{0.002 \times 0.000\ 015}{0.0003 \times 0.04} = \dfrac{2 \times 10^{-3} \times 1.5 \times 10^{-5}}{3 \times 10^{-4} \times 4 \times 10^{-2}} = \dfrac{3 \times 10^{-8}}{12 \times 10^{-6}}$

$= 0.25 \times 10^{-2} = 0.0025$

4. $\dfrac{125\ 000 \times 400}{2\ 000\ 000 \times 10\ 000} = \dfrac{1.25 \times 10^5 \times 4 \times 10^2}{2 \times 10^6 \times 1 \times 10^4} = \dfrac{5 \times 10^7}{2 \times 10^{10}}$

$= 2.5 \times 10^{-3} = 0.0025$

SOLUTIONS/ANSWERS UNIT 3

ANSWERS TO DRILL EXERCISES

1. 1
2. a^{21}
3. 1
4. $32b^5$
5. y^4
6. $(x + y)^5$
7. $32a^5 b^5$ or $(2ab)^5$
8. $3x + 5$
9. 100
10. $(x - 2)^7$
11. $\dfrac{x^{15}}{a^{30}}$
12. $\dfrac{1}{x^8}$
13. $\dfrac{27m^6 n^3}{8s^9 t^9}$
14. 121
15. $x^{12} y^{12}$
16. 23
17. $(y - 3)^3$
18. $144a^{10} b^6$
19. x^7
20. $\dfrac{1}{16}$
21. 729
22. $\dfrac{125}{64}$ or $1 \dfrac{61}{64}$
23. $\dfrac{b^6}{a^4}$
24. $\dfrac{x^3 y^6}{4}$
25. $\dfrac{3}{m^7 n^6}$
26. $\dfrac{1}{p^3 q^2}$
27. $\dfrac{8}{m^3 s^4}$
28. $\dfrac{5z^2}{y^7}$
29. 100
30. $\dfrac{1}{100}$
31. $\dfrac{75}{81}$
32. $\dfrac{9}{56}$
33. $\dfrac{64}{125}$
34. $\dfrac{2025}{4}$ or $506 \dfrac{1}{4}$
35. 3.74×10^{-5}
36. $KmMr^{-2}$
37. $VR^{-1}e - \dfrac{t}{RC}$
38. 3.75×10^{0}
39. 4×10^{-5}
40. 3.1×10^{-1}
41. 6.34×10^{-2}
42. 7.05321×10^{6}
43. 6.1527×10^{2}
44. 203 000
45. 4 000 000
46. 0.03
47. 0.007 102
48. 52 000 000
49. 0.000 09
50. 4 240 000 000
51. 0.000 21
52. 30 000
53. 13.888 889
54. 0.000 045
55. 1200.9628
56. 3.253 655 8
57. 0.000 000 155 000 31
58. 0.256 25

SOLUTIONS/ANSWERS UNIT 4

SOLUTIONS TO CHECK POINTS

Check Point 1, Frame 5

I   1. $5 - x^3 + 2x - 6x^2$; Terms are 5, $-x^3$, $2x$, $-6x^2$.

   2. $2a^3 b^3 - ab - 3a^4 b^4 + 12 - 8a^2 b^2$ Terms are $2a^3 b^3$, $-ab$, $-3a^4 b^4$, 12, $- 8a^2 b^2$.

II 1. $5 - x^3 + 2x - 6x^2 = 5 + (-x^3) + 2x + (-6x^2)$

      $= -x^3 - 6x^2 + 2x + 5$

   2. $2a^3 b^3 - ab - 3a^4 b^4 + 12 - 8a^2 b^2$

      $= 2a^3 b^3 + (-ab) + (-3a^4 b^4) + 12 + (-8a^2 b^2)$

      $= -3a^4 b^4 + 2a^3 b^3 - 8a^2 b^2 - ab + 12$

Check Point 2, Frame 12

I   1. 3 is the numerical coefficient of the term $3m^2 t$.

      -2 is the numerical coefficient of the term $-2mt^2$.

      4 is the numerical coefficient of the term $4m$.

   2. -1 is the numerical coefficient of the term $-n$.

      1 is the numerical coefficient of the term x^2.

      -1 is the numerical coefficient of the term $-xy$.

      -1 is the numerical coefficient of the term $-y^2$.

II 3. ab^2 and $6ab^2$ are the like terms.

   4. x^2, $-3x^2$; $4x^2 y$, $-5x^2 y$

   5. $7mn^2 t^3$, $mn^2 t^3$

SOLUTIONS TO CHECK POINTS

Check Point 3, Frame 19

1. $-xy + 2x^2 + 10xy - y - x^2 + 3y = x^2 + 9xy + 2y$

2. $4a^3 - 3a^4 - 2a^3 + 5ab + 5a^4 - 3ab = 2a^4 + 2a^3 + 2ab$

3. $2m^4nt^3 - 3mn + 6m^4nt^3 - 9mn - n + 2mn = 8m^4nt^3 - 10mn - n$

4. $6a + 2b + 3c$ cannot be simplified.

Check Point 4, Frame 25

1. $(y^2 - y) - (4y + 6) = y^2 - y - 4y - 6$

   $= y^2 - 5y - 6$

2. $-(a^2 - 4a + 1) + (3a^2 - a + 2) - (-a^2 + 3a)$

   $= -a^2 + 4a - 1 + 3a^2 - a + 2 + a^2 - 3a$

   $= 3a^2 + 0a + 1$

   $= 3a^2 + 1$

3. $4pq - [6p + (3pq - 4)] = 4pq - [6p + 3pq - 4]$

   $= 4pq - 6p - 3pq + 4$

   $= pq - 6p + 4$

4. $m^2n^2 - [3m^2n - (-m + n - 1)] = m^2n^2 - [3m^2n + m - n + 1]$

   $= m^2n^2 - 3m^2n - m + n - 1$

5. $x^2 - \{4x - [2x - (6x - 3)] - 3x^2 + 9\} = x^2 - \{4x - [2x - 6x + 3] - 3x^2 + 9\}$

   $= x^2 - \{4x - 2x + 6x - 3 - 3x^2 + 9\}$

   $= x^2 - 4x + 2x - 6x + 3 + 3x^2 - 9$

   $= 4x^2 - 8x - 6$

Check Point 5, Frame 29

1.
$$
\begin{array}{r}
a - 3 \\
4a^2 + 6a \\
3a^2 - 9a - 4 \\
\hline
7a^2 - 2a - 7
\end{array}
$$

2.
$$
\text{Subtract:}\quad
\begin{array}{r}
3y^2 - 6y + 4 \\
y^2 + 2y + 8
\end{array}
\qquad
\text{Add:}\quad
\begin{array}{r}
3y^2 - 6y + 4 \\
-y^2 - 2y - 8 \\
\hline
2y^2 - 8y - 4
\end{array}
$$

ANSWERS TO DRILL EXERCISES

I   1. $8a^2 - 20ab + 10b^2$ 2. $-x^3 - x^2 - 3xy$

    3. $-2x^2yz - xz - x$ 4. $m + 3m^2 - 9p$

II   5. $-r - 3a$ 6. $a^3b^3 + 3ab - 2b + a$

    7. $4x^2 + 2x - 7$ 8. $m^2 - 3m + 2n$

    9. $x - 3xy - x^2 + 7z - y$ 10. $p^3 - q^3 + p^2q^2 + pq - 2p + 7$

III   11. $a^3 + a^2 - 3a + 2$ 12. $5x^2 + 2x - 5$

    13. $-mn + 7n + 2$ 14. $-2x^2y - x^2 - 6x + y - 3$

    15. $-7p^2 - 6p + 12$

IV   16. $6x^2 + 3x + 1$ 17. $-3y^2 + 15y + 2$

    18. $2a^3 + 6a^2 - 4a + 4$ 19. $2a^4 - 5a^3 + a^2 - 7a - 1$

    20. $3x^4 - x^3 - 3x^2 - 2x + 6$

SOLUTIONS TO CHECK POINTS

Check Point 3, Frame 18

1. $3x + 5 = 2x + 12$
   $3x - 2x = 12 - 5$
   $x = 7$

   Check: $3(7) + 5 = 2(7) + 12$
   $21 + 5 = 14 + 12$
   $26 = 26$

2. $18 = 4x + 26$
   $-4x = 26 - 18$
   $-4x = 8$
   $x = -2$

   Check: $18 = 4(-2) + 26$
   $18 = -8 + 26$
   $18 = 18$

3. $-x + 4 = 9$
   $-x = 9 - 4$
   $-x = 5$
   $(-1)(-x) = (-1)(5)$
   $x = -5$

   Check: $-(-5) + 4 = 9$
   $5 + 4 = 9$
   $9 = 9$

4. $3 + 12 = 3x + 2x$
   $3x + 2x = 3 + 12$
   $5x = 15$
   $\dfrac{5x}{5} = \dfrac{15}{5}$
   $x = 3$

   Check: $3 + 12 = 3(3) + 2(3)$
   $15 = 9 + 6$
   $15 = 15$

5. $-3(w - 1) = -2(w - 4)$
   $-3w + 3 = -2w + 8$
   $-3w + 2w = 8 - 3$
   $-w = 5$
   $(-1)(-w) = (-1)(5)$
   $w = -5$

   Check: $-3(w - 1) = -2(w - 4)$
   $-3(-5 - 1) = -2(-5 - 4)$
   $-3(-6) = -2(-9)$
   $18 = 18$

SOLUTIONS/ANSWERS UNIT 5

SOLUTIONS TO CHECK POINTS

Check Point 1, Frame 5

I 1. No, since 2(10) = 30 is false.

   2. No, since 10 + 1 = 9 is false.

   3. Yes, since 2(10 - 10) = 0 is true.

   4. No, since 10[3(10) - 30]= 10 is false.

   5. Yes, since 3(10) + 10 + 6 = 4(10) + 6 is true.   (In fact, it is an identity.)

II 6. Identity. True for every value of y.

   7. Conditional. True only for x = 6.

   8. Identity. True for every value of x.

   9. Conditional. True only for y = 1.

   10. Identity. True for every value of x.

Check Point 2, Frame 8

1. False.

2. True by Addition Property.

3. True by Division Property.

4. True by Subtraction Property.

5. False.

6. True by Multiplication Property.

7. True by Symmetry Property.

SOLUTIONS TO CHECK POINTS

Check Point 3, Frame 18

6. $8y - 6(2y + 5) = 12 + 2(y + 3)$ Check: $8(-8) - 6[2(-8) + 5] = 12 + 2(-8 + 3)$

$$8y - 12y - 30 = 12 + 2y + 6$$
$$-64 - 6[-16 + 5] = 12 + 2(-5)$$
$$-4y - 30 = 18 + 2y$$
$$-64 - 6[-11] = 12 - 10$$
$$-4y - 2y = 18 + 30$$
$$-64 + 66 = 2$$
$$-6y = 48$$
$$2 = 2$$
$$\frac{-6y}{-6} = \frac{48}{-6}$$
$$y = -8$$

Check Point 4, Frame 24

1.   The LCD of $\frac{2}{3}$, $\frac{5}{6}$ and $\frac{1}{18}$ is 18, since 3 and 6 are factors of 18.

2.   The LCD of $\frac{1}{x}$ and $\frac{3}{y}$ is xy.

3.   The LCD of $\frac{7}{n+1}$ and $\frac{a}{m-1}$ is $(n + 1)(m - 1)$.

4.   The LCD of $\frac{z}{3y}$ and $\frac{9a}{5b}$ is $15by$.

5.   The LCD of $\frac{1}{(p+1)^2}$ and $\frac{p}{p+1}$ is $(p + 1)^2$, since $p + 1$ is a factor of $(p + 1)^2$.

6.   The LCD of $\frac{1}{4R}$ and $\frac{a}{8R^2}$ is $8R^2$, since $4R$ is a factor if $8R^2$.

Check Point 5, Frame 31

1.   The LCD is 12.

$$12\left(\frac{2m}{3} - \frac{1}{4}\right) = 12 \cdot \frac{1}{6}$$
$$12 \cdot \frac{2m}{3} - 12 \cdot \frac{1}{4} = 12 \cdot \frac{1}{6}$$
$$8m - 3 = 2$$
$$8m = 5$$
$$m = \frac{5}{8}$$

Check:

$$\text{L.H.S.} = \frac{2 \cdot \frac{5}{8}}{3} - \frac{1}{4} = \frac{5}{4} \cdot \frac{1}{3} - \frac{1}{4} = \frac{5}{12} - \frac{3}{12} = \frac{2}{12} = \frac{1}{6}$$
$$\text{R.H.S.} = \frac{1}{6}$$

Hence, the solution is $m = \frac{5}{8}$.

SOLUTIONS TO CHECK POINTS

Check Point 5, Frame 31

2.   The LCD is 12.

$$12\left(\frac{R}{3} - \frac{R}{4}\right) = 12 \cdot \frac{2}{3}$$
$$12 \cdot \frac{R}{3} - 12 \cdot \frac{R}{4} = 12 \cdot \frac{2}{3}$$
$$4R - 3R = 8$$
$$R = 8$$

Check:

$$\text{L.H.S.} = \frac{8}{3} - \frac{8}{4} = 2\frac{2}{3} - 2 = \frac{2}{3}$$
$$\text{R.H.S.} = \frac{2}{3}$$

Hence, the solution is $R = 8$.

3.   The LCD is 6.

$$6\left(\frac{2M}{3} - \frac{M+1}{2}\right) = 6 \cdot \frac{1}{3}$$
$$6 \cdot \frac{2M}{3} - 6 \cdot \frac{M+1}{2} = 6 \cdot \frac{1}{3}$$
$$4M - 3(M + 1) = 2$$
$$4M - 3M - 3 = 2$$
$$M = 5$$

Check:

$$\text{L.H.S.} = \frac{2(5)}{3} - \frac{5+1}{2} = \frac{10}{3} - 3 = 3\frac{1}{3} - 3 = \frac{1}{3}$$
$$\text{R.H.S.} = \frac{1}{3}$$

Hence, the solution is $M = 5$.

Check Point 6, Frame 40

1.

$$V_1 P_1 = V_2 P_2$$
$$\frac{V_1 P_1}{P_2} = \frac{V_2 \cancel{P_2}}{\cancel{P_2}}$$
$$\text{Therefore, } V_2 = \frac{V_1 P_1}{P_2}.$$

2.

$$E = \frac{MV^2}{R}$$
$$ER = \frac{MV^2}{\cancel{R}} \cdot \cancel{R}$$
$$ER = MV^2$$
$$\frac{\cancel{E}R}{\cancel{E}} = \frac{MV^2}{E}$$
$$\text{Therefore, } R = \frac{MV^2}{E}$$

ANSWERS TO DRILL EXERCISES

1. Identity. True for every value of x.
2. Conditional. True only for x = 3.
3. Conditional. True only for y = 6.
4. Identity. True for every value of x.
5. Identity. True for every value of x.

6. $y = 7$	7. $x = 6$
8. $w = 2$	9. $x = \frac{1}{2}$
10. $w = 8$	11. $y = 1$
12. $y = -1\frac{2}{3}$	13. $m = 2$
14. $x = -4$	15. $t = 3$
16. $m = 28$	17. $p = \frac{3}{4}$
18. $V = \frac{1}{10}$	19. $R = \frac{21}{16}$
20. $x = -\frac{35}{8}$	21. $G = \frac{1}{5}$
22. $F = 10$	23. $V_2 = 36$
24. $S = 1026.67$	25. $C = -6.7^\circ$
26. $d = 0.011$	27. $x = \frac{y - 7}{2}$
28. $y = \frac{15x + 26}{2}$	29. $y = \frac{-3x + 24}{13}$
30. $x = \frac{12y + 8}{9}$	31. $y = \frac{-5x + 2}{8}$
32. $r = \frac{C}{2\pi}$	33. $P_2 = \frac{V P_1}{V_1}$
34. $V = \frac{IbS}{Q}$	35. $P = \frac{A}{(1 + i)^n}$
36. $M = \frac{2T}{g + a}$	37. $m_1 = \frac{r^2 F}{Gm_2}$
38. $C = L - N$	39. $A = \frac{P - D}{F}$
40. $S_2 = P - S_1 - S_3$	41. $m = \frac{y - b}{x}$
42. $Z = N(X - L)$	43. $t = \frac{n}{V_2 - V_1}$
44. $a = \frac{2T - gM}{M}$	45. $D = P - AF$

SOLUTIONS TO CHECK POINTS

Check Point 6, Frame 40

3.
$$Q = C(T - t)$$
$$\frac{Q}{T - t} = \frac{C(T - t)}{T - t}$$

Therefore, $C = \dfrac{Q}{T - t}$.

4.
$$\overline{X} = u + \sigma z$$
$$-\sigma z = u - \overline{X}$$
$$\frac{-\sigma z}{-\sigma} = \frac{u - \overline{X}}{-\sigma}$$
$$z = \frac{u - \overline{X}}{-\sigma}$$
$$z = \frac{(-1)(u - \overline{X})}{(-1)(-\sigma)}$$
$$= \frac{-u + \overline{X}}{\sigma}$$
$$= \frac{\overline{X} - u}{\sigma}$$

Therefore, $z = \dfrac{\overline{X} - u}{\sigma}$.

5.
$$P = 2L + 2W$$
$$-2L = -P + 2W$$
$$\frac{-2L}{-2} = \frac{-P + 2W}{-2}$$
$$L = \frac{-P + 2W}{-2}$$
$$L = \frac{(-1)(-P + 2W)}{(-1)(-2)}$$
$$= \frac{P - 2W}{2}$$

Therefore, $L = \dfrac{P - 2W}{2}$.

6.
$$T = \frac{1}{a} + t$$
$$-\frac{1}{a} = t - T$$
$$a\left(-\frac{1}{a}\right) = a(t - T)$$
$$-1 = a(t - T)$$
$$\frac{-1}{t - T} = \frac{a(t - T)}{t - T}$$
$$a = \frac{-1}{t - T}$$
$$= \frac{(-1)(-1)}{(-1)(t - T)}$$
$$= \frac{1}{-t + T}$$
$$= \frac{1}{T - t}$$

Therefore, $a = \dfrac{1}{T - t}$.

SOLUTIONS TO CHECK POINTS

Check Point 1, Frame 6

1. $n + 10$

2. $12n$

3. $n - 4$

4. $2n + 8$

5. $20 - 5n$

Check Point 2, Frame 13

1. Let n = the number.

    $2n + 25 = 103$
    $2n = 103 - 25$
    $2n = 78$
    $n = 39$

    Check: $2(39) + 25 = 78 + 25 = 103$

    Hence, 39 is correct. ∴ the number is 39.

2. Let n = the number.

    $3n - 75 = -39$
    $3n = -39 + 75$
    $3n = 36$
    $n = 12$

    Check: $3(12) - 75 = -39$
    $36 - 75 = -39$
    $-39 = -39$

    Hence, 12 is correct. ∴ the number is 12.

3. Let n = the number.

    $5n + 10 = 45$
    $5n = 45 - 10$
    $5n = 35$
    $n = 7$

    Check: $5(7) + 10 = 35 + 10 = 45$

    Hence, 7 is correct. ∴ the number is 7.

Check Point 3, Frame 19

1. If we knew the second number, we could evaluate the first number.

    Let x = the second number.
    Then $3x$ = the first number.

2. Let x = the first number.
    Then $x - 6$ = the second number.

3. Let x = the length of the third piece.
    Then $2x - 2$ = the length of the first piece.
    Then $x + 10$ = the length of the second piece.

4. Let x = the second number.
    Then $4x - 8$ = the first number.
    Then $2[(4x - 8) + x]$ = the third number.

SOLUTIONS TO CHECK POINTS

Check Point 4, Frame 25

1. Let n = the number of amperes carried by the first circuit.
    Then $3n + 12$ = the number carried by the second circuit.
    Then $2[n + (3n + 12)]$ = the number carried by the third circuit.

    Then: $n + (3n + 12) + 2[n + (3n + 12)] = 84$
    $n + 3n + 12 + 2[n + 3n + 12] = 84$
    $n + 3n + 12 + 2n + 6n + 24 = 84$
    $12n + 36 = 84$
    $12n = 84 - 36$
    $12n = 48$
    $n = 4$

    $3n + 12 = 3(4) + 12 = 12 + 12 = 24$

    $2[n + (3n + 12)] = 2[4 + 24] = 2(28) = 56$

    Hence, the first circuit carries 4 amperes, the second 24 and the third carries 56 amperes.

    Check: $4 + 24 + 56 = 84$ amperes.

2. Let L = the length of the rectangle.
    $\frac{L}{4} - 3$ = the width of the rectangle.

    "The perimeter is 54 cm." $(P = 2L + 2W)$

    $2L + 2(\frac{L}{4} - 3) = 54$
    $2L + \frac{L}{2} - 6 = 54$
    $2(2L + \frac{L}{2} - 6) = 2 \cdot 54$
    $4L + L - 12 = 108$
    $5L = 120$
    $L = 24$

    If $L = 24$, then $\frac{L}{4} - 3 = \frac{24}{4} - 3 = 6 - 3 = 3$.

    Hence, the length is 24 cm and the width is 3 cm.

    Check: $2(24) + 2(3) = 48 + 6 = 54$

L

$\frac{L}{4} - 3$

SOLUTIONS TO CHECK POINTS

Check Point 4, Frame 25

3.   Let x = the length of the larger piece.

$\frac{1}{3}$ x - 8 = the length of the smaller piece.

"A wire which is 36 m long is cut into two pieces."

$$x + \frac{1}{3} x - 8 = 36$$

$$3(x + \frac{1}{3} x - 8) = 3 \cdot 36$$

$$3x + x - 24 = 108$$

$$4x = 132$$

$$x = 33$$

If the first piece x = 33 m, then the length of the second piece

is $\frac{1}{3}$ x - 8 = $\frac{1}{3}$ (33) - 8 = 11 - 8 = 3m.

Hence, the lengths of the two pieces are 33m and 3m.

Check: 33 + 3 = 36.

Let x = his sales last month.

$$375 + \frac{15x}{100} = \text{his pay last month.}$$

$$375 + \frac{15x}{100} = 4602$$

$$100 \left(375 + \frac{15x}{100}\right) = 100(4602)$$

$$100(375) + 100(\frac{15x}{100}) = 100(4602)$$

$$37\,500 + 15x = 460\,200$$

$$15x = 422\,700$$

$$x = 28\,180$$

If x = 28 180, then 375 + $\frac{15}{100}$ (28 180) = 375 + 4227 = 4602.

Hence, his sales last month were $28 180.

SOLUTIONS TO CHECK POINTS

Check Point 5, Frame 28

1.   If one number is x,
     then the other is 120 - x.

2.   If the length of one piece is x cm,
     then the length of the other piece is 350 - x cm.

3.   If the number of dimes is x,
     then the number of quarters is 75 - x.

4.   If the first parcel has x acres,
     then the second parcel has 240 - x acres.

5.   If the amount of interest is x,
     then the amount of principal is 2350 - x dollars.

Check Point 6, Frame 38

1.   Let x = the number of grams of nickel to be added.

400 + x = the number of grams of the 10% alloy.

Pure silver in original alloy	=	Pure silver in new alloy
80 g	=	10% of (400 + x)

$$80 = \frac{10}{100} (400 + x)$$

$$10 \cdot 80 = 10 \cdot \frac{1}{10} (400 + x)$$

$$800 = 400 + x$$

$$x = 400$$

Hence, we have to add 400 grams.

Check: 10% of 800 = 80.

ANSWERS TO DRILL EXERCISES

1. $x + 17$

2. $6n$

3. $4m - 9$

4. $x + 9x^2$

5. $11z + 3$

6. Lengths: 134 cm, 66 cm

7. Cost of house: $80 000

8. 20 type A batteries
   12 type B batteries

9. 4 nickels, 8 quarters and 5 dimes

10. Number: 3

11. Number: 5

12. Number: 9

13. Number: 75

14. Numbers: 3 and 75

15. Interest: $390

16. Amount earned: $22 000

17. Profit: (a) $33\frac{1}{3}$ %
    (b) 16.67%

18. Interest: $2250
    Percentage: 45%

19. $35 000 at 18%
    $25 000 at 12%

20. Length: 27 cm
    Width: 20 cm

21. Lengths: 5 m, 6 m, 8 m

22. Lengths: 3.5m, 1.5m

23. Water added: 750 mL

24. 50 000 kg of 65% alloy
    100 000 kg of 20% alloy

25. Copper added: 40 kg

SOLUTIONS TO CHECK POINTS

Check Point 6, Frame 38

2. Let x = the number of litres of the 5% solution required.

   $4 - x$ = the number of litres of the 20% solution required.

Pure hydrochloric acid in 5% solution		Pure hydrochloric acid in 20% solution		Pure hydrochloric acid in 10% solution
5% of x	+	20% of $(4 - x)$	=	10% of 4

   $$\frac{5}{100}x + \frac{20}{100}(4 - x) = \frac{10}{100}(4)$$
   $$5x + 20(4 - x) = 10(4)$$
   $$5x + 80 - 20x = 40$$
   $$-15x = -40$$
   $$x = \frac{40}{15} = \frac{8}{3} \text{ or } 2\frac{2}{3}$$

   If $x = \frac{8}{3}$, then $4 - \frac{8}{3} = \frac{4}{3}$ or $1\frac{1}{3}$.

   Hence, to obtain 4 litres of a 10% solution, we require $2\frac{2}{3}$ litres of the 5% solution and $1\frac{1}{3}$ litres of the 20% solution.

   Check: $2\frac{2}{3} + 1\frac{1}{3} = 4$.

3. Let p = the amount invested at 14%.

   $500\ 000 - p$ = the amount invested at 17%.

   "He has a guaranteed income of $74 500 each year."

   $$\frac{14}{100}p + \frac{17}{100}(500\ 000 - p) = 74\ 500$$
   $$100 \cdot \frac{14}{100}p + 100 \cdot \frac{17}{100}(500\ 000 - p) = 100(74\ 500)$$
   $$14p + 8\ 500\ 000 - 17p = 7\ 450\ 000$$
   $$-3p = -1\ 050\ 000$$
   $$p = 350\ 000$$

   If $p = 350\ 000$, then $500\ 000 - p = 500\ 000 - 350\ 000 = 150\ 000$.

   Hence, $350 000 has been invested at 14% and $150 000 has been invested at 17%.

   Check: Interest earned at 14% is $\frac{14}{100}(350\ 000) = $49\ 000$.

   Interest earned at 17% is $\frac{17}{100}(150\ 000) = $ 25\ 500.

   Total interest earned on $500 000 $= $74\ 500$ is correct.

SOLUTIONS TO CHECK POINTS

CHECK POINT 1, FRAME 7

1. trinomial
2. monomial
3. binomial
4. binomial
5. monomial
6. monomial

CHECK POINT 2, FRAME 13

1. $(2x^3y)(4x^2y^2) = 8x^5y^3$

2. $(-a^3b^{-3})(4a^2b^3) = -4a^5b^0$
   $= -4a^5$

3. $(5m^3n^2t)(-2mn^{-1}t^3)(-m^{-2}n^3t^{-2}) = 10m^2n^4t^2$

4. $(p^4q)(-3pqt)(6p^{-5}q^{-2}t^2) = -18p^0q^0t^3$
   $= -18t^3$

CHECK POINT 3, FRAME 16

1. $5(a^2b - c^2) = 5a^2b - 5c^2$

2. $-xy(3x^4z + 2y^2z^3) = (-xy)(3x^4z) + (-xy)(2y^2z^3)$
   $= -3x^5yz - 2xy^3z^3$

3. $2pq(5p - 3q + 9) = (2pq)(5p) - (2pq)(3q) + (2pq)(9)$
   $= 10p^2q - 6pq^2 + 18pq$

4. $-3mn^2(2m - 4n - t + 6) = (-3mn^2)(2m) - (-3mn^2)(4n) - (-3mn^2)(t) + (-3mn^2)(6)$
   $= -6m^3n + 12mn^3 + 3mn^2t - 18mn^2$

CHECK POINT 4, FRAME 20

I 1. $(y + 6)(3y^2 + 4y + 1) = y(3y^2 + 4y + 1) + 6(3y^2 + 4y + 1)$
   $= 3y^3 + 4y^2 + y + 18y^2 + 24y + 6$
   $= 3y^3 + 22y^2 + 25y + 6$

   2. $(m - n)(m^2 - mn + n^2) = m(m^2 - mn + n^2) - n(m^2 - mn + n^2)$
   $= (m^3 - m^2n + mn^2) - (m^2n - mn^2 + n^3)$
   $= m^3 - 2m^2n + 2mn^2 - n^3$

II 3. $(3a + 2)(a^2 - 2a - 3) = (3a + 2)a^2 - (3a + 2)2a - (3a + 2)3$
   $= (3a^3 + 2a^2) - (6a^2 + 4a) - (9a + 6)$
   $= 3a^3 + 2a^2 - 6a^2 - 4a - 9a - 6$
   $= 3a^3 - 4a^2 - 13a - 6$

   4. $(2x^2 + 3x - 1)(5x^2 - 2x - 4) = (2x^2 + 3x - 1)5x^2 - (2x^2 + 3x - 1)2x - (2x^2 + 3x - 1)$
   $= (10x^4 + 15x^3 - 5x^2) - (4x^3 + 6x^2 - 2x) - (8x^2 + 12x - 4)$
   $= 10x^4 + 15x^3 - 5x^2 - 4x^3 - 6x^2 + 2x - 8x^2 - 12x + 4$
   $= 10x^4 + 11x^3 - 19x^2 - 10x + 4$

CHECK POINT 5, FRAME 24

1.
$$\begin{array}{r} 2x^2 - xy + y^2 \\ 3x + y \\ \hline 6x^3 - 3x^2y + 3xy^2 \\ 2x^2y - xy^2 + y^3 \\ \hline 6x^3 - x^2y + 2xy^2 + y^3 \end{array}$$

2.
$$\begin{array}{r} p^2 + 3p - 4 \\ 4p^2 - 2p + 3 \\ \hline 4p^4 + 12p^3 - 16p^2 \\ -2p^3 - 6p^2 + 8p \\ 3p^2 + 9p - 12 \\ \hline 4p^4 + 10p^3 - 19p^2 + 17p - 12 \end{array}$$

3.
$$\begin{array}{r} 3a - 4b + 6 \\ a - 2b \\ \hline 3a^2 - 4ab + 6a \\ -6ab + 8b^2 - 12b \\ \hline 3a^2 - 10ab + 6a + 8b^2 - 12b \end{array}$$

CHECK POINT 8, FRAME 36

1. $(x + 7)^2 = x^2 + 2(7x) + 7^2$

   $= x^2 + 14x + 49$

2. $(p - 3)^2 = p^2 + 2(-3p) + (-3)^2$

   $= p^2 - 6p + 9$

3. $(2m + 3n)^2 = (2m)^2 + 2(6mn) + (3n)^2$

   $= 4m^2 + 12mn + 9n^2$

4. $(3y - 5z)^2 = (3y)^2 + 2(-15yz) + (-5z)^2$

   $= 9y^2 - 30yz + 25z^2$

CHECK POINT 9, FRAME 40

1. $(r + 1)(r^2 - r + 1) = r^3 + 1^3 = r^3 + 1$

2. $(m - 1)(m^2 + m - 1)$ $\overset{(-1)^2 = +1}{\longrightarrow}$ $\therefore$ cannot be simplified without multiplying multinomials by the distributive property

3. $(2a + 3b)(4a^2 - 6ab + 9b^2) = (2a)^3 + (3b)^3 = 8a^3 + 27b^3$

4. $(v - 5p)(v^2 + 5vp + 25p^2) = v^3 - (5p)^3 = v^3 - 125p^3$

CHECK POINT 6, FRAME 30

          F   O   I   L

1. $(x - 2)(x + 3) = x^2 + 3x - 2x - 6$

   $= x^2 + x - 6$

          F   O   I   L

2. $(2a - b)(5a - 4b) = 10a^2 - 8ab - 5ab + 4b^2$

   $= 10a^2 - 13ab + 4b^2$

          F   O   I   L

3. $3x(x + 2)(x - 1) = 3x(x^2 - x + 2x - 2)$

   $= 3x(x^2 + x - 2)$

   $= 3x^3 + 3x^2 - 6x$

          F   O   I   L

4. $(x - 3)(x - 1)(x + 2) = (x^2 - x - 3x + 3)(x + 2)$

   $= (x^2 - 4x + 3)(x + 2)$

   $= (x^2 - 4x + 3)x + (x^2 - 4x + 3)2$

   $= x^3 - 4x^2 + 3x + x^2 - 4x^2 - 8x + 6$

   $= x^3 - 2x^2 - 5x + 6$

CHECK POINT 7, FRAME 33

1. $(y + 3)(y - 3) = y^2 - 9$

4. $(2p^4 + 3q^3)(2p^4 - 3q^3) = (2p^4)^2 - (3q^3)^2$

   $= 4p^8 - 9q^6$

2. $(5x + 2y)(5x - 2y) = (5x)^2 - (2y)^2$

   $= 25x^2 - 4y^2$

3. $(3y^2 + 4z)(3y^2 - 4z) = (3y^2)^2 - (4z)^2$

   $= 9y^4 - 16z^2$

ANSWERS TO DRILL EXERCISES

1. monomial
2. trinomial
3. monomial
4. binomial
5. monomial
6. $6m^3 n^4$
7. $-2y^5$
8. $6y^3 z^2$
9. $3p^3 q + 3r^3$
10. $-5a^4 bc^2 - 4ab^4 c^4$
11. $20m^2 n + 8mn^2 - 24mn$
12. $-10x^5 y - 5x^2 y^2 + 5x^2 yz + 20x^2 y$
13. $2x^3 + 11x^2 + 14x + 8$
14. $12m^2 - 11mn + 2n^2$
15. $2x^3 - 2x^2 - 12x$
16. $x^3 - 7x + 6$
17. $a^3 - 3a^2 b + 3ab^2 - b^3$
18. $3m^3 + m^2 - 7m - 4$
19. $2x^4 + 3x^3 - 13x^2 - 7x + 15$
20. $3x^3 + 14x^2 y + 7xy^2 - 4y^3$
21. $2m^2 - 7mn - 14m + 3n^2 + 7n$
22. $5a^4 - 13a^3 + a^2 + 13a - 6$
23. $y^2 + 16y + 64$
24. $x^2 + 8x + 16$
25. $a^2 - 16$
26. $x^2 - 8x + 16$
27. $9a^2 + 18ab + 9b^2$
28. $4m^2 - 9n^2$
29. $36x^4 - 4y^2$
30. $25y^2 - 20yz + 4z^2$
31. $v^3 + 1$
32. $8m^3 - 1$
33. $27x^3 - 125$
34. $x^3 + 64y^3$

SOLUTIONS TO CHECK POINTS

CHECK POINT 1, FRAME 3

1. $\dfrac{3m^4 n^7}{m^3 n} = 3m^{4-3} n^{7-4} = 3mn^3$

2. $(-6x^2 y^5) \div (3xy^2) = -2x^{2-1} y^{5-2}$
   $= -2xy^3$

3. $\dfrac{-a^4 b^3 c}{7a^4 bc} = \dfrac{-a^{4-4} b^{3-1}}{7c^{2-1}} = \dfrac{-b^2}{7c}$

4. $(-15p^5 t^4) \div (-5p^4 qt^2) = 3p^{5-4} q^{1-1} t^{4-2}$
   $= 3pt^2$

CHECK POINT 2, FRAME 6

1. $\dfrac{4a^3 - 8a^2 b}{4a} = \dfrac{4a^3}{4a} - \dfrac{8a^2 b}{4a}$
   $= a^2 - 2ab$

2. $\dfrac{8y^6 - 3y^4 + 6y^2}{2y^2} = \dfrac{8y^6}{2y^2} - \dfrac{3y^4}{2y^2} + \dfrac{6y^2}{2y^2}$
   $= 4y^4 - \dfrac{3y^2}{2} + 3$

3. $\dfrac{-16m^8 n^2 + 8m^5 n^3 - 4m^3 n^5}{4m^3 n^2} = \dfrac{-16m^8 n^2}{4m^3 n^2} + \dfrac{8m^5 n^3}{4m^3 n^2} - \dfrac{4m^3 n^5}{4m^3 n^2}$
   $= -4m^5 + 2m^2 n - n^3$

CHECK POINT 2, FRAME 6

4. $\dfrac{36x^7y^5z^2 - 9x^5y^3z^3 + 18x^4y^2z^4}{-9x^4y^2z^2} = \dfrac{36x^7y^5z^2}{-9x^4y^2z^2} - \dfrac{9x^5y^3z^3}{-9x^4y^2z^2} + \dfrac{18x^4y^2z^4}{-9x^4y^2z^2}$

$\qquad = -4x^3y^3 + xyz - 2z^2$

CHECK POINT 3, FRAME 9

1. $(6a^2 - a - 2) \div (2a + 1)$

```
              3a - 2
       ┌─────────────────
2a + 1 │ 6a² -  a - 2
         6a² + 3a
         ───────
              - 4a - 2
              - 4a - 2
              ───────
                    0
```

Hence, $(6a^2 - a - 2) \div (2a + 1) = 3a - 2$ R0

Check: $(2a + 1)(3a - 2) = 6a^2 - a - 2$

2. $(-2x - 1 + 3x^2) \div (x - 2)$

```
             3x + 4
       ┌───────────────
x - 2 │ 3x² - 2x - 1
        3x² - 6x
        ───────
              4x - 1
              4x - 8
              ──────
                   7
```

Hence, $3x^2 - 2x - 1 \div (x - 2) = 3x + 4$ R7

or $\dfrac{3x^2 - 2x - 1}{x - 2} = 3x + 4 + \dfrac{7}{x - 2}$

Check: $(x - 2)(3x + 4) + 7 = 3x^2 - 2x - 8 + 7$

$\qquad = 3x^2 - 2x - 1$

CHECK POINT 4, FRAME 14

1. $(6x^3 - 11x^2 - 7x + 15) \div (2x - 3)$

```
              3x² -  x - 5
       ┌───────────────────────
2x - 3 │ 6x³ - 11x² - 7x + 15
         6x³ -  9x²
         ─────────
              - 2x² - 7x
              - 2x² + 3x
              ──────────
                     -10x + 15
                     -10x + 15
                     ─────────
                            0
```

Hence, $(6x^3 - 11x^2 - 7x + 15) \div (2x - 3) = 3x^2 - x - 5$

Check: $(2x - 3)(3x^2 - x - 5) = 6x^3 - 11x^2 - 7x + 15$

2. $(x^3 + 1) \div (x + 1)$

```
              x² - x + 1
       ┌─────────────────
x + 1 │ x³         + 1
        x³ + x²
        ──────
            -x²
            -x² - x
            ───────
                  x + 1
                  x + 1
                  ─────
                      0
```

Hence, $(x^3 + 1) \div (x + 1) = x^2 - x + 1$

Check: $(x + 1)(x^2 - x + 1) = x^3 + 1$

3. $(4m^3 - 31m + 15) \div (2m^2 + m - 5)$

```
                  2m - 1
            ┌─────────────────────────
2m² + m - 5 │ 4m³        - 31m + 15
              4m³ + 2m² - 10m
              ──────────────
                  - 2m² - 21m + 15
                  - 2m² -   m +  5
                  ───────────────
                         - 20m + 10
```

Hence, $(4m^3 - 31m + 15) \div (2m^2 + m - 5) = 2m - 1$ R(-20m + 10)

Check: $(2m^2 + m - 5)(2m - 1) + (-20m + 10) = 4m^3 - 11m + 5 - 20m + 10$

$\qquad = 4m^3 - 31m + 15$

SOLUTIONS/ANSWERS UNIT 8

CHECK POINT 4, FRAME 14

4. $(27m^3 + 125n^3) \div (3m + 5n) = (9m^2 - 15mn + 25n^2)$

   since $(3m + 5n)(9m^2 - 15mn + 25n^2) = 27m^3 + 125n^3$

ANSWERS TO DRILL EXERCISES

1. $6x^2y$

2. $\dfrac{2a^2b^4}{c}$

3. $4y - 3$

4. $\dfrac{-4x^2z^2}{y^3}$

5. $-\dfrac{9s}{2r^2t}$

6. $-2 + 5vt$

7. $2ab + \dfrac{8}{a} - \dfrac{1}{2b}$

8. $-\dfrac{3x}{2yz} - \dfrac{8y}{xz} + 6x^2z^3$

9. $\dfrac{2b}{c} + \dfrac{3a}{2bc^2} + 6a^2c$

10. $v - 1$

11. $x - 2$

12. $m^2 + mn + n^2$

13. $3a^2 - a - 3$

14. $4m^2 + 2m + 4 + \dfrac{5}{2m - 1}$

15. $2a - b$

16. $3t - 2$

17. $4x^2 + 6x + 9$

18. $a^2 + 4a - 8 + \dfrac{24}{a + 2}$

19. $3x + 1$

20. $v^3 + v^2 + v + 1 + \dfrac{2}{v - 1}$

21. $2a^2 - 4a + 2$

22. $3y - 5$

23. $m + 9 + \dfrac{18}{m - 2}$

24. $x^3 + 3x^2 - 2x + 1$

25. $2a^2 - \dfrac{3}{2}a + \dfrac{3}{4} - \dfrac{\frac{3}{4}}{2a + 1}$

26. $5m + n$

27. $v^2 + 5v + 25 + \dfrac{135}{v - 5}$

28. $t^2 - 2t + \dfrac{1}{2} + \dfrac{\frac{3}{2}}{2t - 1}$

SOLUTIONS/ANSWERS UNIT 9

SOLUTIONS TO CHECK POINTS

CHECK POINT 1, FRAME 7

1. 4 is the greatest common factor of $4a - 16b$.

2. $2x$ is the greatest common factor of $6x^3y + 2x$.

3. $2m^2$ is the greatest common factor of $2m^5 - 8m^3 + 10m^2$.

4. $5p^3q$ is the greatest common factor of $15p^5q^2 - 10p^4q^3 - 20p^3q^4$.

5. $6x^2y^3z - 3x^2y + 8z$ has no common factor.

6. $3x^2yz$ is the greatest common factor of $3x^4y^2z - 6x^3y^2z^2 + 15x^2yz^2$.

CHECK POINT 2, FRAME 13

1. $7a - 7b = 7(a - b)$ 2. $3x + 2xy = x(3 + 2y)$

3. $10xy - 3z$ cannot be factored.

4. $8m^2 - 4mn = 4m(2m - n)$

5. $6x^3 + 12x^2 - 3x = 3x(2x^2 + 4x - 1)$

6. $4x^3y - x^2y - 8xy^2 = xy(4x^2 - x - 8y)$

7. $9m^3n - 3m^2t + 2nt^2$ cannot be factored.

8. $5a^4b^2c + 2a^3b^3c^2 - 8a^2b^4c^3$

   $= a^2b^2c(5a^2 + 2abc - 8b^2c^2)$

CHECK POINT 3, FRAME 18

1. $x^2 + 5x + 6 = (x + 2)(x + 3)$

2. $x^2 + 11x + 18 = (x + 2)(x + 9)$

3. $x^2 + 8x + 15 = (x + 3)(x + 5)$

4. $x^2 + 13x + 36 = (x + 4)(x + 9)$

CHECK POINT 4, FRAME 21

1. $x^2 - 10x + 16 = (x - 2)(x - 8)$

2. $x^2 - 11x + 30 = (x - 6)(x - 5)$

CHECK POINT 5, FRAME 25

1. $x^2 - 3x - 18 = (x - 6)(x + 3)$

2. $y^2 + y - 30 = (y + 6)(y - 5)$

3. $a^2 + 6a - 16 = (a + 8)(a - 2)$

4. $x^2 - 5x - 24 = (x - 8)(x + 3)$

5. $p^2 + 9p - 36 = (p + 12)(p - 3)$

CHECK POINT 6, FRAME 28

1. $5x^2 + 40x + 60 = 5(x^2 + 8x + 12)$
   $= 5(x + 2)(x + 6)$

2. $3m^3 - 12m^2 - 36m = 3m(m^2 - 4m - 12)$
   $= 3m(m - 6)(m + 2)$

CHECK POINT 7, FRAME 33

1. $2x^2 + 3x + 1 = (2x + 1)(x + 1)$

2. $6m^2 + m - 5 = (6m - 5)(m + 1)$

3. $5x^2 - 6x + 1 = (5x - 1)(x - 1)$

4. $3a^2 - 7a + 4 = (3a - 4)(a - 1)$

CHECK POINT 8, FRAME 39

1. $x^2 + 6x + 5 = (x + 1)(x + 5)$

2. $m^2 - 2m - 8 = (m - 4)(m + 2)$

3. $t^2 - 9t + 18 = (t - 3)(t - 6)$

4. $a^2 + 4a - 21 = (a + 7)(a - 3)$

5. $y^2 + 12y + 20 = (y + 2)(y + 10)$

6. $n^2 - 5n - 24 = (n + 3)(n - 8)$

CHECK POINT 9, FRAME 43

1. $x^2 - y^2 = (x - y)(x + y)$

2. $4a^2 - 9b^2 = (2a - 3b)(2a + 3b)$

3. $36m^3 - m = m(36m^2 - 1)$
   $= m(6m - 1)(6m + 1)$

4. $100 - 9p^2 = (10 - 3p)(10 + 3p)$

5. $p^4 - q^4 = (p^2 - q^2)(p^2 + q^2)$
   $= (p - q)(p + q)(p^2 + q^2)$

6. $-7x + 63x^3 = -7x(1 - 9x^2)$
   $= -7x(1 - 3x)(1 + 3x)$

   or

   $-7x + 63x^3 = 63x^3 - 7x$
   $= 7x(9x^2 - 1)$
   $= 7x(3x - 1)(3x + 1)$

CHECK POINT 10, FRAME 47

1. $p^3 + 1 = p^3 + 1^3 = (p + 1)(p^2 - p + 1)$

2. $8 - a^3 = 2^3 - a^3 = (2 - a)(4 + 2a + a^2)$

3. $125m^3 + 27 = (5m)^3 + 3^3 = (5m + 3)(25m^2 - 15m + 9)$

4. $24x^3y - 3y^4 = 3y(8x^3 - y^3) = 3y\left[(2x)^3 - y^3\right]$
   $= 3y(2x - y)(4x^2 + 2xy + y^2)$

5. $128pt^3 + 250p^4 = 2p(64t^3 + 125p^3) = 2p\left[(4t)^3 + (5p)^3\right]$
   $= 2p(4t + 5p)(16t^2 - 20pt + 25p^2)$

6. $3b^6 - 81 = 3(b^6 - 27) = 3\left[(b^2)^3 - 3^3\right]$
   $= 3(b^2 - 3)(b^4 + 3b^2 + 9)$

SOLUTIONS TO CHECK POINTS

CHECK POINT 1, FRAME 10

1. $x(x + 6) = 30$

   $x^2 + 6x = 30$

   $x^2 + 6x - 30 = 0$

2. $(y - 5)(y + 1) = -9$

   $y^2 + y - 5y - 5 = -9$

   $y^2 - 4y + 4 = 0$

3. $m(3m + 2) = (m + 1)^2 + 7$

   $3m^2 + 2m = (m + 1)(m + 1) + 7$

   $3m^2 + 2m = m^2 + 2m + 1 + 7$

   $2m^2 - 8 = 0$

4. $6(x + 3) + 6(x - 2) = (x + 2)(x + 3)$

   $6x + 18 + 6x - 12 = x^2 + 5x + 6$

   $12x + 6 = x^2 + 5x + 6$

   $-x^2 + 7x = 0$

   or

   $x^2 - 7x = 0$

5. $4p^2 + p = 2(2p^2 + 3p + 1)$

   $4p^2 + p = 4p^2 + 6p + 2$

   $-5p - 2 = 0$

   or

   $5p + 2 = 0$

CHECK POINT 2, FRAME 17

1. $3y^2 - 12y = 0$

   $3y(y - 4) = 0$

   either $3y = 0$ or $y - 4 = 0$

   $\therefore$ $y = 0$ or $y = 4$

2. If $y = 0$, $3y^2 - 12y = 3(0)^2 - 12(0)$

   $= 0$

   If $y = 4$, $3y^2 - 12y = 3(4)^2 - 12(4)$

   $= 3(16) - 48$

   $= 0$

Check: Hence, the solutions to $3y^2 - 12y = 0$ are $y = 0$ and $y = 4$.

SOLUTIONS/ANSWERS UNIT 9

ANSWERS TO DRILL EXERCISES

I
1. $3(3x + y)$
2. $4b(2ab - 1)$
3. $5n^2(2n^4 + n^3 - 3)$
4. cannot be factored
5. $7pq^3(3p^3q^2 - p^2 - 2q)$
6. $3x^3yz(x^2z + 3xy^2 - 4z)$

II
7. $5x(x + 4)(x - 2)$
8. $(m - 9)(m - 2)$
9. $2(x + 5)(x + 9)$
10. $(x + 3)(x + 3)$
11. $(a + 6)(a - 8)$
12. $(x + 3)(x - 2)$
13. $(x + 4)(x + 5)$
14. $(h + 9)(h - 3)$
15. cannot be factored
16. $(y - 4)(y - 7)$
17. $(y + 2)(y - 12)$
18. $(2x - 3)(x - 5)$
19. $(5a - 3)(2a + 5)$
20.
21. $(4x - 7)(2x + 3)$
22. $(3m + 5)(4m - 3)$
23.
24. $(9a + 4)(6a + 7)$
25. $(x - 7)(11x + 2)$
26.
27. $(4p + 1)(2p - 1)$
28. $(7r + 5)(3r - 2)$
29.
30. $(2x + 1)(2x - 3)$
31. $(3m - 4)(2m + 3)$

III
32. $(a - b)(a + b)$
33. $(3x - 4y)(3x + 4y)$
34. $(t + 1)(t^2 - t + 1)$
35. $(8 - 5m)(8 + 5m)$
36. $(2x^2 - y^2)(2x^2 + y^2)$
37. $(3 - k)(9 + 3k + k^2)$
38. cannot be factored
39. $a(7a - 1)(7a + 1)$
40. $(4d + 3)(16d^2 - 12d + 9)$
41. $3x(x - 2)(x + 2)$
42. $2(t^2 - 2)(t^4 + 2t^2 + 4)$
43. $3n(5n + 2m)(25n^2 - 10mn + 4m^2)$

CHECK POINT 3, FRAME 22

2. $-y^2 + 100 = 0$

$y^2 - 100 = 0$

$(y + 10)(y - 10) = 0$

$y + 10 = 0$ or $y - 10 = 0$

$\therefore$ $y = -10$ or $y = 10$

Check: 1. If $y = -10$, $-y^2 + 100 = -(-10)^2 + 100$

$= -100 + 100$

$= 0$

2. If $y = 10$, $-y^2 + 100 = -(10)^2 + 100$

$= 0$

Hence, $y = -10$ and $y = 10$ are the solutions to $-y^2 + 100 = 0$.

CHECK POINT 4, FRAME 26

1. $y^2 + 5y + 6 = 0$

$(y + 2)(y + 3) = 0$

$y + 2 = 0$ or $y + 3 = 0$

$y = -2$ or $y = -3$

Check: 1. If $y = -2$, $y^2 + 5y + 6 = (-2)^2 + 5(-2) + 6$

$= 4 - 10 + 6$

$= 0$

2. If $y = -3$, $y^2 + 5y + 6 = (-3)^2 + 5(-3) + 6$

$= 9 - 15 + 6$

$= 0$

Hence, $y = -2$ and $y = -3$ are the solutions to $y^2 + 5y + 6 = 0$.

CHECK POINT 2, FRAME 17

2. $-4p^2 + 28p = 0$

$-4p(p - 7) = 0$

either $-4p = 0$ or $p - 7 = 0$

$\therefore$ $p = 0$ or $p = 7$

Check: 1. If $p = 0$, $-4p^2 + 28p = -4(0)^2 + 28(0)$

$= 0$

2. If $p = 7$, $-4p^2 + 28p = -4(7)^2 + 28(7)$

$= -4(49) + 196$

$= 0$

Hence, the solutions to $-4p^2 + 28p = 0$ are $p = 0$ and $p = 7$.

CHECK POINT 3, FRAME 22

1. $m^2 - 49 = 0$

$(m + 7)(m - 7) = 0$

$m + 7 = 0$ or $m - 7 = 0$

$m = -7$ or $m = 7$

Check: 1. If $m = -7$, $m^2 - 49 = (-7)^2 - 49$

$= 49 - 49$

$= 0$

2. If $m = 7$, $m^2 - 49 = 7^2 - 49$

$= 0$

Hence, $m = -7$ and $m = 7$ are the solutions to $m^2 - 49 = 0$.

CHECK POINT 4, FRAME 26

2. $-m^2 + 3m + 18 = 0$

$m^2 - 3m - 18 = 0$

$(m - 6)(m + 3) = 0$

$m - 6 = 0$ or $m + 3 = 0$

$m = 6$ or $m = -3$

Check: 1. If $m = 6$, $-m^2 + 3m + 18 = -(6^2) + 3(6) + 18$

$= -36 + 18 + 18$

$= 0$

2. If $m = -3$, $-m^2 + 3m + 18 = -(-3)^2 + 3(-3) + 18$

$= -9 - 9 + 18$

$= 0$

Hence, $m = 6$ and $m = -3$ are the solutions to $-m^2 + 3m + 18 = 0$.

CHECK POINT 5, FRAME 32

1. $x(x + 7) = 30$

$x^2 + 7x - 30 = 0$

$(x + 10)(x - 3) = 0$

$x + 10 = 0$ or $x - 3 = 0$

$x = -10$ or $x = 3$

Check: 1. Substituting $x = -10$ into $x(x + 7) = 30$

we get $-10(-10 + 7) = 30$

$-10(-3) = 30$

$30 = 30$

2. Substituting $x = 3$ into $x(x + 7) = 30$

we get $3(3 + 7) = 30$

$3(10) = 30$

$30 = 30$

Hence, $x = -10$ and $x = 3$ are the solutions to $x(x + 7) = 30$.

CHECK POINT 5, FRAME 32

2. $6(m + 3) + 6(m - 2) = (m + 2)(m + 3)$

$6m + 18 + 6m - 12 = m^2 + 5m + 6$

$-m^2 + 7m = 0$

$-m(m - 7) = 0$

$-m = 0$ or $m - 7 = 0$

$m = 0$ or $m = 7$

Check: 1. Substituting $m = 0$ into $6(m + 3) + 6(m - 2) = (m + 2)(m + 3)$

we get $6(0 + 3) + 6(0 - 2) = (0 + 2)(0 + 3)$

$18 - 12 = 6$

$6 = 6$

2. Substituting $m = 7$ into $6(m + 3) + 6(m - 2) = (m + 2)(m + 3)$

we get $6(7 + 3) + 6(7 - 2) = (7 + 2)(7 + 3)$

$6(10) + 6(5) = (9)(10)$

$60 + 30 = 90$

$90 = 90$

Hence, $m = 0$ and $m = 7$ are the solutions to $6(m + 3) + 6(m - 2) = (m + 2)(m + 3)$.

3. $a(3a + 2) = (a + 1)^2 + 7$

$3a^2 + 2a = a^2 + 2a + 1 + 7$

$2a^2 - 8 = 0$

$2(a^2 - 4) = 0$

$2(a + 2)(a - 2) = 0$

$a + 2 = 0$ or $a - 2 = 0$

$a = -2$ or $a = 2$

CHECK POINT 5, FRAME 32

3. Check: 1. Substituting $a = -2$ into $a(3a + 2) = (a + 1)^2 + 7$

we get $-2[3(-2) + 2] = (-2 + 1)^2 + 7$

$$-2[-6 + 2] = (-1)^2 + 7$$

$$-2(-4) = 1 + 7$$

$$8 = 8$$

2. Substituting $a = 2$ into $a(3a + 2) = (a + 1)^2 + 7$

we get $2[3(2) + 2] = (2 + 1)^2 + 7$

$$2[6 + 2] = 3^2 + 7$$

$$2(8) = 9 + 7$$

$$16 = 16$$

Hence, $a = -2$ and $a = 2$ are the solutions
to $a(3a + 2) = (a + 1)^2 + 7$

4.
$$7y(y - 1) + 8 = 3y(y - 4) - 4y + 3$$
$$7y^2 - 7y + 8 = 3y^2 - 12y - 4y + 3$$
$$4y^2 + 9y + 5 = 0$$
$$(4y + 5)(y + 1) = 0$$
$$Y = -\frac{5}{4} \quad \text{or} \quad y = -1$$

Check: 1. Substituting $y = -\frac{5}{4}$ into $7y(y - 1) + 8 = 3y(y - 4) - 4y + 3$

we get $7(-\frac{5}{4})[-\frac{5}{4} - 1] + 8 = 3(-\frac{5}{4})[-\frac{5}{4} - 4] - 4(-\frac{5}{4}) + 3$

$$7(-\frac{5}{4})(-\frac{9}{4}) + 8 = 3(-\frac{5}{4})(-\frac{21}{4}) + 8$$

$$\frac{315}{16} + 8 = \frac{315}{16} + 8$$

2. Substituting $y = -1$ into $7y(y - 1) + 8 = 3y(y - 4) - 4y + 3$

we get $7(-1)[-1 - 1] + 8 = 3(-1)[-1 - 4] - 4(-1) + 3$

$$14 + 8 = 15 + 7$$

$$22 = 22$$

Hence, the equation $7y(y - 1) + 8 = 3y(y - 4) - 4y + 3$

Has the solutions $y = -\frac{5}{4}$ and $y = -1$

CHECK POINT 6, FRAME 40

1. Let x = the length in metres.

Then $x - 50$ = the width in metres.

area = length · width

$$5000 = (x)(x - 50)$$

$$x^2 - 50x - 5000 = 0$$

$$(x - 100)(x + 50) = 0$$

$$x - 100 = 0 \quad \text{or} \quad x + 50 = 0$$

$$x = 100 \quad \text{or} \quad x = -50$$

Since the length of the piece of land cannot be negative, we discard the solution $x = -50$.

If $x = 100$, $x - 50 = 50$. Check: Area = 100 m x 50 m = 5000 m²

Hence, the length of piece of land is 100 m and the width of piece of land is 50 m.

5000 m² | x - 50

x

2. Let x represent the length of one side.

Area $= x^2$

Perimeter $= 4x$

Area = 10·perimeter in this case.

$$x^2 = 10(4x)$$

$$x^2 - 40x = 0$$

$$x(x - 40) = 0$$

$$x = 0 \quad \text{or} \quad x - 40 = 0$$

$$x = 40$$

Disregard $x = 0$ because the area is not zero.

$$\therefore x = 40$$

Check: Area = 40 x 40 = 1600   Perimeter = 4(40) = 160

$$\therefore \text{Area} = 10 \times \text{Perimeter}$$

Hence, the length of one side of the metal plate is 40 units.

CHECK POINT 6, FRAME 40

3. Let one of the numbers be x.

   Then, the other number is 19 - x.

   We know $x(19 - x) = 48.$

   $$-x^2 + 19x - 48 = 0$$
   $$x^2 - 19x + 48 = 0$$
   $$(x - 16)(x - 3) = 0$$
   $$x - 16 = 0 \quad \text{or} \quad x - 3 = 0$$
   $$x = 16 \quad \text{or} \quad x = 3$$

   Hence, the numbers are 3 and 16.   Check: $16 + 3 = 19$ $16 \times 3 = 48$

4. Let x be the width and let x + 7 be the length of the window.

   area of window $= x(x + 7)$

   perimeter of window $= 2(x) + 2(x + 7)$

   In our case: area = perimeter + 40

   $$x(x + 7) = 2(x) + 2(x + 7) + 40$$
   $$x^2 + 7x = 2x + 2x + 14 + 40$$
   $$x^2 + 3x - 54 = 0$$
   $$(x + 9)(x - 6) = 0$$
   $$x + 9 = 0 \quad \text{or} \quad x - 6 = 0$$
   $$x = -9 \qquad\qquad x = 6$$

   If $x = -9$, $x + 7 = -2$ and if
   $x = 6$, $x + 7 = 13$.

   Disregard $x = -9$ Check: Area $= 6 \times 13 = 78$ Perimeter $= 2(6) + 2(13) = 38$

   $$\therefore \text{Area} = \text{Perimeter} + 40$$

   Hence, the width is 6 units and the length is 13 units.

CHECK POINT 6, FRAME 40

5. Let x = the length in centimetres of one side of the square piece of sheet metal.

   Then x - 20 = the length of one side of the container.
   The height is 10 cm.

   Volume of container = (width)(length)(height)
   Hence, $9000 = (x - 20)(x - 20)(10)$

   $$10(x - 20)(x - 20) = 9000$$
   $$(x - 20)(x - 20) = 900$$
   $$x^2 - 40x + 400 = 900$$
   $$x^2 - 40x - 500 = 0$$
   $$(x - 50)(x + 10) = 0$$
   $$(x - 50) = 0 \quad \text{or} \quad x + 10 = 0$$
   $$x = 50 \quad \text{or} \quad x = -10$$

   Since the length of the piece cannot be negative, we discard $x = -10$.

   If $x = 50$, $x - 20 = 30$.

   The volume of the container is $30 \cdot 30 \cdot 10 = 9000$ cm^3.

   Hence, the size of the piece of sheet metal required is 50 cm by 50 cm.

SOLUTIONS TO CHECK POINTS

CHECK POINT 1, FRAME 7

I 1. $\frac{3a}{5} = (3a) \div 5$

4. $\frac{3p^2 + 2p - 1}{p^2 - p + 6} = (3p^2 + 2p - 1) \div (p^2 - p + 6)$

2. $\frac{5x - 1}{6} = (5x - 1) \div 6$

5. $\frac{8x}{2x - 1} = (8x) \div (2x - 1)$

3. $\frac{m^2 - 1}{m + 1} = (m^2 - 1) \div (m + 1)$

6. $\frac{1}{a + 1} = 1 \div (a + 1)$

II 7. $(x + 1) \div (2x - 5) = \frac{x + 1}{2x - 5}$

8. $p - 3p^2 + 6r + 1 = p - \frac{3p^2}{6r} + 1$

9. $(y^2 - 2y + 1) \div y - 6 = \frac{y^2 - 2y + 1}{y} - 6$

10. $m^2 - 3mn^2 \div (m - 1) = m^2 - \frac{3mn^2}{m - 1}$

CHECK POINT 2, FRAME 9

1. If a = 3, $\frac{5a - 2}{5} = \frac{5(3) - 2}{5} = \frac{15 - 2}{5} = \frac{13}{5}$ or $2\frac{3}{5}$.

2. If y = 5, $\frac{6y - 4}{9y} = \frac{6(5) - 4}{9(5)} = \frac{30 - 4}{45} = \frac{26}{45}$.

3. If x = 2, $\frac{4x^3}{4x^3} = \frac{4(2^3)}{4(2^3)} = \frac{4(8)}{4(8)} = \frac{32}{32} = 1$.

4. If m = 9, n = 3, $\frac{3m - n^3}{m - n} = \frac{3(9) - 3^3}{9 - 3} = \frac{27 - 27}{6} = \frac{0}{6} = 0$.

5. If p = 3, $\frac{2(p - 1)}{p - 3} = \frac{2(3 - 1)}{3 - 3} = \frac{2(2)}{0} = \frac{4}{0}$ undefined.

6. If x = 0, y = 1, $\frac{5x^2 + 3y}{2x(y - 1)} = \frac{5(0^2) + 3(1)}{2(0)(1 - 1)} = \frac{0 + 3}{0}$ undefined.

ANSWERS TO DRILL EXERCISES

I 1. $y^2 - y - 12 = 0$

2. $x^2 - x + 20 = 0$

3. $9a + 6 = 0$

4. $x^2 - 4x = 0$

5. $x^2 - 5x - 24 = 0$

II 6. x = 0 or x = 4

7. $x = \frac{3}{4}$ or x = 4

8. $p = \frac{2}{3}$ or $p = -\frac{2}{3}$

9. y = 5 or y = -3

10. m = 9 or m = -9

11. y = 0 or y = 6

12. x = 3 or x = -1

13. a = 3 or a = -3

14. y = 0 or y = 12

15. m = 2

III 16. numbers = 16, 5

17. width = 10 m, length = 17 m

18. width = 6 units, length = 8 units

19. numbers = 11, 12

20. number = 6

21. 28 cm x 12 cm x 2 cm

22. width = 5 cm, length = 15 cm

23. walk: 6 units wide

SOLUTIONS TO CHECK POINTS

Check Point 3, Frame 11

1. $\dfrac{2x}{5y}$, $y \neq 0$

2. $\dfrac{7a - 1}{a + 2}$, $a \neq -2$

3. $\dfrac{m^2 - 5n + 3}{9(m-1)(n+2)}$, $m \neq 1$, $n \neq -2$

Check Point 4, Frame 23

1. $\dfrac{2a}{b} = \dfrac{2a(3a^2)}{b(3a^2)} = \dfrac{6a^3}{3a^2 b}$

2. $n^2 + n = n(n + 1)$

   $\therefore \dfrac{2}{n} = \dfrac{2(n+1)}{n(n+1)} = \dfrac{2n+2}{n^2 + n}$

3. $m^2 + 5m + 6 = (m + 3)(m + 2)$

   $\therefore \dfrac{m-2}{m+3} = \dfrac{(m-2)(m+2)}{(m+3)(m+2)} = \dfrac{m^2 - 4}{m^2 + 5m + 6}$

4. $12n + 8 = 4(3n + 2)$

   $\therefore \dfrac{m-5}{3n+2} = \dfrac{(m-5)(4)}{(3n+2)(4)} = \dfrac{4m-20}{12n+8}$

5. $\dfrac{x-1}{y} = \dfrac{(x-1)(x^2 y)}{y(x^2 y)} = \dfrac{x^3 y - x^2 y}{x^2 y^2}$

6. $y^2 - 4 = (y - 2)(y + 2)$

   $\therefore \dfrac{y+4}{y-2} = \dfrac{(y+4)(y+2)}{(y-2)(y+2)} = \dfrac{y^2 + 6y + 8}{y^2 - 4}$

SOLUTIONS TO CHECK POINTS

Check Point 5, Frame 35

1. $\dfrac{4x^2 + x}{5x} = \dfrac{x(4x + 1)}{5x} = \dfrac{4x + 1}{5}$

2. $\dfrac{3a^2 - 6a}{a - 2} = \dfrac{3a(a - 2)}{a - 2} = 3a$

3. $\dfrac{3m}{6m^2 + 9m} = \dfrac{3m}{3m(2m + 3)} = \dfrac{1}{2m + 3}$

4. $\dfrac{2a^2 - 7a + 3}{4a^2 - 1} = \dfrac{(2a - 1)(a - 3)}{(2a - 1)(2a + 1)} = \dfrac{a - 3}{2a + 1}$

5. $\dfrac{x^2 + 6x + 9}{4x^2 + 7x - 15} = \dfrac{(x + 3)(x + 3)}{(x + 3)(4x - 5)} = \dfrac{x + 3}{4x - 5}$

6. $\dfrac{8y^3 + 6y^2}{2y^4 - 4y^5} = \dfrac{2y^2(4y + 3)}{2y^4(1 - 2y)} = \dfrac{2y^2(4y + 3)}{2y^2 y^2(1 - 2y)} = \dfrac{4y + 3}{y^2(1 - 2y)}$ or $\dfrac{4y + 3}{y^2 - 2y^3}$

CHECK POINT 6, FRAME 41

1. $\dfrac{4xy}{4xy} \cdot \dfrac{-2y}{x^2} = \dfrac{(4xy)(-2y)}{(4xy)(x^2)} \neq \dfrac{4xy - 2y}{4xy + x}$

   Hence, $\dfrac{4xy - 2y}{x^2 + 4xy} \neq \dfrac{-2y}{x^2}$; 4xy is incorrectly cancelled.

2. $\dfrac{4m^2}{4m^2} \cdot \dfrac{1}{m - 1} = \dfrac{4m^2}{4m^2(m - 1)}$; $4m^2$ is correctly cancelled.

3. $\dfrac{p}{p} \cdot (-12) = \dfrac{6(p - 2)}{p} = \dfrac{6p - 12}{p}$; p is incorrectly cancelled.

4. $\dfrac{a + 1}{a + 1} \cdot 3x = \dfrac{3x(a + 1)}{a + 1}$; a + 1 is correctly cancelled.

CHECK POINT 7, FRAME 47

1. $-\dfrac{7}{11} = \dfrac{7}{-11}$ is true, because if a fraction has one negative sign, it can be placed in front of either the numerator, the fraction bar, or the denominator.

2. False. $\dfrac{-3xy}{-x^2} = \dfrac{3xy}{x^2}$

3. $\dfrac{-(2x - y)}{-(y - 2x)} = \dfrac{2x - y}{-(y - 2x)}$ is true, because any two of the three signs of a fraction can be interchanged without changing the value of the fraction.

4. False. $\dfrac{-6a^2}{a^2 - 1} = \dfrac{6a^2}{-(a^2 - 1)} = \dfrac{6a^2}{-a^2 + 1}$

5. False. $\dfrac{-m^2}{-n^3} = \dfrac{m^2}{n^3}$

6. $-\dfrac{-(x + 1)}{-(x - 2)} = -\dfrac{x + 1}{x - 2}$ is true, because any two of the three signs of a fraction can be interchanged without changing the value of the fraction.

CHECK POINT 8, FRAME 52

1. Step 1. $4 - a = -(a - 4)$

   Step 2. $\dfrac{4 - a}{a - 4} = \dfrac{-(a - 4)}{a - 4} = -\dfrac{\cancel{a - 4}}{\cancel{a - 4}} = -1$

2. Step 1. $1 - x = -(x - 1)$

   Step 2. $\dfrac{5(x - 1)(x + 2)}{(x - 2)(1 - x)} = \dfrac{5\cancel{(x - 1)}(x + 2)}{-\cancel{(x - 1)}(x - 2)} = -\dfrac{5(x + 2)}{x - 2}$

ANSWERS TO DRILL EXERCISES

I(a) 1. $(4b) \div (3a + 2)$

2. $(2x^2 + x + 1) \div (x^2 - 1)$

3. $7 \div (n - 1)$

(b) 4. $\dfrac{m^2 - 3}{2m + 7}$

5. $y^2 - \dfrac{3x^3y}{2x} + 3y$

6. $p^2 + \dfrac{6p}{2p - 1}$

II 7. $\dfrac{5}{14}$

8. $\dfrac{19}{6}$ or $3\dfrac{1}{6}$

9. $\dfrac{32}{0}$ undefined

III 10. $x \neq 0$

11. $y \neq -3$

12. $b \neq 0$, $a \neq 2$

13. $a \neq -2$, $b \neq 3$

14. $m \neq 2$, $m \neq -2$

IV 15. $\dfrac{28}{32}$

16. $\dfrac{12mn^2}{20n^3}$

17. $\dfrac{a^2b + 2ab}{ab^2}$

18. $\dfrac{5x + 5}{x^2 - x - 2}$

19. $\dfrac{y^2 + 5y + 4}{y^2 - 16}$

20. $\dfrac{6m^2 + 11m - 10}{9m^2 - 12m + 4}$

21. $\dfrac{a - 3}{a^2 - a - 6}$

V 22. $\dfrac{3}{7}$

23. $\dfrac{3}{4x^2y^3}$

24. $\dfrac{3a^2 - 4b}{a^2 + 7b}$

25. $\dfrac{3}{4}$

26. $\dfrac{1}{x(x - 2)}$

27. $\dfrac{b - 3}{b^2(2 + 5b)}$

28. $\dfrac{x + 3}{x + 5}$

29. $\dfrac{m + 2}{m}$

30. $\dfrac{a + 4}{3a - 2}$

ANSWERS TO DRILL EXERCISES

VI 31. True.

  32. False. $\dfrac{-a^2b}{-b^2} = \dfrac{a^2b}{b^2}$

  33. False. $\dfrac{x^2-1}{-3x^2} = \dfrac{-x^2+1}{3x^2}$

  34. True.

VII 35. -1

  36. $-\dfrac{y+3}{3(y-3)}$

SOLUTIONS TO CHECK POINTS

Check Point 1, Frame 3

1. $\dfrac{1}{m^2} \cdot \dfrac{1}{7m} = \dfrac{1 \cdot 1}{m^2 \cdot 7m} = \dfrac{1}{7m^3}$

2. $\left(\dfrac{3x}{y}\right)\left(\dfrac{x}{4y}\right) = \dfrac{(3x)(x)}{(y)(4y)} = \dfrac{3x^2}{4y^2}$

3. $\left(\dfrac{-5}{R}\right)\left(\dfrac{2z}{R^2}\right) = \dfrac{(-5)(2z)}{(R)(R^2)} = \dfrac{-10z}{R^3}$

4. $\dfrac{3a}{4a-2} \cdot \dfrac{2a+3}{4a+2} = \dfrac{(3a)(2a+3)}{(4a-2)(4a+2)} = \dfrac{6a^2+9a}{16a^2-4}$

5. $\dfrac{m-3}{m+3} \cdot \dfrac{m-3}{8m} = \dfrac{(m-3)(m-3)}{(m+3)(8m)} = \dfrac{m^2-6m+9}{8m^2+24m}$

6. $\dfrac{2x+1}{x-5} \cdot \dfrac{3x-2}{4x+5} = \dfrac{(2x+1)(3x-2)}{(x-5)(4x+5)} = \dfrac{6x^2-x-2}{4x^2-15x-25}$

Check Point 2, Frame 5

1. $\left(\dfrac{1}{3}\right)^4 = \dfrac{1^4}{3^4} = \dfrac{1}{81}$

2. $\left(\dfrac{5x^2\,y}{2z^3}\right)^3 = \dfrac{(5x^2\,y)^3}{(2z^3)^3} = \dfrac{125x^6 y^3}{8z^9}$

3. $\left(\dfrac{2a-1}{a+4}\right)^2 = \dfrac{(2a-1)^2}{(a+4)^2} = \dfrac{(2a-1)(2a-1)}{(a+4)(a+4)} = \dfrac{4a^2-4a+1}{a^2+8a+16}$

SOLUTIONS TO CHECK POINTS

Check Point 5, Frame 16

4. $\dfrac{9a^2 - 25}{a^2 + 2a - 8} \cdot \dfrac{a^2 + 5a + 4}{5 - 3a} = \dfrac{(3a-5)(3a+5)(a+1)}{-(3a-5)(a-2)} = \dfrac{(3a+5)(a+1)}{-(a-2)}$ or $-\dfrac{3a^2 + 8a + 5}{a-2}$

5. $\dfrac{4x^3 - 10x^2 + 6x}{x^2 - 4x - 12} \cdot \dfrac{x^2 - 7x + 6}{2x^2 + 5x - 12} = \dfrac{2x(2x-3)(x-1)(x-6)(x-1)}{(x-6)(x+2)(2x-3)(x+4)}$

$= \dfrac{2x(x-1)(x-1)}{(x+2)(x+4)}$ or $\dfrac{2x^3 - 4x^2 + 2x}{x^2 + 6x + 8}$

Check Point 6, Frame 18

1. The reciprocal of $\dfrac{3m^2}{-5}$ is $\dfrac{-5}{3m^2}$ or $-\dfrac{5}{3m^2}$.

2. The reciprocal of $\dfrac{x-5}{x+1}$ is $\dfrac{x+1}{x-5}$.

3. The reciprocal of $\dfrac{6v^2 + 5v}{RV^2}$ is $\dfrac{RV^2}{6v^2 + 5v}$.

4. The reciprocal of $\dfrac{a^2 - 2a + 1}{9a^2 - 16}$ is $\dfrac{9a^2 - 16}{a^2 - 2a + 1}$.

5. The reciprocal of 9 is $\dfrac{1}{9}$.

6. The reciprocal of $3x^2 + 1$ is $\dfrac{1}{3x^2 + 1}$.

SOLUTIONS TO CHECK POINTS

Check Point 3, Frame 8

1. $7 \cdot \dfrac{a}{3b} = \dfrac{7}{1} \cdot \dfrac{a}{3b} = \dfrac{7a}{3b}$

2. $4x\left(\dfrac{3x}{2y}\right) = \dfrac{4x}{1} \cdot \dfrac{3x}{2y} = \dfrac{12x^2}{2y} = \dfrac{6x^2}{y}$

3. $\dfrac{6v^2}{5p} \cdot 3v = \dfrac{6v^2}{5p} \cdot \dfrac{3v}{1} = \dfrac{18v^3}{5p}$

4. $5m^2 \cdot \dfrac{m+1}{2m-3} = \dfrac{5m^2}{1} \cdot \dfrac{m+1}{2m-3} = \dfrac{5m^2(m+1)}{2m-3}$ or $\dfrac{5m^3 + 5m^2}{2m-3}$

5. $\dfrac{a-6}{3a+2} \cdot (a+6) = \dfrac{a-6}{3a+2} \cdot \dfrac{a+6}{1} = \dfrac{(a-6)(a+6)}{3a+2}$ or $\dfrac{a^2 - 36}{3a+2}$

Check Point 4, Frame 10

1. $\dfrac{12}{13} = 12 \cdot \dfrac{1}{13}$

2. $\dfrac{-5m^2}{2n^3} = -5m^2 \cdot \dfrac{1}{2n^3}$

3. $\dfrac{v^2 - 3v}{2p^2} = (v^2 - 3v) \cdot \dfrac{1}{2p^2}$

4. $\dfrac{5a^2 - 2a + 3}{4a^2 - 9} = (5a^2 - 2a + 3) \cdot \dfrac{1}{4a^2 - 9}$

Check Point 5, Frame 16

1. $\dfrac{-2v^3}{p^2} \cdot \dfrac{vp}{6v^2} = \dfrac{-2v^4 p}{6p^2 v^2} = -\dfrac{v^2}{3p}$

2. $\dfrac{-2}{5x - 5} \cdot \dfrac{x-1}{6(x+1)} = \dfrac{-1}{15(x+1)}$ or $\dfrac{-1}{15x + 15}$

3. $\dfrac{-7m}{m^2 - 3m + 2} \cdot \dfrac{4m^2 - 4m}{14m^3} = \dfrac{-2}{m(m-2)}$ or $\dfrac{2}{2m - m^2}$

SOLUTIONS/ANSWERS UNIT 12

SOLUTIONS TO CHECK POINTS

Check Point 8, Frame 26

3. The reciprocal of $\frac{R+1}{1}$ is $\frac{1}{R+1}$.

$\frac{2V^2}{R+1} \div (R+1) = \frac{2V^2}{R+1} \cdot \frac{1}{R+1} = \frac{2V^2}{(R+1)^2}$ or $\frac{2R^2}{R^2+2R+1}$

Check Point 9, Frame 31

1. $\frac{R+1}{3V^2} + \frac{5R+5}{V} = \frac{R+1}{3V^2} \cdot \frac{V}{5R+5} = \frac{\cancel{R+1}}{\cancel{V}(3V)} \cdot \frac{1}{5(\cancel{R+1})}$

$= \frac{1}{(3V)(5)} = \frac{1}{15V}$

2. $\frac{4(a-3)}{ab} \div \frac{8b^2}{2a(a-3)} = \frac{\overset{1}{\cancel{4}}(a-3)}{a\,b} \cdot \frac{\overset{1}{\cancel{2a}}(a-3)}{\underset{b^2}{\cancel{8}}} = \frac{(a-3)^2}{b^3}$

or $\frac{a^2-6a+9}{b^3}$

SOLUTIONS TO CHECK POINTS

Check Point 7, Frame 24

1. The reciprocal of $\frac{P}{V^2}$ is $\frac{V^2}{P}$.

$\frac{2V}{3P^2} \div \frac{P}{V^2} = \frac{2V}{3P^2} \cdot \frac{V^2}{P} = \frac{2V^3}{3P^3}$

2. The reciprocal of $\frac{7n-5}{m^3}$ is $\frac{m^3}{7n-5}$.

$\frac{m+2}{3n} \div \frac{7n-5}{m^3} = \frac{m+2}{3n} \cdot \frac{m^3}{7n-5}$

$= \frac{m^3(m+2)}{3n(7n-5)}$ or $\frac{m^4+2m^3}{21n^2-15n}$

3. The reciprocal of $\frac{3y^2}{x-1}$ is $\frac{x-1}{3y^2}$.

$\frac{\dfrac{10x^2}{7y}}{\dfrac{3y^2}{x-1}} = \frac{10x^2}{7y} \div \frac{3y^2}{x-1}$

$= \frac{10x^2}{7y} \cdot \frac{x-1}{3y^2}$

$= \frac{10x^2(x-1)}{21y^3}$ or $\frac{10x^3-10x^2}{21y^3}$

Check Point 8, Frame 26

1. The reciprocal of $\frac{z^2}{1}$ is $\frac{1}{z^2}$.

$\frac{5x}{y} \div z^2 = \frac{5x}{y} \cdot \frac{1}{z^2} = \frac{5x}{yz^2}$

2. The reciprocal of $\frac{n+1}{1}$ is $\frac{1}{n+1}$.

$\frac{m-1}{n^2} \div (n+1) = \frac{m-1}{n^2} \cdot \frac{1}{n+1}$

$= \frac{m-1}{n^2(n+1)}$ or $\frac{m-1}{n^3+n^2}$

SOLUTIONS/ANSWERS UNIT 12

SOLUTIONS TO CHECK POINTS

Check Point 10, Frame 34

1. The reciprocal of $\dfrac{x^2-25}{2xy}$ is $\dfrac{2xy}{x^2-25}$.

$$\frac{5-x}{4y^2} \div \frac{x^2-25}{2xy} = \frac{5-x}{4y^2} \cdot \frac{2xy}{x^2-25} = \frac{-(x-5)}{(2y)(2y)} \cdot \frac{(2y)x}{(x+5)(x-5)} =$$

$$= -\frac{1}{2y} \cdot \frac{x}{x+5} = -\frac{x}{2y(x+5)} \quad\text{or}\quad -\frac{x}{2xy+10y}$$

2. The reciprocal of $\dfrac{2m+14}{m^2-4}$ is $\dfrac{m^2-4}{2m+14}$.

$$\frac{m^2+5m-14}{3m-6} \div \frac{2m+14}{m^2-4} = \frac{m^2+5m-14}{3m-6} \cdot \frac{m^2-4}{2m+14} =$$

$$= \frac{(m+7)(m-2)}{3(m-2)} \cdot \frac{(m-2)(m+2)}{2(m+7)} = \frac{m-2}{3} \cdot \frac{m+2}{2} =$$

$$= \frac{(m-2)(m+2)}{6} \quad\text{or}\quad \frac{m^2-4}{6}$$

3. The reciprocal of $\dfrac{a^2+5a-6}{a^2-2a-8}$ is $\dfrac{a^2-2a-8}{a^2+5a-6}$.

$$\frac{a^2+9a+18}{a^2+3a+2} \div \frac{a^2+5a-6}{a^2-2a-8} = \frac{a^2+9a+18}{a^2+3a+2} \cdot \frac{a^2-2a-8}{a^2+5a-6} =$$

$$= \frac{(a+6)(a+3)}{(a+2)(a+1)} \cdot \frac{(a-4)(a+2)}{(a+6)(a-1)} = \frac{(a+3)(a-4)}{(a+1)(a-1)} =$$

$$\text{or}\quad \frac{a^2-a-12}{a^2-1}$$

SOLUTIONS/ANSWERS UNIT 12

SOLUTIONS TO CHECK POINTS

Check Point 11, Frame 40

1. The LCD is y.

$$\frac{\dfrac{x}{y}}{\dfrac{x}{y}+1} = \frac{\left(\dfrac{x}{y}\right)(y)}{\left(\dfrac{x}{y}+1\right)(y)} = \frac{\dfrac{xy}{y}}{\dfrac{xy}{y}+y} = \frac{x}{x+y}$$

2. The LCD is ab.

$$\frac{\dfrac{1}{a}-b}{\dfrac{1}{b}+a} = \frac{\left(\dfrac{1}{a}-b\right)(ab)}{\left(\dfrac{1}{b}+a\right)(ab)} = \frac{\dfrac{ab}{a}-ab^2}{\dfrac{ab}{b}+a^2b} = \frac{b-ab^2}{a+a^2b} = \frac{b(1-ab)}{a(1+ab)}$$

3. The LCD is xy^2.

$$\frac{\dfrac{x^2}{y^2}+\dfrac{2x}{y}-8}{\dfrac{x}{y^2}-\dfrac{1}{y}-\dfrac{2}{x}} = \frac{\left(\dfrac{x^2}{y^2}+\dfrac{2x}{y}-8\right)(xy^2)}{\left(\dfrac{x}{y^2}-\dfrac{1}{y}-\dfrac{2}{x}\right)(xy^2)} = \frac{\dfrac{x^2xy^2}{y^2}+\dfrac{2xxy^2}{y}-8xy^2}{\dfrac{x^2y^2}{y^2}-\dfrac{xy^2}{y}-\dfrac{2xy^2}{x}}$$

$$= \frac{x^3+2x^2y-8xy^2}{x^2-xy-2y^2} = \frac{x(x^2+2xy-8y^2)}{x^2-xy-2y^2}$$

$$= \frac{x(x-2y)(x+4y)}{(x-2y)(x+y)} = \frac{x(x+4y)}{(x+y)}$$

411

SOLUTIONS/ANSWERS UNIT 12

ANSWERS TO DRILL EXERCISES

I

1. $\dfrac{5b^2}{6a^5}$

2. $\dfrac{21m^3}{8n^6}$

3. $-\dfrac{6x}{y^6}$

4. $\dfrac{8a^9 b^3}{27c^6}$

5. $\dfrac{10xy^2}{3z}$

6. $\dfrac{3n^3(n-2)}{2n+1}$

II

7. $(2x)\left(-\dfrac{1}{3y^2}\right)$

8. $(a^3 - 5a)\cdot\dfrac{1}{3b^2}$

9. $(m^2 - 3m + 5)\cdot\dfrac{1}{m+1}$

III

10. $\dfrac{5a^2}{4b}$

11. $\dfrac{2}{15x^2}$

12. $\dfrac{x^5 z}{8y^2}$

13. $\dfrac{(x+2)(x-1)}{x}$

14. $\dfrac{-1}{b-7}$

15. $\dfrac{7a(a+2)}{(a+3)(a-6)}$

IV

16. $-\dfrac{7}{2x^3}$

17. $\dfrac{3a^2 + 2b^2}{a^2 b^3}$

18. $\dfrac{1}{m^2 n + mn + 2}$

19. $\dfrac{6x(x-3)}{5y(y+7)}$

20. $\dfrac{m^3(m-1)}{3n^2(n+1)}$

21. $\dfrac{4a}{3bc^3}$

22. $\dfrac{x^2 - 2x}{y^3(y+5)}$

23. $\dfrac{3}{m}$

24. $\dfrac{12(x+3)^2}{x^2 y}$

SOLUTIONS/ANSWERS UNIT 12

ANSWERS TO DRILL EXERCISES

VI

25. $\dfrac{5x^2}{6}$

26. $\dfrac{4a^2(a+5)}{3}$

27. $\dfrac{4y(y-3)}{3}$

28. $\dfrac{-(x-3)(x+4)}{3(x+2)}$

29. $\dfrac{3(x-2)}{5(x+7)}$

30. $\dfrac{6b(b+7)}{-(b-4)}$

31. $\dfrac{3(2y+3)(y-1)}{2y(y+2)(y+1)}$

32. $\dfrac{x+7}{2x-3}$

33. $-3(m-4)$

34. $\dfrac{5(x+3)(x+1)}{x(x+2)(3x-1)}$

35. $\dfrac{a}{a+1}$

36. $\dfrac{x+4}{x-5}$

37. $-\dfrac{(2y-5)(y+1)}{(y+6)^2}$

38. $\dfrac{3x-1}{8x^2}$

39. $\dfrac{(x-1)^2}{x+2}$

40. $\dfrac{3m+9K}{4}$ or $\dfrac{3(m+3K)}{4}$

41. $\dfrac{y-x}{y+x}$

42. $\dfrac{(t+5)(5t-1)}{4t-22}$

43. $\dfrac{p(p-5a)}{p-a}$

44. $\dfrac{-a}{h^2(a+h)}$ or $-\dfrac{a}{h^2(a+h)}$

SOLUTIONS TO CHECK POINTS

Check Point 3, Frame 13

1. The LCD of $\dfrac{m^2}{n-3}$ and $\dfrac{n}{m+1}$ is $(n-3)(m+1)$.

2. The LCD of $\dfrac{6}{R+3}$ and $\dfrac{5R}{(R+3)^2}$ is $(R+3)^2$ since R + 3 is a factor of $(R+3)^2$.

3. The LCD of $\dfrac{1}{4x^2-9}$ and $\dfrac{1}{2x+3}$ is $4x^2 - 9 = (2x+3)(2x-3)$, since 2x + 3 is a factor of $4x^2 - 9$.

Check Point 4, Frame 16

1. $8a^2 + 24a = 8a(a+3) = 2^3 \cdot a(a+3)$

2. $12m^2 - 48n^2 = 12(m^2 - 4n^2) = 2^2 \cdot 3(m-2n)(m+2n)$

3. $6a^3 - 15a^2 - 9a = 3a(2a^2 - 5a - 3) = 3a(2a+1)(a-3)$

Check Point 5, Frame 20

1. $32R^3Z = 2^5 \cdot R^3 \cdot Z$

$18R^3 + 36R^2 = 18R^2(R+2) = 2 \cdot 3^2 \cdot R^2 \cdot (R+2)$

$LCD = 2^5 \cdot 3^2 \cdot R^3 \cdot Z \cdot (R+2)$

$= 288R^3Z(R+2)$

2. $5a - 10 = 5(a-2)$

$a^2 + 2a - 3 = (a+3)(a-1)$

$LCD = 5(a-1)(a-2)(a+3)$

3. $7x^3 = 7 \cdot x^3$

$x^2 - 4 = (x-2)(x+2)$

$x^2 + 5x + 6 = (x+2)(x+3)$

$LCD = 7x^3(x-2)(x+2)(x+3)$

SOLUTIONS TO CHECK POINTS

Check Point 1, Frame 3

1. $\dfrac{2}{T-1} - \dfrac{9V}{T-1} = \dfrac{2-9V}{T-1}$

2. $\dfrac{3ab}{vw^2} + \dfrac{4ab}{vw^2} = \dfrac{7ab}{vw^2}$

3. $\dfrac{-x^2y^3}{2x+3} + \dfrac{x^3y}{2x+3} = \dfrac{-x^2y^3+x^3y}{2x+3}$ or $\dfrac{x^3y - x^2y^3}{2x+3}$

Check Point 2, Frame 8

1. $\dfrac{2a+3}{6b} + \dfrac{4a-1}{6b} = \dfrac{2a+3+4a-1}{6b}$

$= \dfrac{6a+2}{6b}$

$= \dfrac{2(3a+1)}{2(3b)}$

$= \dfrac{3a+1}{3b}$

2. $\dfrac{5m^2-3m}{2n} - \dfrac{m^2-m}{2n} = \dfrac{5m^2-3m}{2n} - \dfrac{(m^2-m)}{2n}$

$= \dfrac{5m^2-3m-(m^2-m)}{2n}$

$= \dfrac{5m^2-3m-m^2+m}{2n}$

$= \dfrac{5m^2-3m-m^2+m}{2n}$

$= \dfrac{4m^2-2m}{2n}$

$= \dfrac{2m(2m-1)}{2n}$

$= \dfrac{m(2m-1)}{n}$ or $\dfrac{2m^2-m}{n}$

413

SOLUTIONS TO CHECK POINTS

Check Point 6, Frame 25

1. Since $15n^3p = 3 \cdot 5n^3 p$ and $24np^2 = 2^3 \cdot 3np^2$,

   the LCD $= 2^3 \cdot 3 \cdot 5 \cdot n^3 \cdot p^2 = 120n^3p^2$,

   $$\frac{2m^3}{15n^3p} - \frac{-5m}{24np^2} = \frac{(2m^3)(8p)}{(15n^3p)(8p)} - \frac{(-5m)(5n^2)}{(24np^2)(5n^2)}$$

   $$= \frac{16m^3p}{120n^3p^2} - \frac{(-25mn^2)}{120n^3p^2}$$

   $$= \frac{16m^3p + 25mn^2}{120n^3p^2}$$

2. The LCD $= (x - 6)(x + 1) = x^2 - 5x - 6$ by inspection.

   $$\frac{9}{x-6} + \frac{3x}{x+1} = \frac{(9)(x+1)}{(x-6)(x+1)} + \frac{(3x)(x-6)}{(x+1)(x-6)}$$

   $$= \frac{9x+9}{(x-6)(x+1)} + \frac{3x^2-18x}{(x+1)(x-6)}$$

   $$= \frac{3x^2 - 9x + 9}{(x+1)(x-6)}$$

3. The LCD $= (4a^2 - 9) = (2a + 3)(2a - 3)$, since $2a + 3$ is a factor of $4a^2 - 9$.

   $$\frac{1}{4a^2-9} - \frac{a}{2a+3} = \frac{1}{(2a+3)(2a-3)} - \frac{a(2a-3)}{(2a+3)(2a-3)}$$

   $$= \frac{1}{4a^2-9} - \frac{(2a^2-3a)}{4a^2-9}$$

   $$= \frac{1-2a^2+3a}{4a^2-9} \quad \text{or} \quad \frac{-2a^2+3a+1}{4a^2-9}$$

CHECK POINT 7, FRAME 30

1. Since $2R^2 + 4R = 2R(R + 2)$ and $8R^2 = 2^3 \cdot R^2$,

   the LCD $= 2^3 \cdot R^2 \cdot R^2(R + 2) = 8R^2(R + 2)$.

   $$\frac{R^2}{2R^2+4R} - \frac{5}{8R^2} = \frac{R^2}{2R(R+2)} - \frac{5}{8R^2}$$

   $$= \frac{R^2(4R)}{2R(R+2)(4R)} - \frac{5(R+2)}{8R^2(R+2)}$$

   $$= \frac{4R^3}{8R^2(R+2)} - \frac{(5R+10)}{8R^2(R+2)}$$

   $$= \frac{4R^3 - (5R+10)}{8R^2(R+2)}$$

   $$= \frac{4R^3 - 5R - 10}{8R^2(R+2)}$$

2. Since $3a - 9 = 3(a - 3)$ and $a^2 - 9 = (a - 3)(a + 3)$,

   the LCD $= 3(a - 3)(a + 3)$.

   $$\frac{2a-1}{3a-9} - \frac{a+2}{a^2-9} = \frac{2a-1}{3(a-3)} - \frac{a+2}{(a-3)(a+3)}$$

   $$= \frac{(2a-1)(a+3)}{3(a-3)(a+3)} - \frac{(a+2)(3)}{3(a-3)(a+3)}$$

   $$= \frac{(2a^2+6a-a-3) - (3a+6)}{3(a^2-9)}$$

   $$= \frac{2a^2+2a-9}{3(a^2-9)}$$

414

CHECK POINT 7, FRAME 30

3. Since $3x^2 - 5x - 2 = (3x + 1)(x - 2)$ and $x^2 + x - 6 = (x + 3)(x - 2)$,

the LCD = $(3x + 1)(x - 2)(x + 3)$.

$$\frac{5x}{3x^2 - 5x - 2} - \frac{3x}{x^2 + x - 6} = \frac{5x}{(3x + 1)(x - 2)} - \frac{3x}{(x + 3)(x - 2)}$$

$$= \frac{5x(x + 3)}{(3x + 1)(x - 2)(x + 3)} - \frac{3x(3x + 1)}{(x + 3)(x - 2)(3x + 1)}$$

$$= \frac{(5x^2 + 15x) - (9x^2 + 3x)}{(3x + 1)(x - 2)(x + 3)}$$

$$= \frac{5x^2 + 15x - 9x^2 - 3x}{(3x + 1)(x - 2)(x + 3)}$$

$$= \frac{-4x^2 + 12x}{(3x + 1)(x - 2)(x + 3)}$$

4. Since $n^3 - 2n^2 = n^2(n - 2)$

$$n^2 + n - 6 = (n + 3)(n - 2)$$

$$n^3 = n^3$$

the LCD = $n^3(n - 2)(n + 3)$

$$\frac{2}{n^3 - 2n^2} + \frac{n}{n^2 + n - 6} - \frac{7}{n^3} = \frac{2}{n^2(n - 2)} + \frac{n}{(n + 3)(n - 2)} - \frac{7}{n^3}$$

$$= \frac{2n(n + 3)}{n^3(n + 3)(n - 2)} + \frac{n \cdot n^3}{n^3(n + 3)(n - 2)} - \frac{7n(n + 3)(n - 2)}{n^3(n + 3)(n - 2)}$$

$$= \frac{2n(n + 3) + n^4 - 7(n + 3)(n - 2)}{n^3(n + 3)(n - 2)}$$

$$= \frac{2n^2 + 6n + n^4 - 7(n^2 + n - 6)}{n^3(n + 3)(n - 2)}$$

$$= \frac{2n^2 + 6n + n^4 - 7n^2 - 7n + 42}{n^3(n + 3)(n - 2)}$$

$$= \frac{n^4 - 5n^2 - n + 42}{n^3(n + 3)(n - 2)}$$

ANSWERS TO DRILL EXERCISES

I   1. $\dfrac{T^2S + TS^2}{2T - 1}$ 2. $\dfrac{7y + 1}{7y^2}$ 3. $\dfrac{8x - 5}{3b}$

4. $\dfrac{4n^3 + 2n}{5m}$

II   5. $3^2y^2(y - 4)$ 6. $2^2(2y + 3z)(2y - 3z)$

7. $2(m - 3)(m - 6)$ 8. $x^3(3x + 2)(x - 1)$

III   9. $(x + 3)(x - 2)$ 10. $6xy$

11. $(x + 3)^2$ 12. $9a^2$

13. $112a^2b^3c^4$ 14. $72x^3y^2(x - 3)$

15. $4(a - 3)(a + 7)$ 16. $3x^2(x + 5)(x - 5)(x - 2)$

IV   17. $\dfrac{2z + 3y + 4x}{xyz}$ 18. $\dfrac{21m^3p^2 + 4np}{70m^4n^2}$

19. $\dfrac{2n + 3m - 1}{(m + 1)(n - 2)}$ 20. $\dfrac{28x^2 - 3x + 6}{16x^2(x - 2)}$

21. $\dfrac{4x^3 + x^2 - 3x - 4}{2x^2(x^2 + 3x + 2)}$ 22. $-\dfrac{x^2 - 7x + 5}{x^2 - 4}$

23. $-\dfrac{x^2 + 25x + 1}{5x(3x - 1)}$ or $-\dfrac{x^2 + 25x + 1}{15x^2 - 5x}$

24. $\dfrac{2x^2 - 3x + 5}{2(3x + 2)(x + 1)}$ or $\dfrac{2x^2 - 3x + 5}{6x^2 + 10x + 4}$

SOLUTIONS TO CHECK POINTS

Check Point 1, Frame 5

1.

$$LCD = 2M$$

$$2M\left(\frac{7}{2} + \frac{2}{M}\right) = 2M \cdot 4$$

$$2M \cdot \frac{7}{2} + 2M \cdot \frac{2}{M} = 2M \cdot 4$$

$$7M + 4 = 8M$$

2.

$$LCD = 60V$$

$$60V\left(\frac{2}{5V} + \frac{3}{10}\right) = 60V \cdot \frac{1}{12}$$

$$60V \cdot \frac{2}{5V} + 60V \cdot \frac{3}{10} = 60V \cdot \frac{1}{12}$$

$$12 \cdot 2 + 6V \cdot 3 = 5V$$

$$24 + 18V = 5V$$

3.

$$LCD = 4(f - 2)$$

$$4(f - 2)\frac{3}{4} = 4(f - 2)\frac{f}{f-2}$$

$$3(f - 2) = 4f$$

$$3f - 6 = 4f$$

SOLUTIONS TO CHECK POINTS

Check Point 1, Frame 5

4.

$$LCD = R(R + 2)$$

$$R(R + 2)\frac{2}{R} = R(R + 2)\frac{3}{R+2}$$

$$2(R + 2) = 3R$$

$$2R + 4 = 3R$$

5.

$$LCD = x + 1$$

$$(x + 1)\frac{2}{x+1} = (x + 1)5 + (x + 1)\frac{2}{x+1}$$

$$2 = 5(x + 1) + 2$$

$$2 = 5x + 5 + 2$$

SOLUTIONS TO CHECK POINTS

Check Point 2, Frame 12

1.
$$LCD = 6$$
$$6\left(\frac{2M}{3} - \frac{M+1}{2}\right) = 6\left(\frac{1}{3}\right)$$
$$6\left(\frac{2M}{3}\right) - 6\left(\frac{M+1}{2}\right) = 6\left(\frac{1}{3}\right)$$
$$4M - 3(M+1) = 2$$
$$4M - 3M - 3 = 2$$
$$M = 5$$

Check the solution.

L.H.S. = $\frac{2(5)}{3} - \frac{5+1}{2} = \frac{10}{3} - \frac{6}{2} = \frac{20}{6} - \frac{18}{6} = \frac{2}{6} = \frac{1}{3}$

R.H.S. = $\frac{1}{3}$

Hence, the solution is M = 5.

2.
$$LCD = x - 3$$
$$(x-3)\left(\frac{2}{x-3} - 5\right) = (x-3) \cdot \frac{2}{x-3}$$
$$(x-3) \cdot \frac{2}{x-3} - (x-3)(5) = (x-3) \cdot \frac{2}{x-3}$$
$$2 - (x-3)(5) = 2$$
$$2 - 5x + 15 = 2$$
$$-5x = -15$$
$$x = 3$$

Check the solution.

L.H.S. = $\frac{2}{3-3} - 5 = \frac{2}{0} - 5$.

Since $\frac{2}{0}$ is undefined, x = 3 cannot be a solution but is extraneous.

Hence, the equation has no solution.

SOLUTIONS TO CHECK POINTS

Check Point 2, Frame 12

3.
$$LCD = a + 6$$
$$(a+6) \cdot \frac{a+10}{a+6} = (a+6)\left(\frac{4}{a+6} + 3\right)$$
$$(a+6) \cdot \frac{a+10}{a+6} = (a+6) \cdot \frac{4}{a+6} + (a+6)(3)$$
$$a + 10 = 4 + (a+6)(3)$$
$$a + 10 = 4 + 3a + 18$$
$$-2a = 12$$
$$a = -6$$

Check the solution.

L.H.S. = $\frac{-6+10}{-6+6} = \frac{4}{0}$

Since $\frac{4}{0}$ is undefined, a = -6 cannot be a solution, but is extraneous.

Hence, the equation has no solution.

4.
$$LCD = y^2 - 4 = (y+2)(y-2)$$
$$(y+2)(y-2)\left(\frac{6}{y^2-4} + \frac{3}{y-2}\right) = (y+2)(y-2) \cdot \frac{4}{y+2}$$
$$(y+2)(y-2)\frac{6}{(y-2)(y+2)} + (y+2)(y-2)\frac{3}{y-2} = (y+2)(y-2)\frac{4}{y+2}$$
$$6 + (y+2)(3) = (y-2)(4)$$
$$6 + 3y + 6 = 4y - 8$$
$$-y = -20$$
$$y = 20$$

Check the solution.

L.H.S. = $\frac{6}{(20)^2 - 4} + \frac{3}{20 - 2} = \frac{6}{396} + \frac{3}{18} = \frac{6}{396} + \frac{66}{396} = \frac{72}{396} = \frac{2}{11}$

R.H.S. = $\frac{4}{20 + 2} = \frac{4}{22} = \frac{2}{11}$

Hence, the solution is y = 20.

417

SOLUTIONS TO CHECK POINTS

Check Point 3, Frame 16

1. $$LCD = p(p + 3)$$

$$p(p+3)\left(\frac{4}{p+3} + \frac{6}{p+3}\right) = p(p+3)\left(1 - \frac{4}{p}\right)$$

$$4p + 6p = p(p+3) - 4(p+3)$$

$$10p = p^2 + 3p - 4p - 12$$

$$0 = p^2 - 11p - 12$$

$$0 = (p-12)(p+1)$$

$$p - 12 = 0 \text{ or } p + 1 = 0$$

$$\therefore \ p = 12 \text{ or } p = -1$$

Check the solution.

If p = 12, $\frac{4}{12+3} + \frac{6}{12+3} = 1 - \frac{4}{12}$ or $\frac{2}{3} = \frac{2}{3}$

If p = -1, $\frac{4}{-1+3} + \frac{6}{-1+3} = 1 - \frac{4}{-1}$ or $5 = 5$

Hence, p = 12 and p = -1 are solutions of the given equation.

2. $$LCD = (a + 4)(a - 4) = a^2 - 16$$

$$(a+4)(a-4)\left(\frac{1}{(a+4)(a-4)} + \frac{a}{a+4}\right) = (a+4)(a-4) \cdot \frac{2}{a-4}$$

$$(a+4)(a-4)\frac{1}{(a+4)(a-4)} + (a+4)(a-4)\frac{a}{a+4} = (a+4)(a-4)\frac{2}{a-4}$$

$$1 + (a-4)(a) = (a+4)(2)$$

$$1 + a^2 - 4a = 2a + 8$$

$$a^2 - 6a - 7 = 0$$

$$(a-7)(a+1) = 0$$

$$a - 7 = 0 \text{ or } a + 1 = 0$$

$$\therefore \ a = 7 \text{ or } a = -1$$

SOLUTIONS TO CHECK POINTS

Check Point 3, Frame 16

2. Check the solution.

If a = 7, $\frac{1}{7^2 - 16} + \frac{7}{7+4} = \frac{2}{7-4}$ or $\frac{2}{3} = \frac{2}{3}$

If a = -1, $\frac{1}{(-1)^2 - 16} + \frac{-1}{-1+4} = \frac{2}{-1-4}$ or $-\frac{2}{5} = -\frac{2}{5}$

Hence, a = 7 and a = -1 are solutions of the original equation.

3. $$LCD = (m^2 + 2)(3m + 2)$$

$$(m^2+2)(3m+2) \cdot \frac{m-1}{(m^2+2)} = (m^2+2)(3m+2) \cdot \frac{2}{(3m+2)}$$

$$(3m+2)(m-1) = (m^2+2)(2)$$

$$3m^2 - 3m + 2m - 2 = 2m^2 + 4$$

$$m^2 - m - 6 = 0$$

$$(m-3)(m+2) = 0$$

$$m - 3 = 0 \text{ or } m + 2 = 0$$

$$\therefore \ m = 3 \text{ or } m = -2$$

Check the solution.

If m = 3, $\frac{3-1}{3^2+2} = \frac{2}{3(3)+2}$ or $\frac{2}{11} = \frac{2}{11}$

If m = -2, $\frac{-2-1}{(-2)^2+2} = \frac{2}{3(-2)+2}$ or $-\frac{3}{6} = -\frac{1}{2}$ or $-\frac{1}{2} = -\frac{1}{2}$

Hence, m = 3 and m = -2 are solutions of the original equation.

SOLUTIONS TO CHECK POINTS

Check Point 3, Frame 16

4.
$$LCD = (t + 3)(t - 5)$$
$$= t^2 - 2t - 15$$

$$(t + 3)(t - 5)\left(\frac{t + 1}{t + 3} + \frac{t + 2}{t - 5}\right) = (t + 3)(t - 5) \cdot \frac{t^2 - t + 4}{t^2 - 2t - 15}$$

$$(t + 3)(t - 5)\frac{t+1}{t+3} + (t + 3)(t - 5)\frac{t+2}{t-5} = (t+3)(t-5)\frac{t^2 - t + 4}{(t+3)(t-5)}$$

$$(t - 5)(t + 1) + (t + 3)(t + 2) = t^2 - t + 4$$
$$t^2 - 4t - 5 + t^2 + 5t + 6 = t^2 - t + 4$$
$$t^2 + 2t - 3 = 0$$
$$(t + 3)(t - 1) = 0$$
$$t + 3 = 0 \quad \text{or} \quad t - 1 = 0$$
$$\therefore \quad t = -3 \quad \text{or} \quad t = 1$$

Check the solution.

If $t = -3$, the L.H.S. is undefined since $\frac{-3 + 1}{-3 + 3} = \frac{-2}{0}$

If $t = 1$, $\frac{1+1}{1+3} + \frac{1+2}{1-5} = \frac{1^2 - 1 + 4}{1^2 - 2(1) - 15}$ or $-\frac{1}{4} = -\frac{1}{4}$

Hence, $t = 1$ is a solution of the original equation.

However, $t = -3$, is not a solution and we call $t = -3$ an extraneous solution.

SOLUTIONS TO CHECK POINTS

Check Point 4, Frame 23

1.
$$W = \frac{2PR}{R - r}$$
$$(R - r)W = \frac{2PR}{R - r}(R - r)$$
$$(R - r)W = 2PR$$
$$RW - rW = 2PR$$
$$RW - 2PR = rW$$
$$R(W - 2P) = rW$$
$$\frac{R(W - 2P)}{W - 2P} = \frac{rW}{W - 2P}$$
$$\therefore \quad R = \frac{rW}{W - 2P}$$

2.
$$S = P(1 + rt)$$
$$S = P + Prt$$
$$-Prt = -S + P$$
$$\frac{-Prt}{-Pt} = \frac{-S + P}{-Pt}$$
$$r = \frac{(-1)(-S + P)}{(-1)(-Pt)}$$
$$= \frac{S - P}{Pt}$$
$$\therefore \quad r = \frac{S - P}{Pt}$$

3.
$$Q = \frac{kAT(t_2 - t_1)}{d}$$
$$dQ = \frac{kAT(t_2 - t_1)}{d} \cdot d$$
$$dQ = kATt_2 - kATt_1$$
$$kATt_1 = kATt_2 - dQ$$
$$\frac{kATt_1}{kAT} = \frac{kATt_2 - dQ}{kAT}$$
$$\therefore \quad t_1 = \frac{kATt_2 - dQ}{kAT}$$

4.
$$A = \frac{m}{t}(p + t)$$
$$At = \frac{m}{t}(p + t)t$$
$$At = mp + mt$$
$$At - mt = mp$$
$$t(A - m) = mp$$
$$\frac{t(A - m)}{A - m} = \frac{mp}{A - m}$$
$$\therefore \quad t = \frac{mp}{A - m}$$

SOLUTIONS TO CHECK POINTS

Check Point 4, Frame 23

5.

$$f = \frac{r_2 - r_1}{r_1(w_1 + w_2)}$$

$$r_1(w_1 + w_2)f = r_2 - r_1$$

$$r_1 w_1 f + r_1 w_2 f + r_1 = r_2$$

$$r_1(w_1 f + w_2 f + 1) = r_2$$

$$\frac{r_1(w_1 f + w_2 f + 1)}{w_1 f + w_2 f + 1} = \frac{r_2}{w_1 f + w_2 f + 1}$$

$$\therefore\ r_1 = \frac{r_2}{w_1 f + w_2 f + 1}$$

Check Point 5, Frame 30

1. Let x = the time in minutes it takes the slower press to print the edition alone.

Then x − 30 = the time in minutes it takes the faster press to print the edition alone.

Hence, $\frac{1}{x}$ = the part of work done by the slower press in one minute.

$\frac{1}{x - 30}$ = the part of the work done by the faster press in one minute.

$\frac{1}{20}$ = the part of the work done by both presses in one minute.

Work done by slower press in 1 minute	+	Work done by faster press in 1 minute	=	Work done by both presses in 1 minute
$\frac{1}{x}$	+	$\frac{1}{x - 30}$	=	$\frac{1}{20}$

Clear the fractions
LCD = 20x(x − 30)

SOLUTIONS TO CHECK POINTS

Check Point 5, Frame 30

$$20(x - 30) + 20x = x(x - 30)$$

$$20x - 600 + 20x = x^2 - 30x$$

$$x^2 - 70x + 600 = 0$$

$$(x - 10)(x - 60) = 0$$

$$\therefore\ x = 10 \text{ or } x = 60$$

Check:

If x = 10, $\frac{1}{x - 30} = \frac{1}{-20}$. The part of work done by the faster press cannot be negative.

Hence, x = 10 is not a solution.

If x = 60, $\frac{1}{x} = \frac{1}{60}$ is the part of work done by the slower press and $\frac{1}{x - 30} = \frac{1}{30}$ is the part of work done by the faster press.

$$\frac{1}{x} + \frac{1}{x - 30} = \frac{1}{60} + \frac{1}{30} = \frac{1}{60} + \frac{2}{60} = \frac{1}{20} = \text{R.H.S.}$$

Hence, it would take 60 minutes (1 h) for the slower press to do the job alone.

2. Let v = the regular average speed in km/h.

v + 10 = the faster average speed in km/h.

Since d = vt, we have $t = \frac{d}{v}$.

We know that d = 100 km.

Hence, $\frac{100}{v}$ = the time in hours required to travel 100 km at the regular speed.

$\frac{100}{v + 10}$ = the time in hours required to travel 100 km at the faster speed.

$$\frac{100}{v} - \frac{100}{v + 10} = \frac{1}{2}$$

Clear the fractions
LCD = 2v(v + 10)

ANSWERS TO DRILL EXERCISES

1.	$m = 10$	2.	$t = 6$
3.	$v = \frac{5}{2}$	4.	$p = \frac{3}{11}$
5.	$M = \frac{3}{11}$	6.	$R = -4$
7.	$p = \frac{17}{4}$	8.	$x = 1$
9.	$x = -\frac{11}{4}$	10.	$x = \frac{30}{11}$
11.	$x = \frac{7}{5}$	12.	$x = \frac{8}{3}$
13.	$t = 20$ or $t = -6$	14.	$t = 10$ or $t = -3$
15.	$v = 9$ or $v = -1$	16.	$v = 30$ or $v = -36$
17.	$a = 3$ or $a = -3$	18.	$x = -1$ is extraneous $\therefore$ no solution
19.	$x = -1$ is extraneous $x = 3$ is the solution	20.	$V = 2$
21.	$x = 4$ or $x = 1$	22.	$d = \frac{1}{2}$
23.	$R = 2$ is extraneous $\therefore$ no solution	24.	$x = \frac{14}{3}$ or $x = -1$
25.	$m = -1$	26.	$R_1 = \frac{C - 2aR_2}{a}$
27.	$V_1 = \frac{P_1 + P_2 - dFV_2}{dF}$	28.	$a = \frac{2D_o + 2vt - D}{t^2}$
29.	$P = \frac{2SEt}{D + 0.2t}$	30.	$Q_w = \frac{Vx - 1440SQx_e}{1440Sx_w - 1440Sx_e}$
31.	$22\frac{2}{9}$ min (22 min, $13\frac{1}{3}$ s)	32.	108 min (1 h, 48 min)
33.	12 min	34.	40 km/h
35.	$4\frac{1}{5}$ km/h		

SOLUTIONS TO CHECK POINTS

Check Point 5, Frame 30

$200v + 2000 - 200v = v(v + 10)$

$v^2 + 10v - 2000 = 0$

$(v + 50)(v - 40) = 0$

$\therefore \quad v = -50$ or $v = 40$

Check:

If $v = -50$, $\frac{100}{v} = -2$h (travel time at regular speed).

Since travel time cannot be negative, $v = -50$ is not a solution.

If $v = 40$, $\frac{100}{v} = \frac{100}{40} = 2\frac{1}{2}$h (travel time at regular speed)

$\frac{100}{v + 10} = \frac{100}{40 + 10} = 2$h (travel time at faster speed)

Travel time saved $= 2\frac{1}{2} - 2$h $= \frac{1}{2}$h.

Hence, the regular average speed is 40 km/h.

421

SOLUTIONS TO CHECK POINTS

Check Point 1, Frame 4

1. The square roots of 49 are +7 and -7.

2. The square roots of 36 are +6 and -6.

3. The square roots of 64 are +8 and -8.

4. The square roots of $\frac{1}{4}$ are $+\frac{1}{2}$ and $-\frac{1}{2}$. (Note that $(+\frac{1}{2})^2 = \frac{1}{2} \times \frac{1}{2} = \frac{1}{4}$. Similarly $(-\frac{1}{2})^2 = (-\frac{1}{2})(-\frac{1}{2}) = \frac{1}{4}$.)

5. The square roots of $\frac{9}{16}$ are $+\frac{3}{4}$ and $-\frac{3}{4}$.

6. The square roots of $\frac{4}{81}$ are $+\frac{2}{9}$ and $-\frac{2}{9}$.

Check Point 2, Frame 8

1. $\sqrt{4} = +2$ or simply 2.

2. $\sqrt{3^2} = \sqrt{9} = +3$ or simply 3.

3. $-\sqrt{36} = -6$

4. $\sqrt{\frac{16}{25}} = \frac{4}{5}$

5. $-\sqrt{8^2} = -\sqrt{64} = -8$

6. $\sqrt{0} = 0$

7. $\sqrt{\frac{64}{81}} = \frac{8}{9}$

8. $-\sqrt{\frac{1}{4}} = -\frac{1}{2}$

Check Point 3, Frame 12

1. $\sqrt[3]{\frac{1}{8}} = \frac{1}{2}$ since $(\frac{1}{2})(\frac{1}{2})(\frac{1}{2}) = \frac{1}{8}$

2. $\sqrt[3]{-64} = -4$ since $(-4)(-4)(-4) = -64$

3. $\sqrt[3]{1000} = 10$

4. $\sqrt[3]{-1} = -1$

5. $\sqrt[3]{\frac{8}{27}} = \frac{2}{3}$

6. $\sqrt[3]{-216} = -6$

SOLUTIONS TO CHECK POINTS

Check Point 4, Frame 18

1. $\sqrt[5]{32} = 2$ since $(2)(2)(2)(2)(2) = 32$

2. $\sqrt[8]{81} = 3$

3. $\sqrt[7]{-1} = -1$

4. $\sqrt[8]{1} = 1$

Check Point 5, Frame 26

I 1. $\sqrt{3} = 3^{1/2}$

   2. $\sqrt[5]{42} = 42^{1/5}$

   3. $\sqrt[6]{9} = 9^{1/6}$

   4. $\sqrt[3]{2^5} = (2^5)^{1/3} = 2^{5/3}$

   5. $\sqrt{7^3} = (7^3)^{1/2} = 7^{3/2}$

   6. $\sqrt[5]{3^4} = (3^4)^{1/5} = 3^{4/5}$

II 7. $16^{1/4} = (2^4)^{1/4} = 2^1 = 2$

    8. $9^{1/2} = (3^2)^{1/2} = 3^1 = 3$

    9. $(125)^{2/3} = (\sqrt[3]{125})^2 = (5)^2 = 25$

    10. $(-32)^{3/5} = (\sqrt[5]{-32})^3 = (-2)^3 = -8$

    11. $(-8)^{5/3} = (\sqrt[3]{-8})^5 = (-2)^5 = -32$

    12. $(81)^{3/4} = (\sqrt[4]{81})^3 = (3)^3 = 27$

SOLUTIONS TO CHECK POINTS

Check Point 6, Frame 31

		Operation	Display	Answer (to 4 decimal places)
1.	$\sqrt{3}$	Enter 3 Press [$\sqrt{\ }$]	3 1.7320508	1.7321
2.	$\sqrt{2.49}$	Enter 2.49 Press [$\sqrt{\ }$]	1.5779734	1.5780
3.	$\sqrt{0.231}$	Enter 0.231 Press [$\sqrt{\ }$]	0.4806246	0.4806
4.	$\sqrt{12396}$	Enter 12396 Press [$\sqrt{\ }$]	111.33733	111.3373
5.	$(139)^{1/2}$	Enter 139 Press [$\sqrt{\ }$]	11.789826	11.7898
6.	$(0.847)^{1/2}$	Enter 0.847 Press [$\sqrt{\ }$]	0.9203260	0.9203
7.	$(0.05)^{1/2}$	Enter 0.05 Press [$\sqrt{\ }$]	0.2236068	0.2236
8.	$(54168)^{1/2}$	Enter 54168 Press [$\sqrt{\ }$]	232.74020	232.7402

Check Point 7, Frame 37

1. $\sqrt[3]{9.84}$

Enter 9.84 Press [$\sqrt[x]{y}$] Enter 3 Press [=] Display 2.14288

Check: Press [y^x] Enter 3 Press [=] Display 9.83996

Hence, $\sqrt[3]{9.84}$ = 2.14288

2. $\sqrt[4]{8963}$

Enter 8963 Press [$\sqrt[x]{y}$] Enter 4 Press [=] Display 9.73001

Check: Press [y^x] Enter 4 Press [=] Display 8962.99

Hence, $\sqrt[4]{8963}$ = 9.73001

SOLUTIONS TO CHECK POINTS

3. $\sqrt[5]{-167}$

Enter 167 Press [$\sqrt[x]{y}$] Enter 5 Press [=] Display 2.78319

Check: Press [y^x] Enter 5 Press [=] Display 166.999

Hence, $\sqrt[5]{-167}$ = -2.78319

4. $\sqrt[8]{0.0596}$

Enter 0.0596 Press [$\sqrt[x]{y}$] Enter 8 Press [=] Display 0.70292

Check: Press [y^x] Enter 8 Press [=] Display 0.0596001

Hence, $\sqrt[8]{0.0596}$ = 0.70292

5. $(12.8)^{2/3}$

Enter 12.8 Press [y^x] Press [( ] Enter 2/3 Press [ )]

Press [=] Display 5.47192

Check: Press [y^x] Press 1.5 Press [=] Display 12.8

Hence, $(12.8)^{2/3}$ = 5.47192

6. $(159.5)^{3/4}$

Enter 159.5 Press [y^x] Enter 0.75 Press [=] Display 44.8818

Check: Press [y^x] Press [( ] Enter 4/3 Press [ )] Press [=]

Display 159.5

Hence, $(159.5)^{3/4}$ = 44.8818

7. $(-8.6)^{3/5}$

Enter 8.6 Press [y^x] Enter 0.6 Press [=] Display 3.63663

Check: Press [y^x] Press [( ] Enter 5/3 Press [ )] Press [=]

Display 8.6

Hence, $(-8.6)^{3/5}$ = -3.63663

SOLUTIONS/ANSWERS UNIT 16

SOLUTIONS TO CHECK POINTS

Check Point 1, Frame 4

		Exponent Laws Used
1.	$15m^{-4} \cdot m^2 = 15m^{(-4)+2} = 15m^{-2} = \dfrac{15}{m^2}$	3.1 and 3.7(a)
2.	$(x^{-1})^8 = x^{-8} = \dfrac{1}{x^8}$	3.3 and 3.7(a)
3.	$4a^8 \div a^{-6} = \dfrac{4a^8}{a^{-6}} = 4a^8 \cdot a^6 = 4a^{14}$ OR $\quad 4a^{8-(-6)} = 4a^{14}$	3.7(b) and 3.1
4.	$(2^{-3} x^{-2} y)^{-2} = 2^{(-3)(-2)} x^{(-2)(-2)} y^{-2}$ $= \dfrac{2^6 x^4}{y^2} = \dfrac{64x^4}{y^2}$	3.2 3.4
5.	$\dfrac{(3^{-1} m^2)^{-2}}{(16m^3 n^2)^0} = \dfrac{3^2 m^{-4}}{1} = \dfrac{9}{m^4}$	3.7(a)
6.	$\left(\dfrac{3x^{-4} y^2}{2^{-1} z^3}\right)^{-2} = \dfrac{3^{-2} x^8 y^{-4}}{2^2 z^{-6}} = \dfrac{x^8 z^6}{2^2 \cdot 3^2 y^4} = \dfrac{x^8 z^6}{36y^4}$	3.4, 3.6 and 3.7(a)

Check Point 2, Frame 10

1.	$(m-n)^{-2} = \dfrac{1}{(m-n)^2}$ OR $\dfrac{1}{m^2 - 2mn + n^2}$	3.5, 3.7(a) and 3.7(b)
2.	$(2a+3b)^{-1} = \dfrac{1}{2a+3b}$	
3.	$(4x)^{-2} + 3y^{-2} = \dfrac{1}{(4x)^2} + \dfrac{3}{y^2} = \dfrac{1}{16x^2} + \dfrac{3}{y^2} = \dfrac{48x^2 + y^2}{16x^2 y^2}$	

SOLUTIONS/ANSWERS UNIT 15

ANSWERS TO DRILL EXERCISES

I
1. 3
2. 1
3. $\frac{5}{7}$
4. $-\frac{1}{4}$
5. $\frac{1}{2}$
6. -1
7. -2
8. 5
9. 2
10. 16
11. -3
12. 125

II
13. 0.0382
14. -6.6494
15. 1.6263
16. -6.1731
17. 0.7035
18. 1.4089
19. 5.2332
20. -0.8607
21. 25.6384
22. 2.0336
23. 0.5612
24. -16.4128

III
25. $16^{1/3}$
26. $3^{5/7}$
27. $-9^{1/5}$
28. $8.5^{3/4}$

424

SOLUTIONS TO CHECK POINTS

Check Point 2, Frame 10

4. $(a^{-1}+2b)^{-2} = \frac{1}{(a^{-1}+2b)^2} = \frac{1}{\left(\frac{1}{a}+2b\right)^2} = \frac{1}{\left(\frac{1+2ab}{a}\right)^2}$

$= \frac{1}{\frac{(1+2ab)^2}{a^2}} = \frac{a^2}{(1+2ab)^2}$

5. $\frac{2m^2 - mn - n^2}{m^2(n^{-2} - m^{-2})} = \frac{(2m+n)(m-n)}{m^2\left(\frac{1}{n^2} - \frac{1}{m^2}\right)} = \frac{(2m+n)(m-n)}{m^2\left(\frac{m^2-n^2}{m^2n^2}\right)}$

$= (2m+n)(m-n) \cdot \frac{n^2}{(m-n)(m+n)}$

$= \frac{n^2(2m+n)}{m+n}$

6. $(4^{-2} - 2^{-3})^2 = \left(\frac{1}{4^2} - \frac{1}{2^3}\right)^2 = \left(\frac{1}{16} - \frac{1}{8}\right)^2 = \left(-\frac{1}{16}\right)^2 = \frac{1}{256}$

7. $\left(\frac{2a^{-1}b^2}{3c^4}\right)^{-3} \cdot \left(\frac{4b}{9a^2c}\right)^2 = \frac{2^{-3}a^3b^{-6}}{3^{-3}c^{-12}} \cdot \frac{4^2b^2}{9^2a^4c^{10}} = \frac{3^3a^3c^{12}}{2^3b^6} \cdot \frac{16b^2}{(3^2)^2a^4c^{10}} = \frac{2c^2}{3ab^4}$

Check Point 3, Frame 14

1. $3x^2(x^2+1)^{-1} - x^3(x^2+1)^{-2}2x$

$= \frac{3x^2}{x^2+1} - \frac{2x^4}{(x^2+1)^2} = \frac{3x^2(x^2+1)}{(x^2+1)^2} - \frac{2x^4}{(x^2+1)^2}$

$= \frac{3x^4 + 3x^2 - 2x^4}{(x^2+1)^2} = \frac{x^4 + 3x^2}{(x^2+1)^2}$ OR $\frac{x^2(x^2+3)}{(x^2+1)^2}$

SOLUTIONS TO CHECK POINTS

Check Point 3, Frame 14

2. $2(2x+1)^{-2} - (2x-1)(2x+1)^{-3}(2) = \frac{2}{(2x+1)^2} - \frac{2(2x-1)}{(2x+1)^3}$

$= \frac{2(2x+1)}{(2x+1)^3} - \frac{2(2x-1)}{(2x+1)^3} = \frac{2(2x+1) - 2(2x-1)}{(2x+1)^3} = \frac{4}{(2x+1)^3}$

Check Point 4, Frame 18

1. $4^{1/3} \cdot 4^{2/3} = 4^1 = 4$

2. $m^{1/2} \cdot m^{-1/4} = m^{1/2 - 1/4} = m^{1/4}$

3. $\frac{x^{3/8}}{x^{1/2}} = x^{3/8 - 1/2} = x^{-1/8} = \frac{1}{x^{1/8}}$

4. $\frac{(R+3)^{3/2}}{(R+3)^{1/2}} = (R+3)^{3/2 - 1/2} = (R+3)^1 = R+3$

Check Point 5, Frame 22

1. $(a^{1/2})^{5/3} = a^{1/2 \cdot 5/3} = a^{5/6}$

2. $(m^{2/3} \cdot n^{1/4})^{3/2} = m^{2/3 \cdot 3/2} \cdot n^{1/4 \cdot 3/2} = mn^{3/8}$

3. $(4x^{4/3} \cdot y^{1/2})^{1/2} = 4^{1/2} \cdot x^{4/3 \cdot 1/2} \cdot y^{1/2 \cdot 1/2} = 2x^{2/3}y^{1/4}$

4. $\left(\frac{1}{32^{6/7}}\right)^{7/10} = \frac{1}{32^{6/7 \cdot 7/10}} = \frac{1}{32^{3/5}} = \frac{1}{(32^{1/5})^3} = \frac{1}{2^3} = \frac{1}{8}$

SOLUTIONS TO CHECK POINTS

Check Point 5, Frame 22

5. $\left(\dfrac{x^{2/3}}{y^{1/4}}\right)^{3/2} = \dfrac{x^{2/3\,\cdot\,3/2}}{y^{1/4\,\cdot\,3/2}} = \dfrac{x}{y^{3/8}}$

6. $\left(\dfrac{16a^{4/3}b^{1/2}}{c^{3/2}}\right)^{1/4} = \dfrac{16^{1/4}\cdot a^{4/3\,\cdot\,1/4}\cdot b^{1/2\,\cdot\,1/4}}{c^{3/2\,\cdot\,1/4}} = \dfrac{2a^{1/3}b^{1/8}}{c^{3/8}}$

Check Point 6, Frame 26

1. $(3^{-1}x^2y^{-2/5})(9x^{-3/4}y^3) = 3^{-1}\cdot 9\cdot x^2\cdot x^{-3/4}\cdot y^{-2/5}+3$

$= \dfrac{1}{3}\cdot 9\cdot x^{5/4}\cdot y^{13/5} = 3x^{5/4}y^{13/5}$

2. $\dfrac{36m^{-1/2}n^{3/5}t^{-1/5}}{9m^{-3/2}n^{-7/5}t^2} = 4m^{-1/2-(-3/2)}\cdot n^{3/5-(-7/5)}\cdot t^{-1/5-2}$

$= 4m^{2/2}n^{10/5}\cdot t^{-11/5} = \dfrac{4mn^2}{t^{11/5}}$

3. $\left(\dfrac{3^2a^{-1/2}}{2b^{-3/4}}\right)^{-2}\cdot\left(\dfrac{9a^{2/3}}{16b}\right)^{1/2} = \dfrac{3^{(2)(-2)}a^{(+1/2)(-2)}}{2^{-2}\cdot b^{(-3/4)(-2)}}\cdot\dfrac{9^{1/2}\cdot a^{(2/3)(1/2)}}{16^{1/2}\cdot b^{1/2}}$

$= \dfrac{3^{-4}\cdot a}{(2^2)^{-1}\cdot b^{3/2}}\cdot\dfrac{3^{1}a^{1/3}}{4b^{1/2}} = \dfrac{a^{4/3}}{4^0 b^{3/2+1/2}} = \dfrac{a^{4/3}}{27b^2}$

4. $\left(\dfrac{1}{8p}\right)^{-2/3}\div\left(\dfrac{16p^2}{t^4}\right)^{3/4} = \dfrac{1p^{2/3}}{(8^{1/3})^{-2}}\div\dfrac{(16^{1/4})^3 p^{3/2}}{t^3}$

$= \dfrac{p^{2/3}}{2^{-2}}\cdot\dfrac{t^3}{2^3 p^{3/2}} = \dfrac{1^{-2/3}}{8^{-2/3}p^{-2/3}}\div\dfrac{16^{3/4}2^{(3/4)}p^{3/2}}{t^{4(3/4)}} = \dfrac{t^3}{2p^{3/2-2/3}} = \dfrac{t^3}{2p^{5/6}}$

Check Point 7, Frame 30

1. $3(x-1)^{1/2} + (3x+1)(x-1)^{-1/2} = 3(x-1)^{1/2} + \dfrac{(3x+1)}{(x-1)^{1/2}}$

$= \dfrac{3(x-1)^{1/2}(x-1)^{1/2}}{(x-1)^{1/2}} + \dfrac{(3x+1)}{(x-1)^{1/2}} = \dfrac{3(x-1)}{(x-1)^{1/2}} + \dfrac{3x+1}{(x-1)^{1/2}}$

$= \dfrac{6x-2}{(x-1)^{1/2}}$ OR $\dfrac{2(3x-1)}{(x-1)^{1/2}}$

SOLUTIONS TO CHECK POINTS

Check Point 7, Frame 30

2. $\dfrac{(2\pi x^2 + 2\pi y^2)^{3/2}}{(x^2+y^2)^{1/2}} = \dfrac{(2\pi(x^2+y^2))^{3/2}}{(x^2+y^2)^{1/2}} = \dfrac{(2\pi)^{3/2}(x^2+y^2)^{3/2}}{(x^2+y^2)^{1/2}}$

$= (2\pi)^{3/2}(x^2+y^2)^{3/2} - 1/2 = (2\pi)^{3/2}(x^2+y^2)$

ANSWERS TO DRILL EXERCISES

1. $3x^3$

2. a^4

3. $\dfrac{4m^6}{n^2}$

4. $\dfrac{125t^6}{r^3}$

5. $\dfrac{z^6}{xy^6}$

6. $\dfrac{81n^{12}}{256m^8z^4}$

7. $\dfrac{2yz^3}{x^3}$

8. $\dfrac{-3}{a^2}$

9. $\dfrac{1}{(a+b)^3}$

10. $\dfrac{n^3 - 6m}{2mn^3}$

11. $\dfrac{y^2}{5xy^2 - 1}$

12. $r(r-2t)$

13. $\dfrac{1}{256}$

14. $\dfrac{y^6}{(2xy^2-1)^3}$

15. -72

16. $\dfrac{3a^2b}{2}$

17. $\dfrac{675t}{8m^2n^9r}$

18. $\dfrac{ab}{b-a}$

19. $-2(1-2t)^2(1+2t)(1+10t)$

20. $\dfrac{2m^5 - 2m - 4m^4}{(m^4-1)^2}$

21. $\dfrac{a^2}{(1-ab)^2}$

22. xy^3

23. 3

24. $a^{1/2}$

SOLUTIONS TO CHECK POINTS

Check Point 1, Frame 7

1. $\sqrt{4^2} = (4^2)^{1/2} = 4$ OR $(4^{1/2})^2 = (\sqrt{4})^2 = 4$

2. $\sqrt[5]{2^5} = (2^5)^{1/5} = 2$ OR $(2^{1/5})^5 = (\sqrt[5]{2})^5 = 2$

3. $\sqrt{81} = \sqrt{9 \cdot 9} = (9 \cdot 9)^{1/2} = 9^{1/2} \cdot 9^{1/2} = \sqrt{9}\sqrt{9}$

4. $\sqrt[3]{216} = \sqrt[3]{8 \cdot 27} = (8 \cdot 27)^{1/3} = 8^{1/3} \cdot 27^{1/3} = \sqrt[3]{8}\ \sqrt[3]{27}$

5. $\sqrt{\dfrac{4}{25}} = \left(\dfrac{4}{25}\right)^{1/2} = \dfrac{4^{1/2}}{25^{1/2}} = \dfrac{\sqrt{4}}{\sqrt{25}}$

6. $\sqrt[3]{\dfrac{27}{64}} = \left(\dfrac{27}{64}\right)^{1/3} = \dfrac{27^{1/3}}{64^{1/3}} = \dfrac{\sqrt[3]{27}}{\sqrt[3]{64}}$

7. $\sqrt[3]{\sqrt{729}} = \left((729)^{1/2}\right)^{1/3} = 729^{1/6} = \sqrt[6]{729}$

Check Point 2, Frame 12

1. $\sqrt{50} = \sqrt{25 \cdot 2} = \sqrt{25}\ \sqrt{2} = 5\sqrt{2}$

2. $\sqrt[3]{24} = \sqrt[3]{8 \cdot 3} = \sqrt[3]{8}\ \sqrt[3]{3} = 2\ \sqrt[3]{3}$

3. $\sqrt[3]{54} = \sqrt[3]{27 \cdot 2} = \sqrt[3]{27}\ \sqrt[3]{2} = 3\ \sqrt[3]{2}$

4. $\sqrt{108} = \sqrt{36 \cdot 3} = \sqrt{36}\ \sqrt{3} = 6\sqrt{3}$

ANSWERS TO DRILL EXERCISES

25. $\dfrac{1}{y^{1/2}}$

26. $(m^2 - 1)^2$

27. $(x + 1)^{5/4}$

28. $t^{1/6}$

29. 2

30. $2^{1/2}$

31. 8

32. $25x^{3/2}y^{2/3}$

33. $\dfrac{m^{5/4}}{n^{1/3}}$

34. $a^{7/2}$

35. $\dfrac{z^{19/5}}{x^{1/6}y^{7/8}}$

36. $t^{13/10}$

37. $\dfrac{1}{x^{1/2}y^{5/2}z^3}$

38. $\dfrac{r^{9/40}t^{1/2}}{8}$

39. $\dfrac{n^{51/8}}{8^{3/2}m^{1/4}}$

40. $\dfrac{4a^{17/6}}{b^{1/6}}$

41. $\dfrac{4x^{7/4}}{y^{1/2}z^{3/4}}$

42. $\dfrac{p^{1/5}}{3v^{1/30}}$

43. $\dfrac{2a^{7/3}}{b^{2/3}}$

44. $x - y$

45. $3^{5/2}(a - b)(a + b)$

46. $\dfrac{29 - 21m}{2(2 - m)^{3/2}}$

47. $\dfrac{15x + 29}{2(5x - 1)^2(x + 3)^{1/2}}$

SOLUTIONS TO CHECK POINTS

Check Point 3, Frame 18

I 1. $\sqrt{12a^3} = \sqrt{4 \cdot 3 \cdot a^2 \cdot a} = \sqrt{4a^2}\sqrt{3a} = 2a\sqrt{3a}$

2. $\sqrt[3]{16x^5y^4} = \sqrt[3]{8 \cdot 2 \cdot x^3 \cdot x^2 \cdot y^3 \cdot y} = \sqrt[3]{8x^3y^3}\sqrt[3]{2x^2y} = 2xy\sqrt[3]{2x^2y}$

3. $\sqrt[6]{m^7n^{13}t^{19}} = \sqrt[6]{m^6 \cdot m \cdot n^{12} \cdot n \cdot t^{18} \cdot t} = \sqrt[6]{m^6\,n^{12}\,t^{18}}\,\sqrt[6]{mnt}$

   $= \sqrt[6]{m^6(n^6)^2(t^6)^3}\,\sqrt[6]{mnt} = mn^2t^3\,\sqrt[6]{mnt}$

4. $\sqrt{8x^5y^9z} = \sqrt{4 \cdot 2 \cdot x^4 \cdot x \cdot y^8 \cdot y \cdot z} = \sqrt{4x^4(y^4)^2}\,\sqrt{2xyz}$

   $= \sqrt{4}\cdot\sqrt{x^4}\cdot\sqrt{(y^4)^2}\cdot\sqrt{2xyz} = 2x^2y^4\sqrt{2xyz}$

II 5. $\sqrt[4]{36} = 36^{1/4} = (6^2)^{1/4} = 6^{1/2} = \sqrt{6}$

6. $\sqrt[6]{16} = 16^{1/6} = (2^4)^{1/6} = 2^{4/6} = 2^{2/3} = \sqrt[3]{2^2} = \sqrt[3]{4}$

7. $\sqrt[9]{64} = (64)^{1/9} = (2^6)^{1/9} = 2^{6/9} = 2^{2/3} = \sqrt[3]{2^2} = \sqrt[3]{4}$

8. $\dfrac{\sqrt[4]{25}}{\sqrt{5}} = \dfrac{(25)^{1/4}}{5^{1/2}} = \dfrac{(5^2)^{1/4}}{5^{1/2}} = \dfrac{5^{1/2}}{5^{1/2}} = 1$

Check Point 4, Frame 28

1. $\sqrt[3]{5} - \sqrt[6]{10}$ is in simplest form. $\sqrt[3]{5}$ and $\sqrt[6]{10}$ are "unlike" radicals.

   You could factor out the 3 common factors and express your final

   answer as $3(\sqrt[3]{5} - 2\sqrt[6]{10})$, but no further simplification is possible.

SOLUTIONS TO CHECK POINTS

Check Point 4, Frame 28

2. $\sqrt{x^3} + 2\sqrt{x} = \sqrt{x^2 \cdot x} + 2\sqrt{x} = \sqrt{x^2}\,\sqrt{x} + 2\sqrt{x} = x\sqrt{x} + 2\sqrt{x} = (x+2)\sqrt{x}$

3. $\sqrt[4]{9} - 2\sqrt{27} + \sqrt{50} = 9^{1/4} - 2\sqrt{9 \cdot 3} + \sqrt{25 \cdot 2}$

   $= (3^2)^{1/4} - 2\sqrt{9}\sqrt{3} + \sqrt{25}\sqrt{2}$

   $= 3^{1/2} - 2(3)\sqrt{3} + 5\sqrt{2}$

   $= \sqrt{3} - 6\sqrt{3} + 5\sqrt{2} = 5\sqrt{2} - 5\sqrt{3} \quad \text{OR} \quad 5(\sqrt{2}-\sqrt{3})$

4. $\sqrt[3]{54} + 2\sqrt[3]{2} - \sqrt[3]{81} = \sqrt[3]{27 \cdot 2} + 2\sqrt[3]{2} - \sqrt[3]{27 \cdot 3}$

   $= \sqrt[3]{27}\,\sqrt[3]{2} + 2\sqrt[3]{2} - \sqrt[3]{27}\,\sqrt[3]{3}$

   $= 3\sqrt[3]{2} + 2\sqrt[3]{2} - 3\sqrt[3]{3}$

   $= 5\sqrt[3]{2} - 3\sqrt[3]{3}$

5. $\sqrt{3x^3y^5} - \sqrt{12\,xy^3} = \sqrt{3x^2 \cdot x \cdot y^4 \cdot y} - \sqrt{4 \cdot 3 \cdot x \cdot y^2 \cdot y}$

   $= \sqrt{x^2 \cdot y^4}\cdot\sqrt{3 \cdot x \cdot y} - \sqrt{4 \cdot y^2}\cdot\sqrt{3 \cdot x \cdot y}$

   $= xy^2\sqrt{3xy} - 2y\sqrt{3xy} = \sqrt{3xy}(xy^2 - 2y)$

   OR $y\sqrt{3xy}(xy - 2)$

6. $\sqrt[3]{2a^4b} - \sqrt[3]{16ab^7} + \sqrt[3]{a^7b^4}$

   $= \sqrt[3]{2 \cdot a^3 \cdot a \cdot b} - \sqrt[3]{8 \cdot 2 \cdot a(b^3)^2 \cdot b} + \sqrt[3]{(a^3)^2 \cdot a \cdot b^3 \cdot b}$

   $= \sqrt[3]{a^3}\,\sqrt[3]{2ab} - \sqrt[3]{8(b^3)^2}\,\sqrt[3]{2ab} + \sqrt[3]{(a^3)^2\,b^3}\,\sqrt[3]{ab}$

   $= a\sqrt[3]{2ab} - 2b^2\sqrt[3]{2ab} + a^2b\sqrt[3]{ab}$

SOLUTIONS TO CHECK POINTS

Check Point 5, Frame 37

1. $\sqrt{32}\sqrt{2} = \sqrt{64} = 8$

2. $\sqrt[3]{5}\,\sqrt[3]{25} = \sqrt[3]{125} = 5$

3. $\sqrt[4]{x}\,\sqrt[4]{ax} = \sqrt[4]{x}\cdot\sqrt[4]{\frac{4}{x}}\cdot a = \sqrt[4]{\frac{4}{x}}\cdot\sqrt[4]{a} = x\sqrt[4]{a}$

4. $\sqrt{\frac{5m}{2t^3}} = \sqrt{\frac{3m}{18t}} = \sqrt{\frac{15m\cdot 6}{36t^4}} = \frac{m^3\sqrt{15}}{6t^2}$

5. $(3\sqrt[5]{4x^2})(6y\sqrt[5]{8x}) = 3\cdot 6y\sqrt[5]{32x^6} = 18y\sqrt[5]{32\cdot x^5\cdot x}$

   $= 18y(2)(x)\sqrt[5]{x}$

   $= 36xy\sqrt[5]{x}$

6. $(4y-3\sqrt{x})(4y+3\sqrt{x}) = 16y^2 - (3\sqrt{x})^2 = 16y^2 - 9x$

7. $(5a+\sqrt{2b})(a+2\sqrt{8b})$

   $= 5a^2 + 10a\sqrt{8b} + a\sqrt{2b} + 2\sqrt{16b^2}$

       F   O   I   L

   $= 5a^2 + 10a\sqrt{4}\sqrt{2b} + a\sqrt{2b} + 2(4b)$

   $= 5a^2 + 20a\sqrt{2b} + a\sqrt{2b} + 8b$

   $= 5a^2 + 21a\sqrt{2b} + 8b$

SOLUTIONS TO CHECK POINTS

Check Point 6, Frame 43

1. $\frac{\sqrt{32}}{\sqrt{2}} = \sqrt{\frac{32}{2}} = \sqrt{16} = 4$

2. $\frac{\sqrt[3]{16x^4y}}{\sqrt[3]{2x}} = \sqrt[3]{\frac{16x^4y}{2x}} = \sqrt[3]{8x^3y} = 2x\sqrt[3]{y}$

3. $\frac{\sqrt[4]{a^2-b^2}}{\sqrt[4]{a-b}} = \sqrt[4]{\frac{a^2-b^2}{a-b}} = \sqrt[4]{\frac{(a-b)(a+b)}{(a-b)}} = \sqrt[4]{a+b}$

4. $\frac{\sqrt{m^2+m-6}}{\sqrt{9m-18}} = \sqrt{\frac{m^2+m-6}{9m-18}} = \sqrt{\frac{(m+3)(m-2)}{9(m-2)}} = \sqrt{\frac{m+3}{9}} = \frac{1}{3}\sqrt{m+3}$

5. $\frac{7}{\sqrt{a}} = \frac{7}{\sqrt{a}}\cdot\frac{\sqrt{a}}{\sqrt{a}} = \frac{7\sqrt{a}}{a}$

6. $\frac{3}{1+2\sqrt{y}} = \frac{3}{1+2\sqrt{y}}\cdot\frac{(1-2\sqrt{y})}{(1-2\sqrt{y})} = \frac{3(1-2\sqrt{y})}{1^2-(2\sqrt{y})^2} = \frac{3(1-2\sqrt{y})}{1-4y}$ OR $\frac{3-6\sqrt{y}}{1-4y}$

7. $\frac{x}{2\sqrt{y}-5} = \frac{x}{(2\sqrt{y}-5)}\cdot\frac{(2\sqrt{y}+5)}{(2\sqrt{y}+5)} = \frac{x(2\sqrt{y}+5)}{(2\sqrt{y})^2-5^2} = \frac{x(2\sqrt{y}+5)}{4y-25}$ OR $\frac{2x\sqrt{y}+5x}{4y-25}$

8. $\frac{2m^2}{\sqrt[3]{4m}} = \frac{2m^2}{\sqrt[3]{4m}}\cdot\frac{\sqrt[3]{2m^2}}{\sqrt[3]{2m^2}} = \frac{2m^2\sqrt[3]{2m^2}}{\sqrt[3]{8m^3}} = \frac{2m^2\sqrt[3]{2m^2}}{2m} = m\sqrt[3]{2m^2}$

SOLUTIONS/ANSWERS UNIT 17

ANSWERS TO DRILL EXERCISES

1. $4\sqrt{2}$

2. $2\,\sqrt[5]{2}$

3. $2\,\sqrt[3]{6}$

4. $5\sqrt{3}$

5. $2\,\sqrt[6]{6}$

6. $2\,\sqrt[3]{2}$

7. $2\sqrt{6}$

8. $3\,\sqrt[4]{2}$

9. $2x\sqrt{2x}$

10. $ab^2\,\sqrt[3]{a^2 b}$

11. $2m^2 nt^2\,\sqrt[5]{4n^3 t^2}$

12. $3xy^3\,\sqrt[3]{2x}$

13. $pr^3 t^2\sqrt{pt}$

14. $2abc\,\sqrt[6]{a^5 b^4 c}$

15. $\sqrt{7}$

16. $\sqrt[3]{5}$

17. $\sqrt[3]{4}$

18. $\sqrt{3}$

19. cannot be simplified

20. $3\sqrt{m}$

21. $6\sqrt{2t}$

22. $(3a^2 + 2)\sqrt{a}$

23. $5\,\sqrt[3]{x} + 3\,\sqrt[3]{xy}$

24. $6\sqrt{3} - 3\sqrt{6}$

25. $3(a - b)\sqrt{3ab}$

26. $(2mn - 1)\,\sqrt[5]{2m^2 n} + 6mn\,\sqrt[5]{m^2 n}$

27. $\dfrac{(1 + 2t)\sqrt{2t}}{5}$

28. $12x^3\sqrt{2xy}$

29. $\dfrac{m}{6n}\sqrt{m}$

30. $5a^2 b\,\sqrt[3]{b} + 4a^2 b\,\sqrt[3]{ab^2}$

31. 6

32. $a^{7/6}$ or $\sqrt[6]{a^7}$

33. $5x$

34. $36t$

35. -72

36. 4

37. $t^{19/15}$ or $\sqrt[15]{t^{19}}$

38. $2\,\sqrt[5]{2}$

39. 36

40. 32

41. $3a^2$

42. $x^5 y$

43. $2m^2$

44. $3x\sqrt{y}$

45. $4a\,\sqrt[3]{a}$

46. $\dfrac{1}{2}$

47. $\dfrac{1}{9}$

48. $\dfrac{6x^2}{5y^2}$

49. $2a$

50. $\dfrac{\sqrt{x - 1}}{6}$

51. $\dfrac{\sqrt{4m^2 - 1}}{2m}$

52. $r^2 t^3\,\sqrt[5]{6rt^2}$

53. $18vp^2\,\sqrt[3]{2v^2}$

54. $2\sqrt{6} - 6$

55. $\sqrt{2x} + 3x\sqrt{y}$

56. $6t\sqrt{2} - 10t\sqrt{2t}$

57. -41

58. $16 - 8\sqrt{x} + x$

59. $m - 9n$

60. $6\sqrt{2y} + 2y - 3\sqrt{2} - \sqrt{y}$

61. $20a + 2b + 12\sqrt{2ab}$

62. $6 + 2\sqrt{7a} - 3\sqrt{6a} - a\sqrt{42}$

63. $v - 2\sqrt{vt} + t$

64. $4\sqrt{3}$

ANSWERS TO DRILL EXERCISES

65. $2\sqrt[3]{a^2}$

66. $3x\sqrt{2y}$

67. $\dfrac{3x\sqrt{xyz}}{2z}$

68. $\dfrac{\sqrt[3]{3mn}}{n}$

69. $\dfrac{ab^2\sqrt{10abc}}{4c^2}$

70. $\sqrt[4]{x-3}$

71. $\dfrac{30\sqrt{5}-20}{41}$

72. $\dfrac{5+5\sqrt{3a}}{1-3a}$

73. $\dfrac{m^2-4m\sqrt{3}+12}{m^2-12}$

74. $\dfrac{2a\sqrt{a}+5\sqrt{ab}}{3a}$

75. $\dfrac{\sqrt{v}+v}{1-v}$

76. $\dfrac{t-2\sqrt{tp}+p}{t-p}$

77. $\dfrac{9t-6\sqrt{t}+1}{9t-1}$

78. $\dfrac{b\sqrt{a}-b+a-b}{b^2-a+b}$

79. $\dfrac{2x+3\sqrt{xy}}{4x-9y}$

80. $m+1+\sqrt{m(m+1)}$

81. $\dfrac{\sqrt{3(x-4)}}{3}$

SOLUTIONS TO CHECK POINTS

Check Point 1, Frame 6

1. $v^2 = 5$

   $v = \pm\sqrt{5}$

   ∴ $v = \sqrt{5}$ or $v = -\sqrt{5}$

2. $R^2 = 121$

   $R = \pm 11$

   ∴ $R = 11$ or $R = -11$

3. $(x-6)^2 = 1$

   $x - 6 = \pm\sqrt{1}$

   $x = 6 \pm 1$

   ∴ $x = 7$ or $x = 5$

4. $(3y+4)^2 = 7$

   $3y + 4 = \pm\sqrt{7}$

   $3y = -4 \pm\sqrt{7}$

   $y = \dfrac{-4 \pm\sqrt{7}}{3}$

   ∴ $y = \dfrac{-4+\sqrt{7}}{3}$ or $y = \dfrac{-4-\sqrt{7}}{3}$

Check Point 2, Frame 10

1. The coefficient of the squared term is 1, and the coefficient of x is 6.

   ∴ $x^2 + 6x + 3^2 = x^2 + 6x + 9 = (x+3)^2$

   $\left(\tfrac{1}{2}\cdot 6\right)^2$

2. The coefficient of the squared term is 1, and the coefficient of t is (-8).

   ∴ $t^2 - 8t + (-4)^2 = t^2 - 8t + 16 = (t-4)^2$

   $\left(\tfrac{1}{2}\cdot(-8)\right)^2$

3. The coefficient of the squared term is 1, and the coefficient of m is 1.

   ∴ $m^2 + m + \left(\tfrac{1}{2}\right)^2 = m^2 + m + \tfrac{1}{4} = \left(m+\tfrac{1}{2}\right)^2$

   $\left(\tfrac{1}{2}\cdot 1\right)^2$

431

SOLUTIONS TO CHECK POINTS

Check Point 2, Frame 10

4. The coefficient of the squared term is 1, and the coefficient of p is $(-\frac{1}{2})$.

$$\therefore p^2 - \frac{p}{2} + (-\frac{1}{4})^2 = p^2 - \frac{p}{2} + \frac{1}{16} = (p - \frac{1}{4})^2$$

$$(\frac{1}{2} \cdot (-\frac{1}{2}))^2$$

Check Point 3, Frame 14

1. $x^2 + 8x - 9 = 0$

$$x^2 + 8x = 9$$

$$x^2 + 8x + 4^2 = 9 + 4^2$$

$$(\frac{1}{2} \cdot 8)^2$$

$$x^2 + 8x + 4^2 = 9 + 16$$

$$(x + 4)^2 = 25$$

$$x + 4 = \pm 5$$

$$x = -4 \pm 5$$

Hence, the solutions of $x^2 + 8x - 9 = 0$ are $x = -9$ or $x = +1$.

2. $t^2 - 4t - 3 = 0$

$$t^2 - 4t = 3$$

$$t^2 - 4t + (-2)^2 = 3 + (-2)^2$$

$$(\frac{1}{2} \cdot (-4))^2$$

$$t^2 - 4t + (-2)^2 = 3 + 4$$

$$(t - 2)^2 = 7$$

$$t - 2 = \pm\sqrt{7}$$

$$t = 2 \pm \sqrt{7}$$

SOLUTIONS TO CHECK POINTS

Check Point 3, Frame 14

Hence, the solutions of $t^2 - 4t - 3 = 0$ are $t = 2 + \sqrt{7}$ or $t = 2 - \sqrt{7}$.

Note:

Using the calculator to evaluate $\sqrt{7}$, we obtain as approximate solutions $t_1 = 4.646$ or $t_2 = -0.646$ (rounded to three decimal places). However, in the remainder of this unit, we shall not use the calculator to evaluate the radicals but rather leave the solutions in simplified reduced form.

3. $p^2 + 18p - 7 = 0$

$$p^2 + 18p = 7$$

$$p^2 + 18p + 9^2 = 7 + 9^2$$

$$(\frac{1}{2} \cdot 18)^2$$

$$p^2 + 18p + 9^2 = 7 + 81$$

$$(p + 9)^2 = 88$$

$$p + 9 = \pm\sqrt{88}$$

$$p = -9 \pm \sqrt{88}$$

$$= -9 \pm 2\sqrt{22}$$

Hence, the solutions of $p^2 + 18p - 7 = 0$ are $p = -9 + 2\sqrt{22}$ or $p = -9 - 2\sqrt{22}$.

Check Point 4, Frame 18

1. $x^2 - 5x + 1 = 0$

$$x^2 - 5x = -1$$

$$x^2 - 5x + (-\frac{5}{2})^2 = -1 + (-\frac{5}{2})^2$$

$$(x - \frac{5}{2})^2 = -\frac{4}{4} + \frac{25}{4}$$

$$(x - \frac{5}{2})^2 = \frac{21}{4}$$

$$x - \frac{5}{2} = \pm\sqrt{\frac{21}{4}}$$

$$x - \frac{5}{2} = \pm\frac{\sqrt{21}}{2}$$

$$x = \frac{5}{2} \pm \frac{\sqrt{21}}{2}$$

Hence, $x = \frac{5 + \sqrt{21}}{2}$ or $x = \frac{5 - \sqrt{21}}{2}$.

SOLUTIONS TO CHECK POINTS

Check Point 4, Frame 18

2. $t^2 - t + 2 = 0$

$$t^2 - t = -2$$

$$t^2 - t + (-\tfrac{1}{2})^2 = -2 + (-\tfrac{1}{2})^2$$

$$(t - \tfrac{1}{2})^2 = \frac{-8}{4} + \frac{1}{4}$$

$$(t - \tfrac{1}{2})^2 = \frac{-7}{4}$$

$$t - \tfrac{1}{2} = \pm\sqrt{\frac{-7}{4}}$$

$$t - \tfrac{1}{2} = \pm \frac{\sqrt{7}}{2} j$$

$$t = \tfrac{1}{2} \pm \frac{\sqrt{7}}{2} j$$

Hence, $t = \dfrac{1 + \sqrt{7}}{2} j$ or $t = \dfrac{1 - \sqrt{7}}{2} j$.

3. $E^2 + 3E + 10 = 0$

$$E^2 + 3E = -10$$

$$E^2 + 3E + (\tfrac{3}{2})^2 = -10 + (\tfrac{3}{2})^2$$

$$(E + \tfrac{3}{2})^2 = \frac{-40}{4} + \frac{9}{4}$$

$$(E + \tfrac{3}{2})^2 = \frac{-31}{4}$$

$$E + \tfrac{3}{2} = \pm\sqrt{\frac{-31}{4}}$$

$$E + \tfrac{3}{2} = \pm \frac{\sqrt{31}}{2} j$$

$$E = \frac{-3}{2} \pm \frac{\sqrt{31}}{2} j$$

Hence, $E = \dfrac{-3 + \sqrt{31}\, j}{2}$ or $\dfrac{-3 - \sqrt{31}\, j}{2}$.

Check Point 5, Frame 22

1. $3x^2 - 6x - 5 = 0$

$$x^2 - 2x - \frac{5}{3} = 0$$

$$x^2 - 2x = \frac{5}{3}$$

$$x^2 - 2x + (-1)^2 = \frac{5}{3} + (-1)^2 \qquad [\tfrac{1}{2} \cdot (-2) = -1]$$

$$(x - 1)^2 = \frac{5}{3} + 1$$

$$(x - 1)^2 = \frac{8}{3}$$

$$x - 1 = \pm\sqrt{\frac{8}{3}}$$

$$x = 1 \pm\sqrt{\frac{8}{3}}$$

SOLUTIONS TO CHECK POINTS

Check Point 5, Frame 22

Note:

When using the calculator to obtain approximate solutions, radicals such as $\sqrt{\dfrac{8}{3}}$ need not be simplified. However, when leaving solutions in radical form, we simplify the expressions as shown in the following solutions.

$$x = 1 \pm \frac{\sqrt{8}}{\sqrt{3}} \times \frac{\sqrt{3}}{\sqrt{3}} \qquad \text{(Rationalizing the denominator.)}$$

$$= 1 \pm \frac{\sqrt{24}}{3}$$

$$= \frac{3}{3} \pm \frac{2\sqrt{6}}{3}$$

$$= \frac{3 \pm 2\sqrt{6}}{3}$$

Hence, $x = \dfrac{3 + 2\sqrt{6}}{3}$ or $x = \dfrac{3 - 2\sqrt{6}}{3}$.

2. $-6t^2 + t + 1 = 0$

$$t^2 - \frac{t}{6} - \frac{1}{6} = 0$$

$$t^2 - \frac{t}{6} = \frac{1}{6}$$

$$t^2 - \frac{t}{6} + (\tfrac{1}{12})^2 = \frac{1}{6} + (-\tfrac{1}{12})^2 + \frac{1}{144} \qquad [\tfrac{1}{2} \cdot (-\tfrac{1}{6}) = \tfrac{-1}{12}]$$

$$(t - \tfrac{1}{12})^2 = \frac{1}{6} + \frac{1}{144}$$

$$(t - \tfrac{1}{12})^2 = \frac{25}{144}$$

$$t - \frac{1}{12} = \pm\sqrt{\frac{25}{144}}$$

$$t - \frac{1}{12} = \pm\frac{5}{12}$$

$$t = \frac{1}{12} \pm \frac{5}{12}$$

Hence, $t = \dfrac{1}{2}$ or $t = -\dfrac{1}{3}$.

SOLUTIONS TO CHECK POINTS

Check Point 5, Frame 22

3. $-m^2 + 2(5m + 1) = 4m - 3(m^2 - 1)$

$-m^2 + 10m + 2 = 4m - 3m^2 + 3$

$2m^2 + 6m - 1 = 0$

$m^2 + 3m - \frac{1}{2} = 0$

$m^2 + 3m = \frac{1}{2}$

$m^2 + 3m + (\frac{3}{2})^2 = \frac{1}{2} + (\frac{3}{2})^2$

$(m + \frac{3}{2})^2 = \frac{1}{2} + \frac{9}{4}$

$(m + \frac{3}{2})^2 = \frac{11}{4}$

$m + \frac{3}{2} = \pm\sqrt{\frac{11}{4}}$

$m = -\frac{3}{2} \pm \frac{\sqrt{11}}{2}$

Hence, $m = \dfrac{-3 + \sqrt{11}}{2}$ or $m = \dfrac{-3 - \sqrt{11}}{2}$.

ANSWERS TO DRILL EXERCISES

I 1. $t = 1$ or -1

2. $x = 2\sqrt{\pi p}$ or $-2\sqrt{\pi p}$

3. $m = \frac{10}{3}$ or $\frac{8}{3}$

4. $y = 3$ or -4

5. $n = \dfrac{1 + \sqrt{7}}{12}$ or $\dfrac{1 - \sqrt{7}}{12}$

6. $R = \dfrac{-2\pi + \sqrt{2h}}{2}$ or $\dfrac{-2\pi - \sqrt{2h}}{2}$

II 7. $(x - 2)^2$

8. $(y + 5)^2$

9. $(m - \frac{1}{3})^2$

10. $(t + \frac{1}{4})^2$

11. $(p - \frac{1}{3})^2$

12. $(x + \frac{a}{2})^2$

13. $(y - \frac{k}{2})^2$

14. $(n + \frac{1}{2})^2$

ANSWERS TO DRILL EXERCISES

III 15. $x = 3$ or -8

16. $y = 17$ or 3

17. $m = \dfrac{-7 + \sqrt{37}}{2}$ or $\dfrac{-7 - \sqrt{37}}{2}$

18. $t = \dfrac{-1 + \sqrt{5}}{2}$ or $\dfrac{-1 - \sqrt{5}}{2}$

19. $E = \dfrac{-2 + \sqrt{2}}{2}$ or $\dfrac{-2 - \sqrt{2}}{2}$

20. $x = \dfrac{3 + \sqrt{19}\, j}{2}$ or $\dfrac{3 - \sqrt{19}\, j}{2}$

21. $y = -3 + \sqrt{7}$ or $-3 - \sqrt{7}$

22. $n = \dfrac{5 + \sqrt{21}}{2}$ or $\dfrac{5 - \sqrt{21}}{2}$

23. $R = \frac{5}{2}$ or -2

24. $x = \dfrac{-7 + \sqrt{167}\, j}{6}$ or $\dfrac{-7 - \sqrt{167}\, j}{6}$

25. $y = \dfrac{3 + \sqrt{3}}{2}$ or $\dfrac{3 - \sqrt{3}}{2}$

26. $t = \frac{1}{2}$ or $\frac{-11}{2}$

27. $m = \dfrac{2 + \sqrt{5}}{3}$ or $\dfrac{2 - \sqrt{5}}{3}$

28. $E = \frac{-3}{5}$ or -1

29. $n = \dfrac{5 + \sqrt{83}\, j}{6}$ or $\dfrac{5 - \sqrt{83}\, j}{6}$

30. $x = \dfrac{5 + \sqrt{137}}{8}$ or $\dfrac{5 - \sqrt{137}}{8}$

31. $y = \dfrac{1 + j}{3}$ or $\dfrac{1 - j}{3}$

32. $t = \dfrac{11 + \sqrt{61}}{6}$ or $\dfrac{11 - \sqrt{61}}{6}$

33. $m = \dfrac{-5 + \sqrt{61}}{6}$ or $\dfrac{-5 - \sqrt{61}}{6}$

34. $x = \dfrac{3 + \sqrt{89}}{4}$ or $\dfrac{3 - \sqrt{89}}{4}$

35. $y = \dfrac{-2 + \sqrt{19}}{5}$ or $\dfrac{-2 - \sqrt{19}}{5}$

36. $x = \frac{3}{2}$ or $-\frac{1}{2}$

37. $p = \dfrac{-13 + \sqrt{217}}{4}$ or $\dfrac{-13 - \sqrt{217}}{4}$

38. $t = \dfrac{-3 + \sqrt{3}\, j}{3}$ or $\dfrac{-3 - \sqrt{3}\, j}{3}$

39. $v = \dfrac{-1 + \sqrt{5}}{2}$ or $\dfrac{-1 - \sqrt{5}}{2}$

SOLUTIONS TO CHECK POINTS

Check Point 1, Frame 7

1. $2m^2 + 5m + 3 = 0$

$a = 2, b = 5, c = 3$

$m = \dfrac{-b \pm \sqrt{b^2 - 4ac}}{2a}$

$= \dfrac{-5 \pm \sqrt{5^2 - 4(2)(3)}}{2(2)}$

$= \dfrac{-5 \pm \sqrt{1}}{4}$

$= \dfrac{-5 \pm 1}{4}$

$\therefore m = -1$ or $m = \dfrac{-3}{2}$

3. $3x = 1 - 10x^2$

$10x^2 + 3x - 1 = 0$

$a = 10, b = 3, c = -1$

$x = \dfrac{-b \pm \sqrt{b^2 - 4ac}}{2a}$

$= \dfrac{-3 \pm \sqrt{3^2 - 4(10)(-1)}}{2(10)}$

$= \dfrac{-3 \pm \sqrt{49}}{20}$

$= \dfrac{-3 \pm 7}{20}$

$\therefore x = \dfrac{1}{5}$ or $x = \dfrac{-1}{2}$

2. $E^2 - E = 1$

$E^2 - E - 1 = 0$

$a = 1, b = -1, c = -1$

$E = \dfrac{-b \pm \sqrt{b^2 - 4ac}}{2a}$

$= \dfrac{-(-1) \pm \sqrt{(-1)^2 - 4(1)(-1)}}{2(1)}$

$= \dfrac{1 \pm \sqrt{5}}{2}$

$\therefore E = \dfrac{1 + \sqrt{5}}{2}$ or $\dfrac{1 - \sqrt{5}}{2}$

SOLUTIONS TO CHECK POINTS

Check Point 2, Frame 11

1. $3R^2 - 50R + 1200 = 0$

$a = 3, b = -50, c = 1200$

$R = \dfrac{-b \pm \sqrt{b^2 - 4ac}}{2a}$

$= \dfrac{-(-50) \pm \sqrt{(-50)^2 - 4(3)(1200)}}{2(3)}$

$= \dfrac{50 \pm \sqrt{-11\,900}}{6}$

$= \dfrac{50 \pm 10\sqrt{119}\,j}{6}$

$= \dfrac{2(25 \pm 5\sqrt{119}\,j)}{2(3)}$

$= \dfrac{25 \pm 5\sqrt{119}\,j}{3}$

$R = \dfrac{25 + 5\sqrt{119}\,j}{3}$

or $R = \dfrac{25 - 5\sqrt{119}\,j}{3}$

Approximate values:

$R = 8.333 + 18.181\,j$

or $R = 8.333 - 18.181\,j$

2. $145 = 100t - 16t^2$

$16t^2 - 100t + 145 = 0$

$a = 16, b = -100, c = 145$

$t = \dfrac{-b \pm \sqrt{b^2 - 4ac}}{2a}$

$= \dfrac{-(-100) \pm \sqrt{(-100)^2 - 4(16)(145)}}{2(16)}$

$= \dfrac{100 \pm \sqrt{720}}{32}$

$= \dfrac{100 \pm 12\sqrt{5}}{32}$

$= \dfrac{4(25 \pm 3\sqrt{5})}{4(8)}$

$= \dfrac{25 \pm 3\sqrt{5}}{8}$

$\therefore t = \dfrac{25 + 3\sqrt{5}}{8}$

or $t = \dfrac{25 - 3\sqrt{5}}{8}$

Approximate values:

$t = 3.964$ or $t = 2.286$

SOLUTIONS TO CHECK POINTS

Check Point 3, Frame 16

1. $0.015 = 2.5V - 12.6V^2$

$12.6V^2 - 2.5V + 0.015 = 0$

$a = 12.6, \ b = -2.5, \ c = 0.015$

$V = \dfrac{-b \pm \sqrt{b^2 - 4ac}}{2a}$

$= \dfrac{-(-2.5) \pm \sqrt{(-2.5)^2 - 4(12.6)(0.015)}}{2(12.6)}$

$= \dfrac{2.5 \pm \sqrt{5.494}}{25.2}$

∴ $V = \dfrac{2.5 + \sqrt{5.494}}{25.2}$

or $V = \dfrac{2.5 - \sqrt{5.494}}{25.2}$

Approximate values:

$V = 0.192 \text{ or } 0.006$

2. $128.6 = 7.5\pi r + 2\pi r^2$

$0 = 2\pi r^2 + 7.5\pi r - 128.6$

$a = 2\pi, \ b = 7.5\pi, \ c = -128.6$

$r = \dfrac{-b \pm \sqrt{b^2 - 4ac}}{2a}$

$= \dfrac{-7.5\pi \pm \sqrt{(7.5\pi)^2 - 4(2\pi)(-128.6)}}{2(2\pi)}$

$= \dfrac{-7.5\pi \pm \sqrt{56.25\pi^2 + 1028.8\pi}}{4\pi}$

∴ $r = \dfrac{-7.5\pi + \sqrt{56.25\pi^2 + 1028.8\pi}}{4\pi}$

or $r = \dfrac{-7.5\pi - \sqrt{56.25\pi^2 + 1028.8\pi}}{4\pi}$

Approximate values:

$r = 3.023 \text{ or } -6.773$

3. $\sqrt{5}\, R^2 - 7 = \sqrt{7}\, R$

$\sqrt{5}\, R^2 - \sqrt{7}\, R - 7 = 0$

$a = \sqrt{5}, \ b = -\sqrt{7}, \ c = -7$

$R = \dfrac{-b \pm \sqrt{b^2 - 4ac}}{2a}$

$= \dfrac{-(-\sqrt{7}) \pm \sqrt{(-\sqrt{7})^2 - 4\sqrt{5}(-7)}}{2\sqrt{5}}$

$= \dfrac{\sqrt{7} \pm \sqrt{7 + 28\sqrt{5}}}{2\sqrt{5}}$

∴ $R = \dfrac{\sqrt{7} + \sqrt{7 + 28\sqrt{5}}}{2\sqrt{5}}$

or $R = \dfrac{\sqrt{7} - \sqrt{7 + 28\sqrt{5}}}{2\sqrt{5}}$

Approximate values:

$R = 2.457 \text{ or } R = -1.274$

4. $d = vt - 16t^2$

$16t^2 - vt + d = 0$

$a = 16, \ b = -v, \ c = d$

$t = \dfrac{-b \pm \sqrt{b^2 - 4ac}}{2a}$

$= \dfrac{-(-v) \pm \sqrt{(-v)^2 - 4(16)(d)}}{2(16)}$

$= \dfrac{v \pm \sqrt{v^2 - 64d}}{32}$

∴ $t = \dfrac{v + \sqrt{v^2 - 64d}}{32}$

or $t = \dfrac{v - \sqrt{v^2 - 64d}}{32}$

SOLUTIONS TO CHECK POINTS

Check Point 2, Frame 11

3. $\dfrac{E}{6} = \dfrac{4}{3} + \dfrac{2E^2}{9}$

LCD: 18

$18\left(\dfrac{E}{6}\right) = 18\left(\dfrac{4}{3} + \dfrac{2E^2}{9}\right)$

$3E = 24 + 4E^2$

$0 = 4E^2 - 3E + 24$

$a = 4, \ b = -3, \ c = 24$

$E = \dfrac{-b \pm \sqrt{b^2 - 4ac}}{2a}$

$= \dfrac{-(-3) \pm \sqrt{(-3)^2 - 4(4)(24)}}{2(4)}$

$= \dfrac{3 \pm \sqrt{-375}}{8}$

$= \dfrac{3 \pm 5\sqrt{15}\, j}{8}$

∴ $E = \dfrac{3 + 5\sqrt{15}\, j}{8}$

or $E = \dfrac{3 - 5\sqrt{15}\, j}{8}$

Approximate values:

$E = 0.375 + 2.421\, j$

or $E = 0.375 - 2.421\, j$

SOLUTIONS TO CHECK POINTS

Check Point 4, Frame 24

1.
$$\sqrt{2v - 7} - 7 = 0$$
$$\sqrt{2v - 7} = 7$$
$$(\sqrt{2v - 7})^2 = 7^2$$
$$2v - 7 = 49$$
$$2v = 56$$
$$v = 28$$
Check: If v = 28,
$$\sqrt{2(28) - 7} - 7 = \sqrt{49} - 7$$
$$= 7 - 7$$
$$= 0$$
∴ v = 28 is the solution.

2.
$$\sqrt[3]{4x + 25} = 5$$
$$(\sqrt[3]{4x + 25})^3 = 5^3$$
$$4x + 25 = 125$$
$$4x = 100$$
$$x = 25$$
Check: If x = 25,
$$\sqrt[3]{4(25) + 25} = \sqrt[3]{125} = 5$$
∴ x = 25 is the solution.

3.
$$1 = 2\sqrt{x + 4} - x$$
$$1 + x = 2\sqrt{x + 4}$$
$$(1 + x)^2 = (2\sqrt{x + 4})^2$$
$$1 + 2x + x^2 = 4(x + 4)$$
$$1 + 2x + x^2 = 4x + 16$$
$$x^2 - 2x - 15 = 0$$
$$(x - 5)(x + 3) = 0$$
∴ x = 5 or x = -3

Check:

If x = 5, $2\sqrt{x + 4} - x = 2\sqrt{5 + 4} - 5$
$$= 2\sqrt{9} - 5$$
$$= 2(3) - 5$$
$$= 1$$

If x = -3, $2\sqrt{x + 4} - x = 2\sqrt{(-3) + 4} - (-3)$
$$= 2\sqrt{1} + 3$$
$$= 2 + 3$$
$$= 5$$
$$\neq 1$$

∴ x = -3 is an extraneous solution and x = 5 is the only solution to $1 = 2\sqrt{x + 4} - x$.

SOLUTIONS TO CHECK POINTS

Check Point 5, Frame 29

1.
$$\sqrt{2x + 3} - \sqrt{x + 8} = 0$$
$$\sqrt{2x + 3} = \sqrt{x + 8}$$
$$(\sqrt{2x + 3})^2 = (\sqrt{x + 8})^2$$
$$2x + 3 = x + 8$$
$$x = 5$$
Check: If x = 5,
$$\sqrt{2x + 3} - \sqrt{x + 8} = \sqrt{10 + 3} - \sqrt{13}$$
$$= \sqrt{13} - \sqrt{13}$$
$$= 0$$
∴ x = 5 is the solution.

2.
$$\sqrt{m + 27} + 2\sqrt{m} = 0$$
$$\sqrt{m + 27} = -2\sqrt{m}$$
$$(\sqrt{m + 27})^2 = (-2\sqrt{m})^2$$
$$m + 27 = 4m$$
$$27 = 3m$$
$$9 = m$$
Check: If m = 9,
$$\sqrt{m + 27} + 2\sqrt{m} = \sqrt{9 + 27} + 2\sqrt{9}$$
$$= \sqrt{36} + 2(3)$$
$$= 6 + 6$$
$$= 12$$
$$\neq 0$$
∴ m = 9 is an extraneous solution and the equation $\sqrt{m + 27} + 2\sqrt{m} = 0$ has no solution.

3.
$$\sqrt{6t + 13} - 3 = t$$
$$\sqrt{6t + 13} = t + 3$$
$$(\sqrt{6t + 13})^2 = (t + 3)^2$$
$$6t + 13 = t^2 + 6t + 9$$
$$0 = t^2 - 4$$
$$0 = (t + 2)(t - 2)$$
∴ t = -2 or t = +2
Check:
If t = -2, $\sqrt{6t + 13} - 3 = \sqrt{-12 + 13} - 3$
$$= \sqrt{1} - 3$$
$$= -2$$
If t = +2, $\sqrt{6t + 13} - 3 = \sqrt{12 + 13} - 3$
$$= \sqrt{25} - 3$$
$$= 2$$
∴ both t = -2 and t = 2 are solutions.

ANSWERS TO DRILL EXERCISES

I   1. $x = 4.56$ or 0.44

2. $t = 1.18$ or -0.85

3. $x = -0.27$ or -3.73

4. $E = 1.08$ or -1.29

5. $p = 1.62$ or -0.62

6. $R = 2.64$ or -1.14

7. $d = 5$ or $\dfrac{-7}{6}$

8. $y = 0.08 + 0.57\,j$ or $y = 0.08 - 0.57\,j$

9. $x = \dfrac{1}{4}$

10. $x = -50 + 22.36\,j$ or $x = -50 - 22.36\,j$

11. $t = 30.74$ or 0.51

12. $d = 0.13 + 0.81\,j$ or $d = 0.13 - 0.81\,j$

13. $p = 0.29 + 1.39\,j$ or $p = 0.29 - 1.39\,j$

14. $x = 0.78$ or -1.61

15. $x = 0.33 + 0.75\,j$ or $x = 0.33 - 0.75\,j$

16. $z = 3.72$ or 0.45

17. $r = 0.28$ or -0.53

18. $R = 1.24$ or -3.24

19. $t = \dfrac{K + \sqrt{Kd}}{K}$ or $\dfrac{K - \sqrt{Kd}}{K}$

20. $x = 2.69$ or -0.69

21. $t = 0.87 + 1.30\,j$ or $t = 0.87 - 1.30\,j$

22. $v = 0.31$ or -4.79

23. $p = 2.5 + 0.87\,j$ or $p = 2.5 - 0.87\,j$

24. $y = 5.26$ or -0.46

25. $r = -4.98$ or 0.53

26. $E = 47.80$ or 0.20

II 27. $x = 4$

28. $y = \dfrac{26}{3}$

29. $p = \dfrac{101}{2}$

30. $t = 126$

SOLUTIONS TO CHECK POINTS

Check Point 5, Frame 29

4. $\sqrt{7-p} - \sqrt{p+13} + 2 = 0$

$$\sqrt{7-p} + 2 = \sqrt{p+13}$$

$$\left(\sqrt{7-p} + 2\right)^2 = \left(\sqrt{p+13}\right)^2$$

$$7 - p + 4\sqrt{7-p} + 4 = p + 13$$

$$4\sqrt{7-p} = 2p + 2$$

$$2\sqrt{7-p} = p + 1$$

$$\left(2\sqrt{7-p}\right)^2 = (p+1)^2$$

$$4(7-p) = p^2 + 2p + 1$$

$$28 - 4p = p^2 + 2p + 1$$

$$0 = p^2 + 6p - 27$$

$$0 = (p+9)(p-3)$$

$$\therefore \quad p = -9 \text{ or } p = 3$$

Check:

If $p = -9$,

$$\sqrt{7-p} - \sqrt{p+13} + 2 = \sqrt{7+9} - \sqrt{-9+13} + 2$$

$$= 4 - 2 + 2$$

$$= 4$$

$$\neq 0$$

If $p = 3$,

$$\sqrt{7-p} - \sqrt{p+13} + 2 = \sqrt{7-3} - \sqrt{3+13} + 2$$

$$= 2 - 4 + 2$$

$$= 0$$

$\therefore \ p = -9$ is an extraneous solution and the only solution to $\sqrt{7-p} - \sqrt{p+13} + 2 = 0$ is $p = 3$.

SOLUTIONS TO CHECK POINTS

Check Point 1, Frame 7

1. $2^4 = 16$ means $\log_2 16 = 4$ (The log to the base 2 of 16 is 4)

2. $100^{1/2} = 10$ means $\log_{100} 10 = \frac{1}{2}$ (The log to the base 100 of 10 is $\frac{1}{2}$)

3. $5^4 = 625$ means $\log_5 625 = 4$ (The log to the base 5 of 625 is 4)

4. $3^{-2} = \frac{1}{9}$ means $\log_3 \frac{1}{9} = -2$ (The log to the base 3 of $\frac{1}{9}$ is -2)

5. $10^{-3}\ 0.001$ means $\log_{10} 0.001 = -3$ (The log to the base 10 of 0.001 is -3)

6. $4^7 = 16\ 384$ means $\log_4 16\ 384 = 7$ (The log to the base 4 of 16 384 is 7)

Check point 2, Frame 9

1. $\log_3 81 = 4$ means $3^4 = 81$ (3 to the exponent 4 is 81)

2. $\log_{10} 100 = 2$ means $10^2 = 100$ (10 to the exponent 2 is 100)

3. $\log_2 128 = 7$ means $2^7 = 128$ (2 to the exponent 7 is 128)

4. $\log_{10} 0.1 = -1$ means $10^{-1} = 0.1$ (10 to the exponent -1 is 0.1)

5. $\log_{81} 9 = \frac{1}{2}$ means $81^{1/2} = 9$ (81 to the exponent $\frac{1}{2}$ is 9)

6. $\log_8 \left(\frac{1}{64}\right) = -2$ means $8^{-2} = \frac{1}{64}$ (8 to the exponent -2 is $\frac{1}{64}$)

ANSWERS TO DRILL EXERCISES

31. $p = \frac{97}{48}$

32. $y = 39$

33. $x = 25$

34. $t = \frac{4}{3}$

35. $x = 2$

36. $p = \frac{11}{6}$ is extraneous.
    No solution.

37. $x = -3$ is extraneous.
    $x = 1$ is the only solution.

38. $t = 6$ or 7

39. $R = \frac{-11}{9}$ is extraneous.
    $R = 5$ is the only solution.

40. $E = \frac{16}{9}$ is extraneous.
    $E = 0$ is the only solution.

41. $p = 2$ is extraneous.
    $p = 9$ is the only solution.

42. $m = \frac{11}{2}$ is extraneous.
    $m = -\frac{1}{2}$ is the only solution.

43. $n = 4$ or 20

44. $y = 25$ is extraneous.
    $y = 1$ is the only solution.

45. $x = -3$ is extraneous.
    No solution.

46. The dimensions are 8.0 cm × 5.5 cm.

47. The sides are 16.0 m and 30.0 m.

48. $R_1 = 35.6\ \Omega$, $R_2 = 45.6\ \Omega$

49. $V = 9.05$, $p = 11.05$

50. (a) $m = -5 \pm 100j$

    (b) $m = \dfrac{-R \pm \sqrt{R^2 - \frac{4L}{C}}}{2L}$

51. $w = 0.28$ ($w = 6.22$ is not a solution, since $2.5 - w$ cannot be negative.)

52. $t = 40.47^{\circ}F$ or $t = -349.80^{\circ}F$

SOLUTIONS TO CHECK POINTS

Check Point 4, Frame 23

1. $\log_{10}(2 \times 5) = \log_{10} 2 + \log_{10} 5$

2. $\log_e \left(\frac{4}{y}\right) = \log_e 4 - \log_e y$

3. $\log_2 5^3 = 3 \log_2 5$

4. $\log_3 3^z = z$

5. $\log_5 (3xy) = \log_5 3 + \log_5 x + \log_5 y$

6. $\log_{10}\left(\frac{x}{10}\right) = \log_{10} x - \log_{10} 10$
   $= \log_{10} x - 1$

7. $\log_e (4e) = \log_e 4 + \log_e e$
   $= \log_e 4 + 1$

8. $\log_8 \left(\frac{1}{x}\right) = \log_8 1 - \log_8 x$
   $= 0 - \log_8 x$
   $= -\log_8 x$

9. $\log_2 6 + \log_2 3 = \log_2 (6 \cdot 3)$
   $= \log_2 18$

10. $\log_5 8x - \log_5 8 = \log_5 \left(\frac{8x}{8}\right)$
    $= \log_5 x$

11. $\log_e 5 + \log_e y = \log_e (5y)$

12. $7 \log_5 5 = 7(1) = 7$

13. $\log_e e^{0.5} = 0.5$

14. $\log_{10} 10^x + 1 = 3$
    $x + 1 = 3$
    $x = 2$

SOLUTIONS TO CHECK POINTS

Check Point 3, Frame 12

1. $\log_{10}(-100) = x$ is not defined since $N = -100$. N cannot be negative.

2. $\log_2 1 = x$ is defined. In fact, $\log_2 1 = 0$ since $2^0 = 1$.

3. $\log_{-5} 125 = x$ is not defined since $a = -5$.

4. $\log_1 10 = x$ is not defined since $a = 1$.

5. $\log_0 1 = x$ is not defined since $a = 0$. a must be > 0.

6. $\log_5 0 = x$ is not defined since $N = 0$. N must be > 0.

7. $\log_2 V = 0.5$
   $V = 2^{0.5}$

8. $\log_e (x + 1) = \pi$
   $x + 1 = e^\pi$

9. $\log_{10} P = 2y$
   $P = 10^{2y}$

10. $2^\pi = R$
    $\log_e R = 2\pi$

11. $10^{2K} = V + 1$
    $\log_{10}(V + 1) = 2K$

12. $2^x + 1 = 64$
    $\log_2 64 = x + 1$

440

SOLUTIONS TO CHECK POINTS

Check Point 5, Frame 30

I 1. $\log_2 (4 \sqrt[3]{x}) = \log_2 4 + \log_2 \sqrt[3]{x}$

$= \log_2 2^2 + \log_2 x^{1/3}$

$= 2 + \log_2 x^{1/3}$

or

$= 2 + \frac{1}{3} \log_2 x$

2. $\log_5 \left(\frac{m+n}{3}\right) = \log_5 (m+n) - \log_5 3$

3. $\log_{10} \left(\frac{a^3 b^{1/2}}{c^4}\right) = \log_{10} a^3 + \log_{10} b^{1/2} - \log_{10} c^4$

or:

$= 3 \log_{10} a + \frac{1}{2} \log_{10} b - 4 \log_{10} c$

4. $\log_e \sqrt{\frac{4x^5 y}{3z^3}} = \log_e \left(\frac{4x^5 y}{3z^3}\right)^{1/2}$

$= \frac{1}{2} \log_e \left(\frac{4x^5 y}{3z^3}\right)$

$= \frac{1}{2} \left[\log_e 2^2 + 5 \log_e x + \log_e y - (\log_e 3 + 3 \log_e z)\right]$

$= \frac{1}{2} \left[2 \log_e 2 + 5 \log_e x + \log_e y - \log_e 3 - 3 \log_e z\right]$

$= \log_e 2 + \frac{5}{2} \log_e x + \frac{1}{2} \log_e y - \frac{1}{2} \log_e 3 - \frac{3}{2} \log_e z$

SOLUTIONS TO CHECKPOINTS

Check Point 5, Frame 30

II 5. $2 \log_3 x + \log_3 (x+1) = \log_3 x^2 + \log_3 (x+1)$

$= \log_3 [x^2 (x+1)]$

or:

$= \log_3 (x^3 + x^2)$

6. $\frac{1}{3} [\log_{10} m + \log_{10} n - 2 \log_{10} t] = \frac{1}{3} [\log_{10} (mn) - \log_{10} t^2]$

$= \frac{1}{3} \log_{10} \left(\frac{mn}{t^2}\right)$

$= \log_{10} \sqrt[3]{\frac{mn}{t^2}}$

7. $\frac{1}{2} [\log_e (x^2 - 2x + 1) - \log_e (x-1)] = \frac{1}{2} \log_e \left[\frac{(x^2 - 2x + 1)}{(x-1)}\right]$

$= \frac{1}{2} \log_e \left[\frac{(x-1)^2}{(x-1)}\right]$

$= \frac{1}{2} \log_e (x-1)$

$= \log_e \sqrt{x-1}$

8. $\log_e (2y - \sqrt{3})$ cannot be simplified.

441

SOLUTIONS TO CHECK POINTS

Check Point 7, Frame 40

1. $10^{2x} = 529$ means $\log_{10} 529 = 2x$

2. $e^{y+1} = \pi + K$ means $\ln (\pi + K) = y + 1$

3. $2^{x+5} = 128$ means $\log_2 128 = x + 5$

4. $\log_{10} V = z - 1$ means $10^{z-1} = V$

5. $-1.2 + x = \ln R$ means $e^{-1.2 + x} = R$

6. $2\pi = \log_2 (p + 2)$ means $2^{2\pi} = p + 2$

Check Point 8, Frame 48

I.

1. $y = \frac{1}{3} \log \left(\frac{x}{a}\right)$ Multiply both sides by 3.

$3y = \log \left(\frac{x}{a}\right)$ Change to exponential form.

$10^{3y} = \frac{x}{a}$

2. $\pi \ln x - \ln a = -1.25 K$

$\ln x^\pi - \ln a = -1.25 K$ Law 20.3

$\ln \frac{x^\pi}{a} = -1.25 K$ Law 20.2

$\frac{x^\pi}{a} = e^{-1.25K}$ Change to exponential form

SOLUTIONS TO CHECK POINTS

Check Point 6, Frame 37

1. $\log \sqrt[3]{x^2} = 10$

$\log x^{2/3} = 10$

$\frac{2}{3} \log x = 10$ Law 20.3

$\log x = 10 \cdot \frac{3}{2}$

$\log x = 15$

2. $2 \log (am) = 5 \log n$

$\log (am) = \frac{5}{2} \log n$

$\log a + \log m = \frac{5}{2} \log n$ Law 20.1

$\log a = \frac{5}{2} \log n - \log m$

or: $= \log \left(\frac{n^{5/2}}{m}\right)$

3. $3 = \ln \left(\pi \sqrt{\frac{P}{R}}\right)$

$3 = \ln \pi + \ln \left(\frac{P}{R}\right)^{1/2}$ Law 20.1

$3 = \ln \pi + \frac{1}{2} \ln \left(\frac{P}{R}\right)$ Law 20.3

$3 = \ln \pi + \frac{1}{2} (\ln P - \ln R)$ Law 20.2

$3 = \ln \pi + \frac{1}{2} \ln P - \frac{1}{2} \ln R$

$\frac{1}{2} \ln R = \ln \pi + \frac{1}{2} \ln P - 3$

$\ln R = 2 \ln \pi + \ln P - 6$ or: $\ln (\pi^2 P) - 6$

4. $\ln (\pi x^2) - \ln (3x) = 2 \ln \pi$

$\ln \pi + \ln x^2 - (\ln 3 + \ln x) = 2 \ln \pi$ Law 20.1

$\ln \pi + 2 \ln x - \ln 3 - \ln x = 2 \ln \pi$ Law 20.3

$2 \ln x - \ln x = 2 \ln \pi - \ln \pi + \ln 3$

$\ln x = \ln \pi + \ln 3$ or: $\ln (3\pi)$

SOLUTIONS TO CHECK POINTS
Check Point 8, Frame 48

3. $\log (u - a) = v + 3$ — Change to exponential form.

   $u - a = 10^{v + 3}$

4. $\ln R + 5 = 2 \ln P$ — Law 20.3

   $\ln R + 5 = \ln P^2$

   $\ln R - \ln P^2 = -5$

   $\ln\left(\dfrac{R}{P^2}\right) = -5$ — Law 20.2

   $\dfrac{R}{P^2} = e^{-5}$ — Change to exponential form.

II.

5. $2\pi e^{-Kt} = e^{0.1x}$ — Change to the form $a^x = N$.

   $2\pi = \dfrac{e^{0.1x}}{e^{-Kt}}$ — Divide both sides by e^{-Kt}.

   $2\pi = e^{0.1x + Kt}$

   $\ln (2\pi) = 0.1x + Kt$ — Change to logarithmic form.

   $\ln (2\pi) - 0.1x = Kt$ — Isolate term containing "t".

   $t = \dfrac{\ln (2\pi) - 0.1x}{K}$ — Solve for t.

SOLUTIONS TO CHECK POINTS
Check Point 8, Frame 48

6. $M - N = 10^{(V + 3)}$ — Change to logarithmic form.

   $\log (M - N) = V + 3$

   $\log (M - N) - 3 = V$

7. $5E = ae^{\pi R}$ — Divide both sides by "a" to obtain the form $N = a^x$.

   $\dfrac{5E}{a} = e^{\pi R}$ — Change to logarithmic form.

   $\ln\left(\dfrac{5E}{a}\right) = \pi R$

   $\dfrac{\ln\left(\dfrac{5E}{a}\right)}{\pi} = R$ — Solve for R.

8. $10^x + 1 = 1000K$ — Change to logarithmic form.

   $\log (1000K) = x + 1$

   $\log 1000 + \log K = x + 1$

   $3 + \log K = x + 1$ — ($\log 1000 = \log_{10} 10^3 = 3$)

   $2 + \log K = x$ — Solve for x.

ANSWERS TO DRILL EXERCISES

I
1. $\log_2 8 = 3$
2. $\log_4 (\frac{1}{16}) = -2$
3. $\log_6 1296 = 4$
4. $\log_{36} 6 = \frac{1}{2}$
5. $\log_8 (\frac{1}{2}) = -\frac{1}{3}$
6. $\log_{3/5} (\frac{9}{25}) = 2$

II
7. $2^6 = 64$
8. $13^2 = 169$
9. $16^{1/4} = 2$
10. $5^{-3} = \frac{1}{125}$
11. $10^{-4} = 0.0001$
12. $(\frac{1}{2})^{-5} = 32$

III
13. $\log_{-3} x = 2$ is not defined since a = -3.
14. $\log_2 (-5) = x$ is not defined since N = -5.
15. $\log_7 (\frac{1}{49}) = x$ is defined.
16. $\log_0 9 = x$ is not defined since a = 0.
17. $\log_{0.8} x = 2$ is defined.
18. $\log_1 25 = x$ is not defined since a = 1.
19. $\log_{12} 0 = x$ is not defined since N = 0.

IV
20. 7
21. 4
22. $\frac{1}{2}$
23. 4
24. x = 3
25. y = 1

ANSWERS TO DRILL EXERCISES

V
26. $\log_4 2 + \log_4 (x^5)$ or $\log_4 2 + 5 \log_4 x$
27. $\ln a + \ln b - \ln 10$
28. $\log_3 (m - n) - \log_3 n - \log_3 t$
29. $\frac{1}{2}\ln (x + y) - \frac{1}{2}\ln a$
30. $\log_2 a + \log_2 b + \log_2 c$
31. $\log m^{1/2} + \log n^4 - \log t^2$ or $\frac{1}{2} \log m + 4 \log n - 2 \log t$
32. $\log_5 \sqrt{v} - \log_5 b^3$ or $\log_5 v^{1/2} - \log_5 b^3$ or $\frac{1}{2} \log_5 v - 3 \log_5 b$
33. $\frac{1}{5} \log_7 x + \frac{1}{5} \log_7 y$
34. $\frac{1}{3} \log 5 + \frac{2}{3} \log x + \frac{1}{3} \log y - \frac{1}{3} \log 2 - \frac{1}{3} \log z$

VI
35. $\log_2 32$
36. $\log \left(\frac{5}{x^2}\right)$
37. $\log_2 (ab)^{1/2}$ or $\log_2 \sqrt{ab}$
38. $\log (\frac{1}{4})$
39. $\log_5 a$

444

SOLUTIONS TO CHECK POINTS

Check Point 1, Frame 5

x	log x
1.00	0
1.589	0.201124
2.00	0.301030
2.36	0.372912
2.8412	0.453502
3.00	0.477121
3.756	0.574726
4.00	0.602060
4.4531	0.648662
4.98	0.697229

x	log x
5.00	0.698970
5.645	0.751664
6.00	0.778151
6.5983	0.819432
7.00	0.845098
7.814	0.892873
8.00	0.903090
8.32092	0.920171
9.00	0.954243
9.75	0.989005

Check Point 2, Frame 10

1. $2503 = 2.503 \times 10^3$

   $\log 2503 = \log 2.503 + \log 10^3$

   $= (\text{larger than } .2) + 3$

   Hence, a rough estimate of log 2503 is 3.3

2. $0.094 = 9.4 \times 10^{-2}$

   $\log 0.094 = \log 9.4 + \log 10^{-2}$

   $= (\text{larger than } .9) + (-2)$

   Hence, a rough estimate of log 0.094 is $-2 + 1.0 = -1.0$

3. $194.56 = 1.9456 \times 10^2$

   $\log 194.56 = \log 1.9456 + \log 10^2$

   $= (\text{larger than } 0.1) + (2)$

   Hence, a rough estimate of log 194.56 is $2 + 0.2 = 2.2$

ANSWERS TO DRILL EXERCISES

VI 40. $\log_3 \left[c(a - b)^{1/2} \right]$ or $\log_3 c\sqrt{a - b}$

41. $\ell n \left[a(x - 1) \right]^{2/3}$

VII 42. $\log m = 10$

43. $\log b = \log \left(\frac{a}{c} \right)$

44. $\ell n\, t = \ell n \left(\frac{116}{K} \right)$

45. $\log x = \frac{45 - \log 2}{2}$

46. $\ell n\, x = -2 \ell n\, y - c$

47. $\log a = \log b$

48. $\ell n\, a = \ell n \left(\frac{K}{\pi b} \right)$

49. $\log x = \log y - 5y + 1$

50. $\log V = \log R + \log a - 50$ or

    $\log V = \log (Ra) - 50$

VIII 51. $10^x = mt$

52. $e^m = \frac{x + b}{5}$

53. $e^{Kt} = \frac{0.6}{y}$

54. $10^x - 5 = y + \pi$

55. $(1 + i)^{10} = 2$

56. $2 = e^{0.4t}$

57. $0.05\, R^{2/3} = P$

58. $ae^{yt/2} = x$

59. $\log_{10} 259 = x$

60. $\frac{\ell n\, \pi - 0.5t}{0.09} = x$

61. $x + y = \ell n\, Kt$

62. $\log_{10} (a + b) = t - 2$

63. $\ell n \left(\frac{i}{I} \right) = -Kt$

64. $t = -2 \ell n \left(1 - \frac{7R}{K} \right)$ or

    $t = \ell n \left(\frac{K - 7R}{K} \right)^{-2}$

65. $x = y - 1 + \log_{10} (2a + 1)$

SOLUTIONS TO CHECK POINTS

Check Point 4, Frame 17

x	$\ln x = y$	Check: $e^y = x$
29.54	3.3857	$e^{3.3857} = 29.54$
18075	9.8023	$e^{9.8023} = 18075$
0.00243	-6.0199	$e^{-6.0199} = 0.00243$
519.6	6.2531	$e^{6.2531} = 519.6$
0.82	-0.1985	$e^{-0.1985} = 0.82$

Check Point 5, Frame 21

log x	x
4.7521	56506.70713
0.2518	1.7856651
-3.496	0.0003192
-1.28	0.0524807
0.5	3.1622777

$\ln x$	x
2.80	16.444647
-3.8675	0.0209106
0.75	2.1170000
-0.5	0.6065307
6.29248	540.49209

Check Point 6, Frame 24

1. $2^{\log_2 10} = 10$

2. $10^{\log(v-1)} = v - 1$

3. $e^{2\ln x} = e^{\ln x^2} = x^2$

4. $8^{1/2 \log_8 (R+3)} = 8^{\log_8 (R+3)^{1/2}} = (R + 3)^{1/2} = \sqrt{R + 3}$

5. $10^{2/3 \log (2y+1)} = 10^{\log (2y+1)^{2/3}} = (2y + 1)^{2/3} = \sqrt[3]{(2y + 1)^2}$

SOLUTIONS TO CHECK POINTS

Check Point 2, Frame 10

4. $0.000\,84 = 8.4 \times 10^{-4}$

$\log 0.000\,84 = \log 8.4 + \log 10^{-4}$

$= (\text{larger than } 0.8) + (-4)$

Hence, a rough estimate of log 0.000 84 is $-4 + 0.9 = -3.1$

Check Point 3, Frame 12

x	log x	Estimate
17.52	1.243534	$17.52 = 1.752 \times 10^1$ $\log 17.52 \doteq 1 + 0.2$ $= 1.2$
0.984	-0.007005	$0.984 = 9.84 \times 10^{-1}$ $\log 0.984 \doteq -1 + 1.0$ $= 0$
640317.8	5.806396	$640317.8 = 6.403178 \times 10^5$ $\log 640317.8 \doteq 5 + 0.7$ $= 5.7$
849.726	2.929279	$849.726 = 8.49726 \times 10^2$ $\log 849.726 \doteq 2 + 0.9$ $= 2.9$
5000	3.698970	$5000 = 5.0 \times 10^3$ $\log 5000 \doteq 3 + 0.6$ $= 3.6$
0.00293	-2.533132	$0.00293 = 2.93 \times 10^{-3}$ $\log 0.00293 \doteq -3 + 0.3$ $= -2.7$
16428	4.215585	$16428 = 1.6428 \times 10^4$ $\log 16428 \doteq 4 + 0.2$ $= 4.2$
0.000075	-4.124939	$0.000075 = 7.5 \times 10^{-5}$ $\log 0.000075 \doteq -5 + 0.8$ $= -4.2$

SOLUTIONS TO CHECK POINTS

Check Point 7, Frame 31

1. $\log_5 16.9 = \dfrac{\log_{10} 16.9}{\log_{10} 5} = \dfrac{\log 16.9}{\log 5} = \dfrac{1.22789}{0.69897} = 1.7567$

2. $\log_{12} 0.598 = \dfrac{\log_{10} 0.598}{\log_{10} 12} = \dfrac{\log 0.598}{\log 12} = \dfrac{-0.22330}{1.07918} = -0.2069$

3. $\log_8 7.5 = \dfrac{\log_{10} 7.5}{\log_{10} 8} = \dfrac{\log 7.5}{\log 8} = \dfrac{0.87506}{0.90309} = 0.9690$

4. $\ln 0.06 = \dfrac{\log 0.06}{\log e} = 2.3026 \,(\log 0.06)$
$= 2.3026 \,(-1.22185)$
$= -2.8134$

5. $\log_2 45 = \dfrac{\log_{10} 45}{\log_{10} 2} = \dfrac{\log 45}{\log 2} = \dfrac{1.65321}{0.30103} = 5.4919$

6. $\ln 19.8 = \dfrac{\log 19.8}{\log e} = 2.3026 \,(\log 19.8)$
$= 2.3026 \,(1.29667)$
$= 2.9857$

7. $\log_4 0.012 = \dfrac{\log_e 0.012}{\log_e 4} = \dfrac{\ln 0.012}{\ln 4} = \dfrac{-4.42285}{1.38629} = -3.1904$

SOLUTIONS TO CHECK POINTS

Check Point 7, Frame 31

8. $\log_{16} 11.5 = \dfrac{\log_e 11.5}{\log_e 16} = \dfrac{\ln 11.5}{\ln 16} = \dfrac{2.44235}{2.77259} = 0.8809$

9. $\log 100 = \dfrac{\log_e 100}{\log_e 10} = \dfrac{\ln 100}{\ln 10} = \dfrac{4.60517}{2.30259} = 2.0000$

10. $\log_8 594.6 = \dfrac{\log_e 594.6}{\log_e 8} = \dfrac{\ln 594.6}{\ln 8} = \dfrac{6.38789}{2.07944} = 3.0719$

11. $\log 0.008 = \dfrac{\log_e 0.008}{\log_e 10} = \dfrac{\ln 0.008}{\ln 10} = \dfrac{-4.82831}{2.30259} = -2.0969$

12. $\log_7 7269 = \dfrac{\log_e 7269}{\log_e 7} = \dfrac{\ln 7269}{\ln 7} = \dfrac{8.89137}{1.94591} = 4.5693$

Check Point 8, Frame 37

1. $10^{x+3} = 21$
$(x + 3) = \log 21$
$(x + 3) = 1.3222$
$x = 1.3222 - 3$
$x = -1.6778$

2. $4(3^p) = 5$
$3^p = 1.25$
$\log 3^p = \log 1.25$
$p \log 3 = 0.09691$
$p = 0.2031$

447

ANSWERS TO DRILL EXERCISES

1.	-1.23657	2.	3.10206
3.	1.93997	4.	-3.30103
5.	2.80804	6.	-0.60206
7.	5.75446	8.	1.80734
9.	-1.74297	10.	8.51719
11.	2.53290	12.	6.66191
13.	4.54416	14.	-6.50229
15.	1029.2005	16.	5.28932
17.	0.00245	18.	15.10776
19.	0.31623	20.	5.62341
21.	36.59823	22.	0.15009
23.	1.64872	24.	15.64263
25.	0.91393	26.	201.64324
27.	2.7927	28.	-0.9418
29.	0.9613	30.	3.2189
31.	4.6439	32.	-1.6523
33.	3.2710	34.	-1.0000
35.	3.0000	36.	4.0827
37.	3.6780	38.	5.1699
39.	3.02531	40.	1.88897
41.	1.52372	42.	-0.85258
43.	-5.258 029	44.	-24.29820
45.	54.59815	46.	0.63515
47.	0.53710	48.	2
49.	-0.70444	50.	4

Check Point 8, Frame 37 (continued)

3.
$$12e^{1.5y} = 612$$
$$e^{1.5y} = 51$$
$$1.5y = \ell n\ 51$$
$$1.5y = 3.9318$$
$$y = 2.6212$$

4.
$$2\pi e^{-1.5} = e^{0.2x}$$
$$2\pi = e^{0.2x + 1.5}$$
$$\ell n\ (2\pi) = 0.2x + 1.5$$
$$1.8379 = 0.2x + 1.5$$
$$0.3379 = 0.2x$$
$$1.6894 = x$$

Check Point 9, Frame 43

1.
$$0.5\ \log\ (2a) = 3.75$$
$$\log\ (2a) = \frac{3.75}{0.5}$$
$$\log\ (2a) = 7.5$$
$$\log\ 2 + \log\ a = 7.5$$
$$\log\ a = 7.5 - \log\ 2$$
$$\log\ a = 7.19897$$
$$a = 10^{7.19897}$$
$$a = 15811388$$

2.
$$1 = \frac{1}{2}\ \log\ (3m + 10)$$
$$\log\ (3m + 10) = 2$$
$$3m + 10 = 10^2$$
$$3m = 100 - 10$$
$$3m = 90$$
$$m = 30$$

3.
$$\ell n\ \sqrt{\frac{5}{V}} = 1.4$$
$$1.4 = \ell n\ \left(\frac{5}{V}\right)^{1/2}$$
$$1.4 = \frac{1}{2}\ (\ \ell n\ 5 - \ell n\ V\ )$$
$$1.4 = \frac{1}{2}\ \ell n\ 5 - \frac{1}{2}\ \ell n\ V$$
$$\frac{1}{2}\ \ell n\ V = \frac{1}{2}\ \ell n\ 5 - 1.4$$
$$= 0.8047189 - 1.4$$
$$\ell n\ V = -1.19056$$
$$V = e^{-1.19056}$$
$$V = 0.30405$$

4.
$$\ell n\ (x - 2) = \frac{1}{2}\ \ell n\ x$$
$$\ell n\ (x - 2) = \ell n\ x^{1/2}$$
$$x - 2 = x^{1/2}$$
$$(x - 2)^2 = x$$
$$x^2 - 4x + 4 = x$$
$$x^2 - 5x + 4 = 0$$
$$(x - 4)(x - 1) = 0$$
$$x = 4\ \text{or}\ x = 1$$

Substitution of x = 1 into the original equation leads to the logarithm of a negative number which is not defined. Hence, we have x = 4 as the only solution.

Unit 22
Solutions to Check Points

Check Point 1, Frame 7 (continued)

4. $y = -x^2$

x	-3	-2	-1	0	1	2	3
y	-9	-4	-1	0	-1	-4	-9

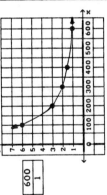

Check Point 2, Frame 14

1. $f(0) = 5(0) - 3,$ $f(-4) = 5(-4) - 3$ $f(3) = 5(3) - 3$
          $= -3$ $= -23$ $= 12$

2. $g(-2) = 2(-2)^2 + 4(-2) - 3,$ $g(0) = 2(0)^2 + 4(0) - 3$
           $= 8 - 8 - 3$ $= -3$
           $= -3$

   $g(4) = 2(4)^2 + 4(4) - 3$
           $= 32 + 16 - 3$
           $= 45$

3. $h(-1) = 3(-1) + 7,$ $h(0) = 3(0) + 7$ $h(3) = 3(3) + 7$
           $= 4$ $= 7$ $= 16$

Unit 22
Solutions to Check Points

Check Point 1, Frame 7

1. $y = 2x - 1$

x	-3	-2	-1	0	1	2	3
y	-7	-5	-3	-1	1	3	5

2. Note: Since s represents the side of a square, we will not choose negative values of s.

   $A = s^2$

s	0	1	2	3	4
A	0	1	4	9	16

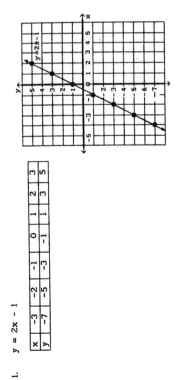
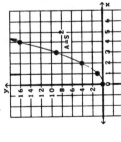

3. Note: $\dfrac{600}{w}$ is not defined when w = 0.

   Also, since L and w represent the length and width, we will not choose negative values of w.

   $\dfrac{600}{w}$

s	100	200	300	400	500	600
w	6	3	2	1.5	1.2	1

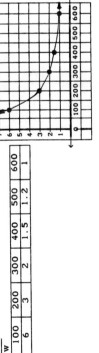

Check Point 3, Frame 25

1. f(x) = 4x - 5
   Make up a table of values.

x	0	5/4	1
f(x)	-5	0	-1

   The x-intercept is $\left(\frac{5}{4}, 0\right)$.

   The y-intercept is (0, -5).
   The slope is 4.

2. g(x) = -3x
   Make up a table of values.

x	0	2	-2
g(x)	0	-6	6

   The x-intercept is (0, 0).
   The y-intercept is (0, 0).
   The slope is -3.

3. $y = \frac{3}{2}$.
   There is not x-intercept.
   The y-intercept is $\left(0, \frac{3}{2}\right)$.
   The slope is 0.

4. $h(x) = -\frac{x}{3} + 1$.
   Make up a table of values.

x	0	3	-3
h(x)	1	0	2

   The x-intercept is (3,0).
   The y-intercept is (0,1).
   The slope is $-\frac{1}{3}$.

5. y = x
   The function y = x is the identity function.
   The x-intercept is (0,0).
   The y-intercept is (0,0).
   The slope is 1.

6. f(x) = -3.
   There is no x-intercept.
   The y-intercept is (0,-3).
   The slope is 0.

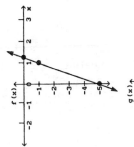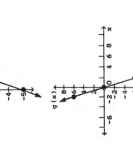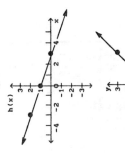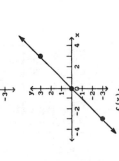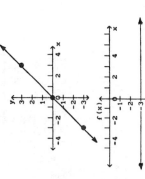

Check Point 4, Frame 33

1. $f(x) = 2x^2 - 3x + 1$.   The graph opens upward.
   Find the x-coordinate of the vertex using $x = -\frac{b}{2a}$.

   a = 2   b = -3 $\therefore x = \frac{-(-3)}{2(2)} = \frac{3}{4}$. $f(x) = 2x^2 - 3x + 1$

   Find the y-coordinate if $x = \frac{3}{4}$.

   $f\left(\frac{3}{4}\right) = 2\left(\frac{3}{4}\right)^2 - 3\left(\frac{3}{4}\right) + 1$

   Hence, the coordinates of the vertex are $\left(\frac{3}{4}, -\frac{1}{8}\right)$.

2. $y = -x^2 + 2$.   The graph opens downward.
   Find the x-coordinate of the vertex using $x = -\frac{b}{2a}$.

   a = -1   b = 0 $\therefore x = -\frac{0}{2(-1)} = 0$.

   Find the y-coordinate if x = 0. $y = -x^2 + 2$
   $y = -0 + 2$
   $y = 2$

   Hence, the coordinates of the vertex are (0, 2).

3. $g(x) = 3x^2 - x$.   The graph opens upward.
   Find the x-coordinates of the vertex using $x = -\frac{b}{2a}$.

   a = 3   b = -1 $\therefore x = -\frac{(-1)}{2(3)} = \frac{1}{6}$

   Find the y-coordinate if $x = \frac{1}{6}$. $g(x) = 3x^2 - x$

   $g\left(\frac{1}{6}\right) = 3\left(\frac{1}{6}\right)^2 - \frac{1}{6}$
   $= -\frac{1}{12}$

   Hence, the coordinates of the vertex are $\left(\frac{1}{6}, -\frac{1}{12}\right)$.

4. $h(x) = -9x^2 - 1$.   The graph opens downward.
   Find the x-coordinate of the vertex using $x = -\frac{b}{2a}$.

   a = -9   b = 0 $\therefore x = -\frac{0}{2(-9)} = 0$.

   Find the y-coordinate if x = 0. $h(x) = -9x^2 - 1$
   $h(0) = -9(0)^2 - 1 = -1$

   Hence, the coordinates of the vertex are (0, -1).

Check Point 4, Frame 33 (continued)

5. $y = x^2$.   The graph opens upward.
Find the x-coordinate of the vertex using $x = -\frac{b}{2a}$

$a = 1$ $b = 0$ $\therefore$ $x = -\frac{0}{2(1)} = 0$
Find the y-coordinate if $x = 0$. $y = x^2$
                                                $y = 0$
Hence, the coordinates of the vertex are $(0, 0)$.

6. $f(x) = -\frac{x^2}{2} - x + 5$ The graph opens downward.
Find the x-coordinate of the vertex using $x = -\frac{b}{2a}$

$a = -\frac{1}{2}$ $b = -1$ $\therefore$ $x = -\frac{(-1)}{2\left(-\frac{1}{2}\right)} = -1$

Find the y-coordinate if $x = -1$.
$f(x) = -\frac{x^2}{2} - x + 5$
$f(-1) = -\frac{(-1)^2}{2} - (-1) + 5 = 5\frac{1}{2}$
Hence, the coordinates of the vertex are $\left(-1, 5\frac{1}{2}\right)$.

7. $y = -4x^2 + 12x - 9$.   The graph opens downward.
Find the x-coordinate of the vertex using $x = -\frac{b}{2a}$

$a = -4$ $b = 12$ $\therefore$ $x = -\frac{12}{2(-4)} = \frac{3}{2}$
Find the y-coordinate if $x = \frac{3}{2}$.
$y = -4x^2 + 12x - 9$
$y = -4\left(\frac{3}{2}\right)^2 + 12\left(\frac{3}{2}\right) - 9 = 0$
Hence, the coordinates of the vertex are $\left(\frac{3}{2}, 0\right)$.

Check Point 5, Frame 36

1. $f(x) = x^2$
To find the zeros of the function $f(x) = x^2$, we solve $0 = x^2$ to find that $x = 0$ is a zero of the function.

2. $y = -x^2 + 4$
To find the zeros of the function $y = -x^2 + 4$, we solve $-x^2 + 4 = 0$ or $x^2 - 4 = 0$ to find that $(x - 2)(x + 2) = 0$, that is $x = 2$ and $x = -2$ are zeros of the given function.

3. To find the zeros of the function $g(x) = x^2 - 4x + 3$, we solve $x^2 - 4x + 3 = 0$ to find that $(x - 3)(x - 1) = 0$; that is $x = 3$ and $x = 1$ are zeros of the given function.

Check Point 5, Frame 36 (continued)

4. To find the zeros of the function $h(x) = -3x^2 + 10x - 4$, we solve $-3x^2 + 10x - 4 = 0$ or $3x^2 - 10x + 4 = 0$.

$x = \frac{-(-10) \pm \sqrt{(-10)^2 - 4(3)4}}{2(3)} = \frac{10 \pm \sqrt{52}}{6}$

$x = 2.87$ and $x = 0.46$ are zeros of the given function.

5. To find the zeros of the function $y = 4x^2 - 12x + 9$, we solve $4x^2 - 12x + 9 = 0$.

$x = \frac{-(-12) \pm \sqrt{(-12)^2 - 4(4)9}}{2(4)} = \frac{12 \pm \sqrt{0}}{8} = \frac{3}{2}$

$x = \frac{3}{2}$ is a zero of the given function.

6. To find the zeros of the function $f(x) = 2x^2 + 8x - 5$, we solve $2x^2 + 8x - 5 = 0$.

$x = \frac{-8 \pm \sqrt{(8)^2 - 4(2)(-5)}}{2(2)} = \frac{-8 \pm \sqrt{104}}{4}$

$x = 0.55$ and $x = -4.55$ are zeros of the given function.

Check Point 6, Frame 42

1. $f(x) = -x^2$
Step 1   Find the coordinates of the vertex.
$a = -1$, $b = 0$ and $x_v = -\frac{0}{2(-1)} = 0$.
$f(0) = 0$
Hence, the coordinates of the vertex are $(0,0)$.

Step 2   Since $a = -1$ is negative, the graph opens downward.

Step 3   Since the vertex is on the x-axis, no other zeros exist. Hence, we make up a table of values choosing x-values to the right and left of the vertex.

x	0	1	-1	2	-2
f(x)	0	-1	-1	-4	-4

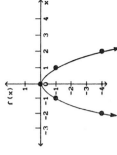

2. $y = x^2 - 4$
Step 1   Find the coordinates of the vertex.
$a = 1$, $b = 0$ and $x_v = -\frac{0}{2(1)} = 0$.
$f(0) = (0)^2 - 4 = -4$
Hence, the coordinates of the vertex are $(0, -4)$.

Step 2   Since $a = 1$ is positive the graph opens upward and will cross the x-axis since the vertex is below the x-axis.

Check Point 6, Frame 42 (continued)

Step 4 Find the zeros of f(x).
$$x^2 - 4 = 0$$
$$(x - 2)(x + 2) = 0$$
Hence, x = 2 and x = -2 are zeros of the function.

Step 4 Plot the coordinates of the vertex and the zeros and then sketch the graph.

3. $g(x) = x^2 - 6x + 2$

Step 1 Find the coordinates of the vertex.
a = 1, b = -6 and $x_V = -\dfrac{-6}{2(1)} = 3$
$$g(3) = (3)^2 - 6(3) + 2 = -7$$
Hence, the coordinates of the vertex are (3, -7).

Step 2 Since a = 1 is positive the graph opens upward and will cross the x-axis since the vertex is below the x-axis.

Step 3 Find the zeros of g(x).
$$x^2 - 6x + 2 = 0$$
$$x = \frac{6 \pm \sqrt{36 - 4(1)(2)}}{2(1)} = \frac{6 \pm \sqrt{28}}{2}$$
Hence, x = 5.65 and x = 0.35 are zeros of the function.

Step 4 Plot the coordinates of the vertex and the zeros and then sketch the graph.

4. $f(x) = - 2x^2 + x - 3$

Step 1 Find the coordinates of the vertex.
a = -2, b = 1 and $x_v = -\dfrac{1}{2(-2)} = \dfrac{1}{4}$.
$$f\left(\frac{1}{4}\right) = -2\left(\frac{1}{4}\right)^2 + \frac{1}{4} - 3 = -2\frac{7}{8}.$$
Hence, the coordinates of the vertex are $\left(\dfrac{1}{4},\ -2\dfrac{7}{8}\right)$.

Step 2 Since a = -2 is negative the graph opens downward. The vertex is below the x-axis and the graph opens downward. Hence, the graph does not cross the x-axis; that is, the function has no zeros that are real numbers.

Step 3 Since no real zeros exist, we make up a table of values.
$f(x) = -2x^2 + x - 3$

x	1/4	0	1	-1
f(x)	-2 7/8	-3	-4	-6

Step 4 Plot the points and sketch the graph.

(continued)

Check Point 6, Frame 42 (continued)

5. $y = 6x - x^2$

Step 1 Find the coordinates of the vertex.
a = -1, b = 6 and $x_V = -\dfrac{6}{2(-1)} = 3$.
$$f(3) = 6(3) - (3)^2 = 9$$
Hence, the coordinates of the vertex are (3, 9).

Step 2 Since a = -1 is negative the graph opens downward, and will cross the x-axis since the vertex is above the x-axis.

Step 3 Find the zeros of f(x).
$$6x - x^2 = 0 \qquad x(6-x) = 0$$
Hence, x = 0 and x = 6 are zeros of the function.

Step 4 Plot the vertex and the zeros of the function and sketch the graph.

6. $f(x) = - 2x^2 - 4x - 5$

Step 1 Find the coordinates of the vertex.
a = -2, b = -4 and $x_V = -\dfrac{-4}{2(-2)} = -1$.
$$f(-1) = -2(-1)^2 - 4(-1) - 5 = -3$$
Hence, the coordinates of the vertex are (-1, -3).

Step 2 Since a = -2 is negative the graph opens downward. The vertex is below the x-axis and the graph opens downward.
Hence, the graph does not cross the x-axis; that is, the function has no zeros that are real numbers.

Step 3 Since no real zeros exist, we make up a table of values.
$f(x) = -2x^2 - 4x - 5.$

x	-1	-2	0	1	-3
f(x)	-3	-5	-5	-11	-11

Step 4 Plot the points and sketch the graph.

5

Check Point 7, Frame 50

1. $y = (2.5)^x$

x	-5	-3	-1	0	1	2	3		5
y	0.010	0.064	0.4	1	2.5	6.25	15.625		97.656

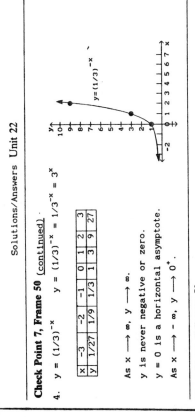

As x is getting larger, y is getting rapidly larger; that is, as $x \longrightarrow \infty$ we have that $y \longrightarrow \infty$. As x gets smaller, y gets smaller. For any negative value x = - a, we have that $y = (2.5)^{-a} = \dfrac{1}{(2.5)^{-a}}$ is

a positive fraction. Hence, y can never be negative or zero. As $x \longrightarrow -\infty$, $y \longrightarrow 0^+$. The line y = 0 is a horizontal asymptote.

2. $g(x) = (1.5)^x$

x	-5	-3	-1	0	1	2	3	5	10
g(x)	0.13	0.30	0.67	1	1.5	2.25	3.38	7.6	57.67

As $x \longrightarrow \infty$, $y \longrightarrow \infty$.
y is never negative or 0.
y = 0 is a horizontal asymptote.
As $x \longrightarrow -\infty$, $y \longrightarrow 0^+$.

3. $f(x) = 3^{-x}$

$f(x) = 3^{-x} = \dfrac{1}{3^x} = \left(\dfrac{1}{3}\right)^x$

x	-3	-2	-1	0	1	2	3
f(x)	27	9	3	1	1/3	1/9	1/27

As $x \longrightarrow -\infty$, $y \longrightarrow \infty$.
y is never negative or 0.
y = 0 is a horizontal asymptote.
As $x \longrightarrow \infty$, $y \longrightarrow 0^+$.

Check Point 7, Frame 50 (continued)

4. $y = (1/3)^{-x}$ $y = (1/3)^{-x} = 1/3^{-x} = 3^x$

x	-3	-2	-1	0	1	2	3
y	1/27	1/9	1/3	1	3	9	27

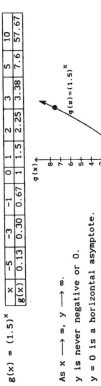

As $x \longrightarrow \infty$, $y \longrightarrow \infty$.
y is never negative or zero.
y = 0 is a horizontal asymptote.
As $x \longrightarrow -\infty$, $y \longrightarrow 0^+$.

Check Point 8, Frame 52

1. $f(x) = e^x$

x	-3	-2	-1	0	1	2	3
f(x)	0.05	0.14	0.37	1	2.7	7.4	20.1

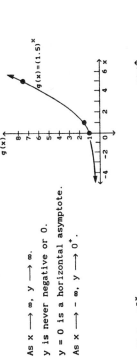

2. $g(x) = e^{-x}$

x	-3	-2	-1	0	1	2	3
g(x)	20.1	7.4	2.7	1	0.37	0.14	0.05

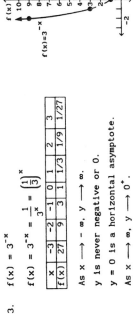

Check Point 9, Frame 57

1. We make up a table of values using the corresponding exponential functions. $y = 3^x$ and $y = (1/3)^x$

$y = 3^x$

x	-3	-2	-1	0	1	2	3
y	1/27	1/9	1/3	1	3	9	27

$y = (1/3)^x$

x	-3	-2	-1	0	1	2	3
y	27	9	3	1	1/3	1/9	1/27

Since $y = \log_3 x$ is the inverse function of $y = 3^x$, we interchange the x and y values in the table for $y = 3^x$ to obtain a table of values for $y = \log_3 x$. We do the same with

Check Point 10, Frame 65 (continued)

2. Let the original amount $A_o = 1$. Then $A(t) = \frac{1}{2}$ when t is the half-life of this substance.

$$A(t) = A_o e^{-kt}$$
$$0.5 = e^{-0.0283t}$$
$$\ln(0.5) = -0.0283t$$
$$t = 24.49$$

Hence, the half-life of this substance is about 24 days.

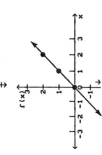

Drill Exercises

1. The x-intercept is (1,0).
   The y-intercept is (0,1).
   The slope is -1.

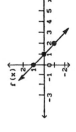

2. There is no x-intercept.
   The y-intercept is (0, -3).
   The slope is 0.

3. The x-intercept is (0,0).
   The y-intercept is (0,0).
   The slope is 1.

Check Point 9, Frame 57 (continued)

$y = (1/3)^x$ and its inverse $y = \log_{(1/3)}x$.

$y = \log_3 x$

x	1/27	1/9	1/3	1	3	9	27
y	-3	-2	-1	0	1	2	3

$y = \log_{(1/3)}x$

x	27	9	3	1	1/3	1/9	1/27
y	-3	-2	-1	0	1	2	3

Plot the points and sketch the graphs.

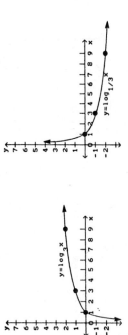

2. $y = \ln(2x)$, y is defined only for $x > 0$.

x	0.125	0.25	0.5	1	2	3	5
2x	0.25	0.5	1.0	2	4	6	10
y	-1.4	-0.7	0	0.7	1.4	1.8	2.3

Check Point 10, Frame 65

1. Let the original amount $A_o = 1$. Then $A(t) = \frac{1}{2}$ when $t = 22$.

Substituting into $A(t) = A_o e^{-kt}$, we have

$$0.5 = e^{-22k} \qquad \text{and} \quad \ln 0.5 = -22k$$
$$k = 0.03151 \text{ (rounded to 5 decimal places)}$$

Hence, the decay constant is $k = 0.03151$.

$$A(t) = A_o e^{-kt}$$
$$5 = 1000e^{-0.03151t}$$
$$0.005 = e^{-0.03151t}$$
$$\ln(0.005) = -0.03151t$$
$$t = \frac{\ln(0.005)}{-0.03151}$$
$$= 168.1648362$$

Hence, it will take about 168 years for 1000 mg of this substance to decay to 5 mg.

Drill Exercises (continued)

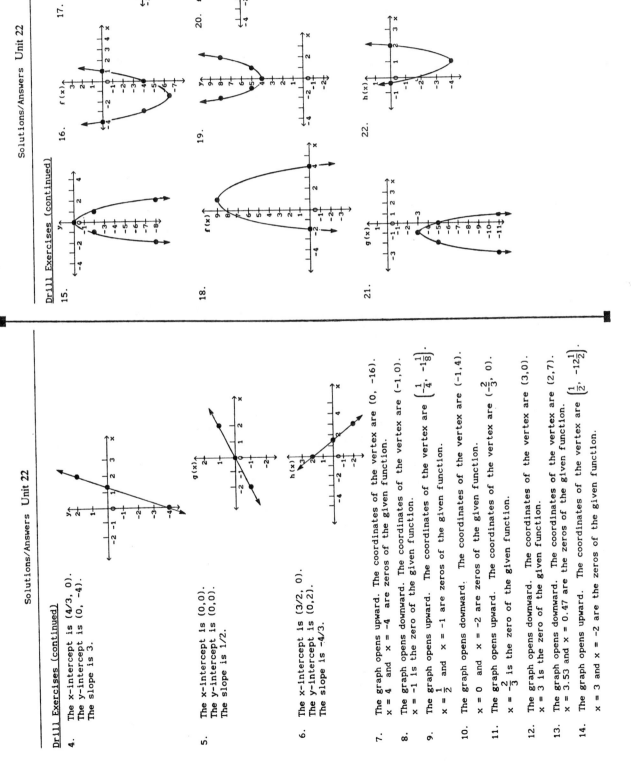

4.   The x-intercept is (4/3, 0).
     The y-intercept is (0, -4).
     The slope is 3.

5.   The x-intercept is (0,0).
     The y-intercept is (0,0).
     The slope is 1/2.

6.   The x-intercept is (3/2, 0).
     The y-intercept is (0,2).
     The slope is -4/3.

7.   The graph opens upward. The coordinates of the vertex are (0, -16).
     x = 4 and x = -4 are zeros of the given function.

8.   The graph opens downward. The coordinates of the vertex are (-1,0).
     x = -1 is the zero of the given function.

9.   The graph opens upward.   The coordinates of the vertex are $\left(-\frac{1}{4}, -1\frac{1}{8}\right)$.
     x = $\frac{1}{2}$ and x = -1 are zeros of the given function.

10. The graph opens downward.   The coordinates of the vertex are (-1, 4).
     x = 0 and x = -2 are zeros of the given function.

11. The graph opens upward. The coordinates of the vertex are $\left(-\frac{2}{3}, 0\right)$.
     x = $-\frac{2}{3}$ is the zero of the given function.

12. The graph opens downward.   The coordinates of the vertex are (3,0).
     x = 3 is the zero of the given function.

13. The graph opens downward.   The coordinates of the vertex are (2,7).
     x = 3.53 and x = 0.47 are the zeros of the given function.

14. The graph opens upward.   The coordinates of the vertex are $\left(\frac{1}{2}, -12\frac{1}{2}\right)$.
     x = 3 and x = -2 are the zeros of the given function.

Drill Exercises (continued)

15.

16.

17.

18.

19.

20.

21.

22.

Dr1ll Exercises

23. $-0.75e^{0.8x}$

24. $\dfrac{A_0}{B_0} 2^{(a-\pi)x}$

25. $\ln\left(\sqrt{\dfrac{x-1}{x+1}}\right), \ x > 1$

26. $\dfrac{A_0}{A_1} e^{-(c+k)t}$

27. $p \ln(x), \ x > 0$

28. $\sqrt{x-1}$

29. $\dfrac{1}{x^2}$

30. x^2

31. πx

32.

$f(x)=\log_3 x$

33.

$f(x)=\log_{(1/2)} x$

34.

$f(x)=\log_{(1.5)} x$

35.

$f(x)=\log x$

36. 167 721 600 bacteria, 3.32 hours

37. Half-life in years is 27.95 years. Decay takes 278.5 years.

38. K = 0.00043. 1000 mg will decay to 5 mg in about 12 322 years.

39. 75% carbon lost: 11 552 years old,
80% carbon lost: 13 412 years old,
90% carbon lost: 19 188 years old,
95% carbon lost: 24 964 years old.

40. 254 years ago.

INDEX